# Emergence, Complexity and Computation

## Volume 25

*About this Series*

The Emergence, Complexity and Computation (ECC) series publishes new developments, advancements and selected topics in the fields of complexity, computation and emergence. The series focuses on all aspects of reality-based computation approaches from an interdisciplinary point of view especially from applied sciences, biology, physics, or chemistry. It presents new ideas and inter-disciplinary insight on the mutual intersection of subareas of computation, com-plexity and emergence and its impact and limits to any computing based on physical limits (thermodynamic and quantum limits, Bremermann's limit, Seth Lloyd limits...) as well as algorithmic limits (Gödel's proof and its impact on calculation, algorithmic complexity, the Chaitin's Omega number and Kolmogorov complexity, non-traditional calculations like Turing machine process and its con-sequences,...) and limitations arising in artificial intelligence field. The topics are (but not limited to) membrane computing, DNA computing, immune computing, quantum computing, swarm computing, analogic computing, chaos computing and computing on the edge of chaos, computational aspects of dynamics of complex systems (systems with self-organization, multiagent systems, cellular automata, artificial life,...), emergence of complex systems and its computational aspects, and agent based computation. The main aim of this series it to discuss the above mentioned topics from an interdisciplinary point of view and present new ideas coming from mutual intersection of classical as well as modern methods of com-putation. Within the scope of the series are monographs, lecture notes, selected contributions from specialized conferences and workshops, special contribution from international experts.

More information about this series at http://www.springer.com/series/10624

Gexiang Zhang · Mario J. Pérez-Jiménez
Marian Gheorghe

# Real-life Applications
# with Membrane Computing

Gexiang Zhang
Key Laboratory of Fluid and Power
  Machinery, Ministry of Education
Xihua University
Chengdu
China

and

Robotics Research Center
Xihua University
Chengdu
China

Mario J. Pérez-Jiménez
Department of Computer Science
  and Artificial Intelligence
University of Seville
Sevilla
Spain

Marian Gheorghe
School of Electrical Engineering
  and Computer Science
University of Bradford
Bradford
UK

ISSN 2194-7287          ISSN 2194-7295  (electronic)
Emergence, Complexity and Computation
ISBN 978-3-319-85798-5          ISBN 978-3-319-55989-6  (eBook)
DOI 10.1007/978-3-319-55989-6

Printed on acid-free paper

This Springer imprint is published by Springer Nature
The registered company is Springer International Publishing AG
The registered company address is: Gewerbestrasse 11, 6330 Cham, Switzerland

# Preface

Membrane computing is a vibrant and fast-growing research area of natural computation which covers the study of computing models, called membrane systems or P systems, inspired by the organization of the living cell and the biochemical reactions and phenomena occurring therein. The results obtained in this field have been published in comprehensive monographs covering theoretical aspects [4, 5] and applications in linguistics, graphics, computer science [1] and systems and synthetic biology [2]. The key research topics and further developments are overviewed in [3].

This book presents for the first time to the international community real-life complex and challenging applications modeled and analyzed with a membrane computing apparatus. These applications require different approaches and tools than those already investigated and described in the above-mentioned publications. They rely on a combination of different membrane systems and evolutionary or fuzzy reasoning methods, applied to a wide range of applications from various engineering areas, including engineering optimization, power systems fault diagnosis, mobile robots controller design, or complex biological systems involving data modeling and process interactions. This book goes far beyond the content of the Chinese textbook focusing mainly on applications of membrane computing [6] by addressing the basis of merging membrane computing concepts and evolutionary computing algorithms, by presenting a broader spectrum of real-life applications and the solid foundation of assessing the results and consequently targeting a much wider and diverse international audience.

The chapters covered in this monograph provide a clear image of the depth and breadth of the real-world applications of membrane systems.

- In Chap. 1, *Membrane Computing—Key Concepts and Definitions*: Basic membrane computing concepts that are used in the models presented in the next chapters are introduced. The most significant references to the research textbooks and overview papers are also provided.

– In Chap. 2, *Fundamentals of Evolutionary Computation*: Fundamental concepts and principles of evolutionary computation are addressed. Five variants of evolutionary computation techniques, including genetic algorithms, quantum-inspired evolutionary algorithms, ant colony optimization, particle swarm optimization and differential evolution, are discussed.

– In Chap. 3, *Membrane Algorithms*: Hybrid approximate optimization algorithms, called membrane algorithms or membrane-inspired evolutionary algorithms, integrating the hierarchical/network structure of P systems with meta-heuristic approaches are introduced and the design principles, their developments with key instances and examples are discussed. In addition, the impact of different variants of P systems with respect to membrane algorithms is analyzed.

– In Chap. 4, *Engineering Optimization with Membrane Algorithms*: A wide range of engineering applications of membrane algorithms with cell-like, tissue-like and neural-like P systems are discussed. The engineering problems include radar emitters signal analysis, digital image processing, controllers design, mobile robots path planning, constrained manufacturing parameters optimization problems, and distribution networks reconfiguration.

– In Chap. 5, *Electric Power System Fault Diagnosis with Membrane Systems*: Spiking neural P systems incorporating fuzzy logics are utilized to solve fault diagnosis problems of electric power systems. Definitions, reasoning algorithms and examples of fuzzy reasoning spiking neural P systems are presented.

– In Chap. 6, *Robot Control with Membrane Systems*: Numerical P systems and enzymatic numerical P systems are employed for designing membrane controllers for mobile robots. Simulators for numerical P systems and enzymatic numerical P systems modeling the robots' behavior are described and the results are analyzed.

– In Chap. 7, *Data Modeling with Membrane Systems: Applications to Real Ecosystems*: A bioinspired computing modeling paradigm within membrane computing, multienvironment P systems, is presented. This paradigm provides two different approaches (multicompartmental P systems and population dynamics P systems). The last approach is used to model population dynamics of real-world ecosystems. Ad hoc algorithms and simulators are introduced to simulate, analyze and (experimentally) validate population dynamics P systems.

This book will be of particular interest to researchers looking for applications of membrane computing, studying the interplay between membrane systems and other computational approaches and methods such as meta-heuristic optimization, fuzzy set theory and control theory. More generally, the book will be of interest to anyone studying bioinspired computing, engineering optimization, electric power systems fault diagnosis, robotics and ecosystems.

Finally, we would like to thank Gheorghe Păun for his continuous support in writing the book, as well as for many insightful comments and suggestions made. We are also very grateful to many colleagues, collaborators, Ph.D. students and friends for their helpful comments and discussions, especially to Jixiang Cheng,

Luis F. Macías-Ramos, Miguel A. Martínez-del-Amor, Agustín Riscos-Núñez, Luis Valencia-Cabrera, Tao Wang and Xueyuan Wang. Gexiang Zhang also acknowledges the support of his research activities provided by the National Natural Science Foundation of China (61170016, 61373047 and 61672437), the Research Project of Key Laboratory of Fluid and Power Machinery (Xihua University), Ministry of Education, P.R. China (JYBFX-YQ-1). We also thank the publisher for a friendly and efficient collaboration.

Chengdu, China                                                                    Gexiang Zhang
Sevilla, Spain                                                                Mario J. Pérez-Jiménez
Bradford, UK                                                                      Marian Gheorghe

# References

1. G. Ciobanu, M.J. Pérez-Jiménez, Gh. Păun (eds.), *Applications of Membrane Computing*, in Natural Computing Series (Springer, 2006)
2. P. Frisco, M. Gheorghe, M.J. Pérez-Jiménez (eds.), *Applications of Membrane Computing in Systems and Synthetic Biology*, in Emergence, Complexity and Computation Series (Springer, 2014)
3. M. Gheorghe, Gh. Păun, M.J. Pérez-Jiménez, G. Rozenberg, Research frontiers of membrane computing: open problems and research topics. *International Journal of Foundations of Computer Science* **24**, 5 (2013), 547–624.
4. Gh. Păun, *Membrane Computing—An Introduction* (Springer, 2002)
5. Gh. Păun, G. Rozenberg, A. Salomaa (eds.), *The Oxford Handbook of Membrane Computing* (Oxford University Press, 2010)
6. G. Zhang, J. Cheng, T. Wang, X. Wang, J. Zhu, *Membrane Computing: Theory & Applications* (Science Press, Beijing, China, 2015)

# Contents

# Chapter 1
# Membrane Computing - Key Concepts and Definitions

**Abstract** The basic membrane computing concepts used in the models presented in the next chapters are introduced. A basic transition membrane system, membrane systems with active membranes, a neural-like network of membranes and a spiking neural membrane system are defined and some simple examples are provided.

## 1.1 Introduction

In this chapter we introduce the basic membrane computing concepts that are used in the models presented in the next chapters. Some examples illustrate the way these concepts are used in various definitions. The presentation is meant to be as intuitive as possible, but also rigorous enough to serve the purpose of the models introduced. We also provide the most significant references to the research text books and overview papers. For a comprehensive and up-to-date list of publications we point to the membrane computing main website [6].

The first membrane computing text book presents the basic membrane computing definitions and variants of the key systems (called *membrane systems* or *P systems*), together with the main theoretical developments of the field in the early 2000 [7]. A later text book reports results on some special classes of membrane systems - symport/antiport P systems, P systems with catalysts, spiking, splicing and conformon P systems -, as well as relationships between membrane systems, on the one hand, and Petri nets and brane calculi, on the other hand [3]. The theory of a special class of deterministic P systems inspired by the cell metabolism, called *metabolic P systems*, and its applications in modeling various biological system are presented in [5]. Applications of membrane computing in modeling protocol communication, sorting, graphics, linguistics and biology problems, but also in describing computationally hard problems and approximate algorithms, are presented in [1]. Specific applications in systems and synthetic biology are discussed in [2]. A more recent text book [9] presents a summary of some theoretical results and applications of evolutionary membrane computing, a type of evolutionary computation algorithms using various membrane computing concepts, such as different membrane structures and rules. A thorough presentation of both theoretical developments in membrane computing and

© Springer International Publishing AG 2017

G. Zhang et al., *Real-life Applications with Membrane Computing*,

Emergence, Complexity and Computation 25, DOI 10.1007/978-3-319-55989-6_1

key applications of this computational model is given in [8]. An overview of the main developments in membrane computing and new research challenges are presented in [4] and an overview of evolutionary membrane computing developments and new challenges are discussed in [10].

## 1.2 Origins of Membrane Computing

*Membrane computing* is a computational paradigm motivated by the structure and functioning of the living cells. The role of membranes in delimiting compartments of the living cells is essential to this approach. They localize the interactions of the objects within compartments and allow exchanges between neighbor compartments. Some of the models are based on a *cell-like* structure, i.e., a hierarchical arrangement of membranes, whereas others, *tissue-like models*, consider membranes linked through communication channels and arranged as a network, i.e., a graph structure. The more recent developments are *neural-like membrane systems* which are motivated by spiking neural networks. Most of the features of these models are inspired by biological phenomena, but there are some derived from computer science or mathematics. The computational models are called either *membrane systems* or *P systems*.

## 1.3 Preliminary Concepts and Notations

Given an *alphabet* $V$ (a nonempty set of symbols), a *string* (or *word*) over $V$ is an ordered finite sequence of elements from $V$. For instance, for $V = \{a, b, c\}$, $aab$, $ccbbb$ and $baabbbc$ are words over $V$; these are also denoted by $a^2b$, $c^2b^3$ and $ba^2b^3c$, respectively. The *empty string* is denoted by $\lambda$. The set of all the strings over $V$ is denoted by $V^*$ and by $V^+$ we denote $V^* \setminus \{\lambda\}$. A *multiset* $M$ over an alphabet $V$ is a mapping, $M : V \longrightarrow \mathbb{N}$, where $\mathbb{N}$ is the set of nonnegative integer numbers. For $a \in V$, $M(a)$ is the multiplicity of $a$ in the multiset $M$. We denote by $supp(M)$, the *support* of $M$, the set $\{a \in V \mid M(a) \neq 0\}$. A multiset is finite (respectively, empty) if its support is a finite (respectively, empty) set. We denote by $\lambda$ the empty multiset and by $M_f(O)$ the set of all finite multisets over $O$. When the set $V$ is finite, i.e. $V = \{a_1, \ldots, a_n\}$, then the multiset $M$ can be explicitly written as $\{(a_1, M(a_1)), \ldots (a_n, M(a_n))\}$ and we use the notation $w = a_1^{M(a_1)} \ldots a_n^{M(a_n)}$. We call $M(a_1) + \cdots + M(a_n)$ the *cardinality* of the multiset and we denote it by $|M|$.

For two multisets $M_1, M_2 : V \longrightarrow \mathbb{N}$, the *union* of $M_1$ and $M_2$ is $M_1 \cup M_2 : V \longrightarrow \mathbb{N}$, where $(M_1 \cup M_2)(a) = M_1(a) + M_2(a)$, for all $a \in V$. The *intersection* of $M_1$ and $M_2$ is $M_1 \cap M_2 : V \longrightarrow N$, where $(M_1 \cap M_2)(a) = min\{M_1(a), M_2(a)\}$, for all $a \in V$. We say that $M_1$ is *included* in $M_2$ ($M_1 \subseteq M_2$) if $M_1(a) \leq M_2(a)$, for all $a \in V$. It follows that $M_{w_1} \cup M_{w_2} = \{(a, 1), (b, 8), (c, 2)\}$ and $M_{w_1} \cap M_{w_2} = \{(b, 3)\}$.

## 1.4 Membrane Computing Concepts

*Membranes* represent a fundamental concept in membrane computing. They delimit *compartments* (or *regions*) of a *membrane system* (also called *P system*). In our approaches we consider that these compartments are arranged in a hierarchical structure (formally, a rooted tree), like in a living cell, and these systems are called *cell-like* membrane systems. A *membrane structure* defines the way the compartments are arranged by providing links between compartments and their neighbors. Its notation will be informally introduced below.

The compartments will be identified by some labels over an alphabet, very often a finite set of nonnegative numbers. A compartment identified by a label $l$ will be denoted by a pair of ordered parentheses $[\ ]_l$. A cell-like membrane structure will be formally represented by a sequence of ordered labeled parentheses, each of them defining the position of a compartment in this hierarchical arrangement. For instance, $\mu = [\ [\ [\ ]_5\ ]_2\ [\ [\ ]_6\ [\ ]_7\ ]_3\ [\ ]_4\ ]_1$, defines a membrane structure with the compartment labeled 1 being the root of the tree and containing compartments 2, 3 and 4. Compartment 2 has inside it compartment 5, whereas compartment 3 contains compartments 6 and 7. Membrane labeled 1 is called *skin membrane*, whereas membranes 5, 6, 7 and 4 delimiting compartments without any compartment inside are called *elementary membranes*.

Each compartment of a membrane system contains multisets of *objects* over a certain alphabet $V$. These objects are abstractions of the biochemical elements that appear inside compartments or are attached to the membranes. The objects are transformed or moved between neighboring compartments by the use of *rewriting* and *communication* rules. These rules represent abstractions of biochemical reactions taking place in the compartments of a cell accounting for both interactions between biochemicals inside compartments as well as the transfer of them across membranes.

The rules are written in the form $u \rightarrow v$, where $u$ is a finite multiset over $V$ and $v$ is a finite multiset over $V \times \{here, out, in\}$. The rules are associated with membranes. For instance, when $V = \{a, b, c\}$ one can have rules $a^2b \rightarrow (c, here)(a, in)(a, here)$, $ac \rightarrow (a, here)(b, out)(b, out)$. These rules are applied only to multisets $w$ that include the multisets on the left hand side of the rules, i.e., $a^2b \subseteq w$ and $ac \subseteq w$. Also the first rule is applied only when the compartment where the rule appears has at least a child compartment, whereas for the second one it is necessary to exist a parent compartment. Given a rule associated with a membrane $i$, the *here*, *in* and *out* are called *target indicators* and their meaning is as follows: *here* means that the object stays in the same compartment; *in* means that the object is sent to one of the child compartments; and *out* means that the object is sent to the parent compartment. For instance, if we consider that compartment 3, of the above defined membrane structure $\mu$, contains the multiset $w = a^2b^3c^2$ and the two rules introduced, then by applying the first rule the multiset $a^2b$ is removed from $w$, $a$ and $c$ are added to it (hence, finally one gets $ab^2c^3$ in 3) and an $a$ is sent to one of the children (let us assume this is 6). When the second rule is applied to the same multiset $w$, then $ac$ is removed from $w$, an $a$ is returned to the same compartment (hence, finally the multiset $a^2b^3c$ is obtained in 3) and two $b'$s are sent to the parent compartment 1. Very often the tag *here* is ignored and the rules are written without it. For instance, the rule $a^2b \rightarrow (c, here)(a, in)(a, here)$ is written as $a^2b \rightarrow c(a, in)a$.

So far we have considered that a single rule is applied to a multiset belonging to a compartment. One can consider that more rules are applied to a multiset in each compartment. For instance, given a compartment with the rules $r_1 : ab \rightarrow c$ and $r_2 : a^2 \rightarrow bc$ and the multiset $w = a^4b$, one can apply $r_1$ or $r_2$ or $r_1, r_2$ or $r_2$ twice (chosen in a *non-deterministic manner*). Applying $r_1$ means that $ab$ is removed from $w$ and $c$ added to the multiset; consequently $w_1 = a^3c$ is obtained. Similarly, applying the other rules or a combination of rules, one can get $w_2 = a^2b^2c$ (when $r_2$ is used), $w_{1,2} = abc^2$ (using $r_1$ and $r_2$) and $w_{2,2} = b^3c^2$ (when $r_2$ is used twice). One can note that when either $r_1, r_2$ are selected or $r_2$ is selected twice then no other rule can be applied in the same time with them. Indeed, in these cases after extracting the corresponding multisets from $w$ the remaining multisets $a$ and $b$, respectively, cannot be associated with any rule. These two cases correspond to maximal parallelism mode. In general, for a given compartment containing a multiset $w$ and a set of rules, $R$, we say that these are applied to $w$ in accordance with the *maximal parallelism* mode, if anytime a multiset of rules is selected to be applied, any submultiset of objects of $w$ that remain unassigned does not include the left hand side of any of the rule of $R$.

We can now formally define a cell-like membrane (P) system.

**Definition 1**  A basic *transition P system* of degree $n \geq 1$ is a tuple

$$\Pi = (O, H, \mu, w_1, \ldots, w_n, R_1, \ldots, R_n, i_0),$$

where:

- $O$ is the alphabet of *objects* of the system;
- $H$ is the alphabet of *membrane labels*;
- $\mu$ is the *membrane structure* consisting of $n$ compartments/membranes injectively labeled by elements from $H$;
- $w_1, \ldots, w_n$ are finite multisets of objects associated with the compartments of $\mu$, called *initial multisets of objects*;
- $R_1, \ldots, R_n$ are the finite sets of *multiset rewriting and communication rules* associated with the $n$ compartments;
- $i_0 \in H \cup \{e\}$ defines the *output compartment*; $e$ is a label, not in $H$.

We assume that $H$ is an ordered set of labels. Very often this is the set $\{1, \ldots, n\}$ and in this case it will be omitted from the system definition. The membrane structure $\mu$ (a rooted tree) will be described as a parentheses expression over $H$. The initial multisets of objects, $w_1, \ldots, w_n$, appear in the compartments 1 to $n$ at the start of the computation (this will be introduced later on in this chapter). Each compartment, $i$, labeled by $l_i \in H$, has its own set of rules, $R_i$, applied only to the objects of this compartment. The label $i_0$ is associated with the output compartment where the result of the computation is obtained. The symbol $e$ is the label of the *environment*. In some sense, the environment can be considered as another compartment containing the entire system.

An *instantaneous description* or a *configuration* $C_t$, at an instant $t$ of a P system, is described by the following elements: (a) the membrane structure at instant $t$; (b) all multisets of objects present in each compartment of the system at that moment. The initial configuration of $\Pi$ is $C_0 = (\mu, w_1, \ldots, w_n)$. Starting from $C_0$ and using rules from $R_1, \ldots, R_n$ in a maximal parallel manner in every compartment one gets a sequence of configurations. The process of getting a configuration from another one is called *transition*. A *computation* of $\Pi$ is a (finite or infinite) sequence of configurations such that: (a) one starts from the initial configuration of the system; (b) for each $n \geq 2$, the $n$-th configuration of the sequence is obtained from the previous configuration in one transition step; and (c) if the sequence is finite (called *halting computation*) then the last configuration is a *halting configuration* (a configuration where no rule of the system is applicable to it). In applications one can consider sometimes only partial computations with a restricted number of steps.

It is worth pointing out that in cell-like P systems the environment plays a "passive" role in the following sense: it only receives objects, but cannot contribute objects to the system.

The P system introduced in Definition 1 is a basic system. Many P system variants have been considered. A great variety of such systems comes from various types of rules that one may consider. Also the execution strategy might not be the maximal parallel mode.

*Remark 1* Sometimes rewriting and communication aspects of a rule are separated into distinct rules. For instance, $a^2b \to ca$ is a multiset rewriting rule that replaces the multiset $a^2b$ by $ca$ and $a \to (a, in)$ is a communication rule sending $a$ to a child compartment of the current one. In a similar way one can consider communication rules sending objects to parent compartments by using the *out* tag. More specific rules utilised in the next chapters are presented below.

Before discussing other types of rules, let us make a note regarding the alphabet used by a P system.

*Remark 2* In certain P systems one might need to distinguish a special alphabet, $T \subseteq O$, of *terminal objects* used for indicating that the results produced by the system are obtained only with symbols from this set.

*Remark 3* Some P systems allow the membrane structure to be changed during the computation through some special rules. These systems are called *P systems with active membranes*. Such a system has the same syntax with a basic transition P system introduced in Definition 1, but the rules are of the following types (Chap. 11 in [8]) listed below, where each membrane is polarised, i.e., has an electrical charge of $+$, $-$ or $0$:

- *Object evolution rules*: $[a \to v]_h^\alpha$, for $h \in H$, $\alpha \in \{+, -, 0\}$, $a \in O$, $v \in M_f(O)$; in membrane $h$ with polarisation $e$, the object $a$ is replaced by the multiset $v$ leaving the membrane with the same polarisation $e$.

- *Communication rules of type in*: $a[\ ]_h^{\alpha_1} \rightarrow [b]_h^{\alpha_2}$, for $h \in H$, $\alpha_1, \alpha_2 \in \{+, -, 0\}$, $a, b \in O$; object $a$ enters membrane $h$ when polarised $e_1$ and changes to $b$ leaving the membrane polarised $e_2$.
- *Communication rules of type out*: $[a]_h^{\alpha_1} \rightarrow [\ ]_h^{\alpha_2} b$, for $h \in H$, $\alpha_1, \alpha_2 \in \{+, -, 0\}$, $a, b \in O$; object $a$ exits membrane $h$ when polarised $e_1$ and changes to $b$ leaving the membrane polarised $e_2$.
- *Dissolution rules*: $[a]_h^{\alpha} \rightarrow b$, for $h \in H$, $\alpha \in \{+, -, 0\}$, $a, b \in O$; the membrane $h$ is dissolved when it contains $a$ which is transformed into $b$ and together with the other objects are made available in the surrounding compartment.
- *Division rules for elementary membranes*: $[a]_h^{\alpha_1} \rightarrow [b]_h^{\alpha_2} [c]_h^{\alpha_3}$, for $h \in H$, $\alpha_1, \alpha_2, \alpha_3 \in \{+, -, 0\}$, $a, b, c \in O$; membrane $h$ polarised $\alpha_1$ and containing $a$ is divided into two membranes, also labelled $h$, but polarised $\alpha_2, \alpha_3$ and containing the same objects as the initial membrane, with the exception of $a$, which becomes $b$ in the membrane polarised $\alpha_2$ and $c$ in the one with polarisation $\alpha_3$.

Some other membrane structure related rules are presented below.

- *Merging rules*: $[x]_h [y]_k \rightarrow [z]_f$; two compartments, labeled $h$ and $k$ and containing $x$, $y$, respectively, and having the same parent compartment, are merged (there contents are united, $x$ and $y$ are removed and $z$ is added) and a new compartment with label $f$ having the same parent compartment with the previous compartments $h$ and $k$ is replacing them.
- *Cell separation rules*: $[x]_h \rightarrow [U]_h [V]_h$, where $x$ is a multiset over $O$, $U, V \subseteq O$ and $U$ and $V$ are complementary subsets. When it is applied, the rule replaces a compartment labeled $h$ with two compartments where the multiset $x$ is removed and the rest of the objects over $U$ are moved into one of them and those over $V$ in the other one.

Apart from considering a special terminal alphabet and some variation of rules one might consider other types of membrane structures than the tree - these models come sometimes with specific rules as well. One of the most utilised membrane structure is the graph with the compartments being vertexes and edges defining communication channels between compartments. This graph structure is associated with a class of systems called *tissue-like* P systems, by analogy with living tissues where cells bump into each other and communicate through pores or other membrane mechanisms.

Now we consider a special class of tissue-like P systems, called *neural-like networks of membranes*, which are inspired by the way neurons work together in order to process impulses in a network of neural cells linked through synapses.

**Definition 2** A neural-like network of membranes is a tuple

$$\Pi = (O, \sigma_1, \dots, \sigma_n, ch, i_0),$$

where

- $O$ and $i_0$ are as in Definition 1;

- $\sigma_1, \ldots, \sigma_n$ denote the *cells* of the system; each of them having the form $\sigma_i = (Q_i, s_{i,0}, w_i, R_i)$, $1 \leq i \leq n$, where $Q_i$ is a *set of states*, $s_{i,0}$ an *initial state*, $w_i$ an *initial multiset of objects* and $R_i$ a set of *rewriting and communication rules*;
- $ch \subseteq \{(i, j)|i, j \in \{0, 1, \ldots, n\}, i \neq j\}$ is the set of *channels* between cells (those labeled 1 to $n$) and cells and environment (labeled by 0).

The rules occurring in the above definition are a bit different from the rewriting and communication rules introduced by Definition 1. In the case of neural-like network of membranes, the rules have the form $s_1 w \rightarrow s_2 x y_{go} z_{out}$, where $s_1$ and $s_2$ are states and $w, x, y, z$ are multisets of objects. When the rule, from cell $i$, is applied to a multiset $u$ containing $w$, then $w$ is extracted from $u$ and replaced by $x$, staying in cell $i$, whereas $y$ is sent to neighboring cells (cells $j$ such that $(i, j) \in ch$) and $z$ is removed from the system, being sent out into the environment.

**Remark 4** Another special case of a tissue-like P system is called *population P system*. In this case the membrane structure is changing all the time. Some special rules, called *bond making rules*, dynamically create communication channels. The format of them is $(i, x; y, j)$, where $i, j$ are membrane labels and $x$ and $y$ are multisets of objects from $i$ and $j$, respectively. When the rules is applied, if $i$ contains the multiset of objects $x$ and $j$ contains $y$ then a bond (a communication channel) is created between these cells.

Neuronal activities have inspired developments in membrane computing. The way spikes are processed and communicated represents the key computational aspect considered by these models. The formal definition of these computational models is given below.

**Definition 3** A spiking neural P system (SNP system, for short) of degree $n \geq 1$ is a tuple $\Pi = (O, \sigma_1, \ldots, \sigma_n, syn, in, out)$, where

- $O = \{a\}$ is the singleton alphabet ($a$ is called *spike*);
- $\sigma_1, \ldots, \sigma_n$ are *neurons*, of the form $\sigma_i = (m_i, R_i)$, $1 \leq i \leq n$, where

1. $m_i \geq 0$ is the *initial number of spikes* contained in $\sigma_i$;
2. $R_i$ is a finite set of *rules* of the following two forms:
   a. *firing* (also called *spiking*) rules, of the form $E/a^c \rightarrow a; d$, where $E$ is a regular expression over $a$, and $c \geq 1$, $d \geq 0$ are integer numbers. If $E = a^c$ then it is usually written in a simpler way $a^c \rightarrow a; d$. If $d = 0$ then it is omitted from the rule;
   b. *forgetting* rules, of the form $a^s \rightarrow \lambda$, $s \geq 1$, with the restriction that for each rule $E/a^c \rightarrow a; d \in R_i$, we have $a^s \notin L(E)$ (where $L(E)$ is the regular language defined by $E$);

- $syn \subseteq \{1, \ldots, n\} \times \{1, \ldots, n\}$, with $(i, i) \notin syn$, $1 \leq i \leq n$, is the directed graph of *synapses* between neurons;
- $in, out \in \{1, \ldots, n\}$ indicate the *input* and the *output* neurons of $\Pi$.

*Remark 5* An SNP system, as defined above, has the following behavior.

1. An *instantaneous description* or a *configuration* $C_t$, at an instant $t$ of a SNP system $\Pi$ is an $n$-tuple $(r_1/t_1, \ldots, r_n/t_n)$, where the neuron $\sigma_i$ contains $r_i \geq 0$ spikes at instant $t$ and will become available (open) after $t_i \geq 0$ steps, $1 \leq i \leq n$. A *computation* starts in the initial configuration, $(m_1/0, \ldots, m_n/0)$. In order to compute a function, a positive integer number is specified in an *input neuron* - this is used in the *acceptance mode*. The result of the computation is captured through the spikes emitted by the *output neurons* into the environment. Output neurons are used in the *generative mode*.

2. A firing rule $E/a^c \rightarrow a; d \in R_i$ is applicable to a configuration $C_t$ if $\sigma_i$ contains $k \geq c$ spikes in $C_t$ and $a^k \in L(E)$. The execution of the rule removes $c$ spikes from $\sigma_i$, leaving $k - c$ spikes in the cell, and sends one spike to every neuron $\sigma_j$ such that $(i, j) \in syn$. The spike is effectively delivered after $d$ computation steps, or immediately when $d = 0$. During the next $d$ computation steps the neuron cannot receive or send any spikes. The neuron is *closed* and does not interact in any way with the system; after that it will *open* again for further interactions.

3. An extended version for firing rules is considered by using more than a spike on the right hand side, i.e., rules of the type $E/a^c \rightarrow a^h; d \in R_i$, with $h \geq 1$.

4. A forgetting rule $a^s \rightarrow \lambda$ can be applied to a configuration $C_t$ if the cell contains exactly $s$ spikes in $C_t$ and no firing rules are applicable. By the execution of the rule $s$ spikes are deleted from $\sigma_i$.

5. In every computation step in each neuron one rule, from those applicable, is selected and executed. **Please note that this is different from the usual maximal parallelism** utilised by other P system variants.

We illustrate now how an SNP system is working in the case of a degree 3 system. Let the SNP system be given by the following tuple $\Pi_1 = (O, \sigma_1, \sigma_2, \sigma_3, syn, out)$, where

- $O = \{a\}$;
- $\sigma_1 = (1, \{r_{1,1} : a \rightarrow a; 1\})$;
- $\sigma_2 = (1, \{r_{2,1} : a \rightarrow a\})$;
- $\sigma_3 = (2, \{r_{3,1} : a^2 \rightarrow a, r_{3,2} : a \rightarrow \lambda\})$;
- $syn = \{(1, 2), (2, 1), (1, 3), (2, 3)\}$;
- $out = 3$.

The neuron $\sigma_1$ has one spike and a rule $r_{1,1}$ which consumes a spike and sends a spike to each of the $\sigma_2$ and $\sigma_3$ neurons, but only after 1 step and this closes $\sigma_1$ for a step. The neuron $\sigma_2$ has also one spike and a rule, $r_{2,1}$, similar to $\sigma_1$, but this one is immediately applied; when the rule is applied a spike is consumed and a spike is sent to $\sigma_1$, when it is open, and another spike to $\sigma_3$. The output neuron, $\sigma_3$, has two spikes and two rules; the first rule, $r_{3,1}$, consumes two spikes and sends one into the environment, whereas the second rule (a forgetting rule), $r_{3,2}$, deletes a spike when only one appears in $\sigma_3$.

The initial configuration of the system is $C_0 = (1/0, 1/0, 2/0)$. In this case the rules $r_{1,1}, r_{2,1}$ and $r_{3,1}$ are applied in $\sigma_1, \sigma_2$ and $\sigma_3$, respectively. As a result, one gets the configuration $C_1 = (0/1, 0/0, 1/0)$. Please note that $\sigma_1$ is closed and it does not send spikes to its connected neurons, nor does it receive any. In the next step the spikes from $\sigma_1$ are released to $\sigma_2$ and $\sigma_3$ and the forgetting rule $r_{3,2}$ is applied in $\sigma_3$, hence the configuration $C_2 = (0/0, 1/0, 1/0)$ is obtained. Now the rules $r_{2,1}$ and $r_{3,2}$ can be used in $\sigma_2$ and $\sigma_3$, respectively, and one gets $C_3 = (1/0, 0/0, 1/0)$. Now the rules $r_{1,1}$ and $r_{3,2}$ are applied and the configuration $C_4 = (0/1, 0/0, 0/0)$ is obtained. This configuration leads to $C_5 = C_2$. The process restarts from this configuration and continues in a periodic manner.

## 1.5   Summary

The definitions introduced in this chapter use the most basic membrane computing concepts. In the following chapters some more specific concepts will be added to these definitions in order to make them suitable for the problems we aim to model.

## References

1. Ciobanu, G., M.J. Pérez-Jiménez, and Gh. Păun (eds.). 2006. *Applications of Membrane Computing*. Natural Computing Series. Berlin: Springer.
2. Frisco, P., M. Gheorghe, and M.J. Pérez-Jiménez (eds.). 2014. *Applications of Membrane Computing in Systems and Synthetic Biology*. Emergence, Complexity and Computation Series. Berlin: Springer.
3. Frisco, P. 2009. *Computing with Cells - Advances in Membrane Computing*. Oxford: Oxford University Press.
4. Gheorghe, M., Gh. Păun, M.J. Pérez-Jiménez, and G. Rozenberg. 2013. Research Frontiers of Membrane Computing: Open Problems and Research Topics. *International Journal of Foundations of Computer Science* 24 (5): 547–624.
5. Manca, V. 2013. *Infobiotics - Information in Biotic Systems*. Emergence, Complexity and Computation Series. Berlin: Springer.
6. P Systems website. http://ppage.psystems.eu.
7. Păun, Gh. 2002. *Membrane Computing - An Introduction*. Berlin: Springer.
8. Păun, Gh, G. Rozenberg, and A. Salomaa (eds.). 2010. *The Oxford Handbook of Membrane Computing*. Oxford: Oxford University Press.
9. Zhang, G., J. Cheng, T. Wang, X. Wang, and J. Zhu. 2015. *Membrane Computing: Theory and Applications*. Beijing: Science Press.
10. Zhang, G., M. Gheorghe, L. Pan, and M.J. Pérez-Jiménez. 2014. Evolutionary Membrane Computing: A Comprehensive Survey and New Results. *Information Sciences* 279: 528–551.

# Chapter 2
# Fundamentals of Evolutionary Computation

**Abstract** The key evolutionary approaches used in the next chapters, including genetic algorithms, quantum-inspired evolutionary algorithms, ant colony optimization, particle swarm optimization and differential evolution are presented.

## 2.1 Introduction

Evolutionary algorithms (EAs) refer to a generic metaheuristic optimization algorithms characterized by implementations looking at a guided random search of an iterative process [49, 59, 111, 122]. EAs include a family of heuristic algorithms, called metaheuristics [10–12]. As a branch of soft computing referring to less exact calculations [115], EAs has become a well-known research area in computer science [82, 111].

Inspired by natural selection [26] and molecular genetics [18], EAs started with three research topics in the 1950s and 1960s: genetic algorithms (GAs) developed by Holland [58], evolution strategies (ES) invented by Rechenberg [102] and Schwefel [106, 107] and evolutionary programming (EP) introduced by Fogel et al. [40], and has a tremendous growth in the past three decades as witnessed by the increasing number of international conferences, workshops, papers, books and dissertations as well as more and more journals dedicated to the field. Historically, one can divide the EAs research into two groups: classic EAs and recently developed EAs. The former consists of GAs, ES, EP, and genetic programming (GP) [71–73], which were developed in the 1990s. The latter is still in a stage of rapid development and includes quantum-inspired evolutionary algorithms (QIEAs) [54, 119], simulated annealing (SA) [20, 70], ant colony optimization (ACO) [30], particle swarm optimization (PSO) [66, 90, 109], differential evolution (DE) [27, 98], estimation of distribution algorithms (EDAs) [74, 94], biogeography-based optimization (BBO) [110, 111], cultural algorithms (CA) [103], tabu search (TS) [48], artificial fish swarm algorithm (AFSA) [76, 87], artificial bee colony algorithm (ABC) [61], firefly algorithm (FA) [118], bacterial foraging optimization algorithm (BFOA) [93], teaching learning based optimization algorithm (TLBO) [100], shuffled frog leaping algorithm (SFL) [37].

© Springer International Publishing AG 2017
G. Zhang et al., *Real-life Applications with Membrane Computing*,
Emergence, Complexity and Computation 25, DOI 10.1007/978-3-319-55989-6_2

Underlying the different variants of EAs, there are several common features: a fundamental algorithm structure show in Algorithm 1 [2, 119], a single solution or a population of tentative solutions, guided random search by an evaluation function called fitness function, iterative progress toward better solutions to the problem [36, 111]. In Algorithm 1, an EA starts from a single or a population of candidate solution(s) $P(t)$ (also called individual(s)), where $t$ represents the number of evolutionary generations, to a problem, and then goes to an iterative search process, and finally stops when a single or a set of satisfactory solution(s) is found. The search process consists of the evaluation of candidate solution(s), the variation of individual(s) from $P(t)$ to $Q(t)$ by using various evolutionary mechanisms and the generation of the offspring individual(s) $P(t+1)$. Defining practical and robust optimization methodologies, EAs have shown outstanding characteristics, such as global search capabilities, flexibility, robust performance and adaptability, in the process of solving complex problems with combinations, discontinuities, constrains, multiple or many objectives, uncertainties or dynamics [1, 2, 36, 111]. So EAs have been increasingly widely applied to problems ranging from practical applications in industry and commerce to leading-edge scientific research [36, 60].

---

**Algorithm 1** EA Fundamental algorithm structure

**Require:** A single or a population of initial solution(s) $P(t), t = 0$
1: Evaluate $P(t)$
2: **while** (not termination condition) **do**
3:      $Q(t) \leftarrow$ Vary $P(t)$
4:      Evaluate $Q(t)$
5:      Produce $P(t+1)$ from $Q(t)$
6:      $t \leftarrow t + 1$
7: **end while**

---

This chapter is devoted to five EA variants, GAs, QIEAs, ACO, PSO and DE, which are the foundation of membrane algorithms and their engineering applications described in the next two chapters. Thus, the following sections focus on the presentation of fundamental concepts and principles, rather than on demonstrating experiments and results. One can note that the EAs discussed in this chapter are used to solve optimization problems.

## 2.2  Genetic Algorithms

As genetic algorithms (GAs) are the earliest, most well-known, and most widely used EAs, there are numerous publications describing them [2, 36, 68, 111]. The aim of this section is to briefly highlight the fundamental concepts and evolution principle so as to make it easy for readers to understand the following four EA variants, QIEAs, ACO, PSO and DE, and membrane algorithms in the next chapter because the notions, operations and procedure underlying GAs are the basics of other types of EAs.

Inspired by natural selection [26] and molecular genetics [18], Holland introduced GAs in the mid-1970s [58], on the basis of the influential works of Fraser [41], Box [14], Friedberg [42], Friedberg et al. [43] and Bremermann [16] in the late 1950s. In a GA, a potential or candidate solution to an optimization problem is called an *individual*; the encoding (binary, numeric, or others) of an individual is known as its *genome* or *chromosome*. A *population* is a set consisting of a certain number of individuals. A chromosome is composed of a sequence of *genes*; specific genes are known as *genotypes*, and the problem-specific parameter representing by a genotype is termed a *phenotype*; the value of a gene is called an *allele*. The individuals at the current generation are named *parents* and correspondingly the new individuals produced by them are called *children* or *offspring*. The function used to evaluate an individual is called *fitness function* and correspondingly the function value with respect to the individual is called its *fitness*, which indicates the quality of the solution in the context of a given problem. The whole process of searching for an optimal solution to a problem is called evolution [68].

Underlying various variants of GAs, there is a common algorithm structure shown in Algorithm 2, where each step is detailed as follows:

---

**Algorithm 2** GA Algorithm structure

---

**Require:** A group of random generated initial solutions $P(t), t = 0$
1: Evaluate $P(t)$
2: **while** (not termination condition) **do**
3:     Select individuals from $P(t)$ to form $Q_1(t)$ for crossover
4:     Crossover $Q_1(t)$ to form $Q_2(t)$
5:     Mutate $Q_2(t)$ to form $Q_3(t)$
6:     Evaluate $Q_3(t)$
7:     $P(t+1) \leftarrow Q_3(t)$
8:     $t \leftarrow t + 1$
9: **end while**

---

Step 0: An initial population $P(t), t = 0$, consisting of a certain number of individuals is randomly generated. Each individual is composed of a sequence of codes such as binary, numeric and permutation codes.

Step 1: Each individual in $P(t)$ is evaluated by using the fitness function associated with the optimization problem. Thus, each individual has assigned a fitness value.

Step 2: The termination criterion may be the prescribed maximal number of evolutionary generations or the preset difference between the best solution searched and the optimal/desired solution of the optimization problem.

Step 3: Determine each pair of individuals in $P(t)$ to perform crossover operation. As usual the roulette-wheel selection with respect to fitness, which is also called fitness-proportional selection or fitness-proportionate selection, is used. The selected population is represented as $Q_1(t)$.

Step 4: Swap partial genes of each pair of selected individuals in $Q_1(t)$ with each other by a probabilistic value called crossover probability. The crossovered population is denoted as $Q_2(t)$. It is well known that the crossover probability should be assigned a bigger value.

Step 5: Each gene of each individual in $Q_2(t)$ is mutated by a probabilistic value called mutation probability, which is usually set to a smaller value. The mutated population is denoted as $Q_3(t)$. The uniform mutation is a popular approach.

Step 6: This step is similar to Step 1, i.e., each individual in $Q_3(t)$ is evaluated.

Step 7: The individuals in $Q_3(t)$ are assigned to $P(t+1)$ as the offspring individuals.

Step 8: The evolutionary generation $t$ increases by 1.

The GA research was mainly developed in the past decades with respect to encoding techniques, selection, crossover operators, mutation operators, fitness functions, hybridization with other techniques and theoretical analysis. As usual the individual representation, selection methods, crossover and mutation operators, and fitness functions depend on the optimization problem. The individuals in GAs could be represented by using various types of codes, such as binary and m-ary codes, numeric values, permutation codes and quantum-inspired bits. The analysis of various representations can be found in [13, 54, 95, 104]. Most of the researches on GAs are related to the modification of selection, crossover and mutation operators. So numerous variants of selection, crossover and mutation operators and their effect on the GA performance were reported in literature [15, 44, 55, 62, 63, 81, 89, 92]. Moreover, the influence of crossover and mutation probabilities were also investigated [3, 4, 79]. To select a suitable fitness function to a real-world application problem is also an important issue. In [60], a comprehensive survey of the research on fitness approximation in GAs was reviewed with respect to approximation levels, approximate model management schemes and model construction techniques. Recent research on GAs principally focused on the hybridization with other techniques such as tabu search, simulated annealing, quantum computing, rough set, fuzzy logic theory and other types of EAs. These investigations mentioned above are more concerned with the question of *whether* GAs work. Actually, the theoretical analysis of GAs answers satisfactorily the questions of *how* or *why* GAs work, which are important and challenging issues in the further advance of GAs, even of EAs [111]. Some methods like schema theory, Markov models and Fourier and Walsh transforms have been applied to analyze the GAs behavior [2, 59, 111].

## 2.3  Quantum-Inspired Evolutionary Algorithms

The past three decades have witnessed the use of various properties from quantum physics to devise a new kind of computers, quantum computers [46, 88]. In contrast with classical computers processing binary digits (*bits*), quantum computers work by handling quantum bits (*qubits*), which are the smallest information units that can be stored in a two-state quantum computer [56]. A qubit can be in a superposition of the usual '0' and '1' states other than themselves. Thus, a quantum particle could simultaneously be in many incompatible states [88]. Each superposition, $|\psi\rangle$ can be represented as a linear sum of the basis states, $|\psi\rangle = \alpha|0\rangle + \beta|1\rangle$, where $\alpha$ and $\beta$ are numbers that denote the corresponding states' probability amplitudes. The values $|\alpha|^2$ and $|\beta|^2$ are the probabilities that the observation of a qubit in state $|\psi\rangle$ will render a '0' or '1' state, respectively [47], and normalization property requires that $|\alpha|^2 + |\beta|^2 = 1$. A quantum gate can be used to modify the state of a qubit [56]. A quantum system $|\psi_n\rangle$ with $n$ qubits can represent $2^n$ states simultaneously [5, 50] as

$$|\psi_n\rangle = \sum_{j=1}^{2^n} C_j |S_j\rangle, \tag{2.1}$$

where $C_j$ is the probability amplitude of the $j$th state $S_j$ described by the binary string $(x_1 x_2 \cdots x_n)$, where $x_i$, $i = 1, 2, \cdots, n$, is either 0 or 1. Nonetheless, the system will "collapse" to a single state if a quantum state is observed.

Inspired by quantum computing, a computational method called quantum-inspired computation is designed to solve various problems in the context of a classical computing paradigm [83]. Amongst the quantum-inspired computation topics, a quantum-inspired evolutionary algorithms (QIEA) is receiving renewed attention. A QIEA is a novel EA *for a classical computer* rather than for a quantum machine (or computer). Generally speaking, a QIEA is designed by integrating the EA framework with quantum-inspired bits (Q-bits), quantum-inspired gates (Q-gates) and probabilistic observation.

Conventional EAs use several different representations to encode solutions onto chromosomes, such as symbolic, binary, and numeric representations [57]. While in a QIEA, a novel probabilistic description, Q-bit representation, of Q-bit individuals is used. A Q-bit individual is represented as a string of Q-bits. The basic computing unit in a QIEA, Q-bit, is defined as a column vector

$$[\alpha \ \beta]^T, \tag{2.2}$$

where $\alpha$ and $\beta$ are real numbers satisfying the normalization condition $|\alpha|^2 + |\beta|^2 = 1$. Equation (2.2) is usually written as $\alpha|0\rangle + \beta|1\rangle$ in quantum mechanics ket-notation. The values $|\alpha|^2$ and $|\beta|^2$ are the probabilities that the Q-bit will be found in the '0' or '1' state, respectively, in quantum theory [54]. By using a probabilistic observation, each Q-bit can be rendered into one binary bit. Algorithm 3 shows the observation process, where $x$ is the observed value of the Q-bit shown in (2.2). Differing from the binary representation that uses 0 or 1 to deterministically represent a bit, the Q-bit representation uses a Q-bit to describe a probabilistic linear superposition of 0 and 1. The Q-bit representation can be easily extended to multi-Q-bit systems.

---

**Algorithm 3** Observation process in the QIEA [54]

---

**Require:** A Q-bit $[\alpha \quad \beta]$
1: **if** $random[0, 1) < |\alpha|^2$ **then**
2:     $x \leftarrow 0$
3: **else**
4:     $x \leftarrow 1$
5: **end if**

---

In what follows, the QIEA in [54] is taken as an example to detail the QIEA algorithm. Algorithm 4 shows the pseudocode QIEA algorithm, where each step is described below.

---

**Algorithm 4** Pseudocode algorithm of the QIEA in [54]

---

**Require:** An initial population $Q(t)$, $t = 0$
1: Make $P(t)$ by observing the states of $Q(t)$
2: Evaluate $P(t)$
3: Store all solutions in $P(t)$, into $B(t)$ and the best solution **b** in $B(t)$
4: **while** (not termination condition) **do**
5:     $t \leftarrow t + 1$
6:     Make $P(t)$ by observing the states of $Q(t - 1)$
7:     Evaluate $P(t)$
8:     Update $Q(t)$ using Q-gates
9:     Store all solutions in $P(t)$, into $B(t - 1)$ and the best solution **b** in $B(t)$
10:     **if** (migration condition) **then**
11:         Migrate **b** or $\mathbf{b}'_j$ to $B(t)$ globally or locally, respectively
12:     **end if**
13: **end while**

---

Step 0: In the step of "initialize $Q(t)$", a population $Q(0)$ with $n$ multi-Q-bit individuals is produced, $Q(t) = \{\boldsymbol{q}_1^t, \boldsymbol{q}_2^t, \cdots, \boldsymbol{q}_n^t\}$, at the generation moment $t = 0$, where $\boldsymbol{q}_i^t$ ($i = 1, 2, \cdots, n$) is an arbitrary individual in $Q(t)$, denoted as

$$\boldsymbol{q}_i^t = \begin{bmatrix} \alpha_{i1}^t | \alpha_{i2}^t | \cdots | \alpha_{im}^t \\ \beta_{i1}^t | \beta_{i2}^t | \cdots | \beta_{im}^t \end{bmatrix}, \tag{2.3}$$

where $m$ is the string length of the Q-bit individual, that is, the number of Q-bits used in each individual's representation. The values $\alpha_{ij}^t$ and $\beta_{ij}^t$, $j = 1, 2, \cdots, m$, $t = 0$, are initialized by the same probability amplitude $1/\sqrt{2}$, which guarantees that all possible states are superposed with the same probability at the beginning.

Step 1: By independently observing each Q-bit of $Q(t)$ (where at this stage $t = 0$), using the process described in Algorithm 3, binary solutions in $P(t)$, $P(t) = \{\boldsymbol{x}_1^t, \boldsymbol{x}_2^t, \cdots, \boldsymbol{x}_n^t\}$, are obtained, where each $\boldsymbol{x}_i^t$ ($i = 1, 2, \cdots, n$) is a binary solution with $m$ bits. Each bit '0' or '1' is the observed value of a Q-bit $[\alpha_{ij}^t \quad \beta_{ij}^t]^T$ in $\boldsymbol{q}_i^t$, respectively, $j = 1, 2, \cdots, m$.

Step 2: The binary solution $x_i^t$ $(i = 1, 2, \cdots, n)$ in $P(t)$ is evaluated thus obtaining its fitness.

Step 3: In this step, all solutions in $P(t)$ are stored into $B(t)$, where $B(t) = \{b_1^t, b_2^t, \cdots, b_n^t\}$ and $b_i^t = x_i^t$ $(i = 1, 2, \cdots, n)$ (again, at this stage, $t = 0$). Furthermore, the best binary solution $b$ in $B(t)$ is also stored.

Step 4: The termination criterion may be the prescribed maximal number of evolutionary generations or the preset difference between the best solution searched and the optimal/desired solution of the optimization problem.

Step 5: The evolutionary generation $t$ increases by 1.

Step 6: This step is similar to Step 1. Observation of the states of $Q(t-1)$ produces the binary solutions in $P(t)$.

Step 7: This step is similar to Step 2.

Step 8: In this step, all the individuals in $Q(t)$ are modified by applying Q-gates. The QIEA uses a *quantum rotation gate* as a Q-gate. To be specific, the $j$th Q-bit in the $i$th Q-bit individual $q_i^t$, $j = 1, 2, \cdots, m$, $i = 1, 2, \cdots, n$, is updated by applying the current Q-gate $G_{ij}^t(\theta)$

$$G_{ij}^t(\theta) = \begin{bmatrix} \cos\theta_{ij}^t & -\sin\theta_{ij}^t \\ \sin\theta_{ij}^t & \cos\theta_{ij}^t \end{bmatrix}, \tag{2.4}$$

where $\theta_{ij}^t$ is an adjustable Q-gate rotation angle. Thus, the update procedure for the Q-bit $[\alpha_{ij}^t \ \beta_{ij}^t]^T$ can be described as

$$\begin{bmatrix} \alpha_{ij}^{t+1} \\ \beta_{ij}^{t+1} \end{bmatrix} = G_{ij}^t(\theta) \begin{bmatrix} \alpha_{ij}^t \\ \beta_{ij}^t \end{bmatrix}, \tag{2.5}$$

where $\theta_{ij}^t$ is defined as

$$\theta_{ij}^t = s(\alpha_{ij}^t, \beta_{ij}^t)\Delta\theta_{ij}^t, \tag{2.6}$$

and $s(\alpha_{ij}^t, \beta_{ij}^t)$ and $\Delta\theta_{ij}^t$ are the sign and the value of $\theta_{ij}^t$, respectively. The particular values used in the QIEA in [54] are illustrated in Table 2.1, in which $f(\cdot)$ is the fitness function, $s(\alpha_{ij}^t, \beta_{ij}^t)$ depends on the sign of $\alpha_{ij}^t\beta_{ij}^t$, and $b$ and $x$ are certain bits of the searched best solution $b$ and the current solution $x$, respectively. It is worth pointing out that Table 2.1 was derived from a maximization problem and hence the condition $f(x) \geq f(b)$ should be replaced by $f(x) \leq f(b)$ if a minimization problem is to be considered.

Step 9: This step is similar to Step 3. The better candidate between $x_i^t$ in $P(t)$ and $b_i^{t-1}$ in $B(t-1)$, $i = 1, 2, \cdots, n$, is selected and stored into $B(t)$. Simultaneously, the best candidate $b$ in $B(t)$ is also stored.

Steps 10–11: This step includes local and global migrations, where a *migration* in this algorithm is defined as the process of copying $b_j^t$ in $B(t)$ or $b$ to $B(t)$. A global migration is realized by substituting $b$ for all the solutions in $B(t)$, and a local migration is realized between each pair of neighboring solutions in $B(t)$, i.e., by

**Table 2.1** Lookup table of $\theta_{ij}^t$, where $f(\cdot)$ is the fitness, $s(\alpha_{ij}^t, \beta_{ij}^t)$ is the sign of $\theta_{ij}^t$, and $b$ and $x$ are certain bits of the searched best solution $b$ and the current solution $x$, respectively [54]

| $x$ | $b$ | $f(x) \geq f(b)$ | $\Delta\theta_{ij}^t$ | $s(\alpha_{ij}^t, \beta_{ij}^t)$ | |
|---|---|---|---|---|---|
| | | | | $\alpha_{ij}^t \beta_{ij}^t \geq 0$ | $\alpha_{ij}^t \beta_{ij}^t < 0$ |
| 0 | 0 | false | 0 | $\pm 1$ | $\pm 1$ |
| 0 | 0 | true | 0 | $\pm 1$ | $\pm 1$ |
| 0 | 1 | false | $0.01\pi$ | $+1$ | $-1$ |
| 0 | 1 | true | 0 | $\pm 1$ | $\pm 1$ |
| 1 | 0 | false | $0.01\pi$ | $-1$ | $+1$ |
| 1 | 0 | true | 0 | $\pm 1$ | $\pm 1$ |
| 1 | 1 | false | 0 | $\pm 1$ | $\pm 1$ |
| 1 | 1 | true | 0 | $\pm 1$ | $\pm 1$ |

substituting the better one of two neighboring solutions for the other solution. For more information about the migrations, see [54].

In summary, in QIEA, Q-bits are applied to represent genotype individuals; Q-gates are employed to operate on Q-bits to generate offspring; and the genotypes and phenotypes are linked by a probabilistic observation process. QIEAs were firstly introduced by Narayanan and Moore in the 1990s to solve the traveling salesman problem [84], in which the crossover operation was performed based on the concept of interference. The contribution of Narayanan and Moore signaled the potential advantage of introducing quantum computational parallelism into the evolutionary algorithm framework. No further attention was paid to QIEAs until a practical algorithm was proposed by [53, 54], but they are now viewed as an emergent theme in evolutionary computation. In the last sixteen years have been considered various variants of QIEAs to solve a large number of problems (for a comprehensive survey see [119]). The main characteristics of QIEAs can be summarized as follows:

- A QIEA uses a novel representation, *Q-bit representation*, to describe individuals of a population. Q-bit representation provides probabilistically a linear superposition of multiple states.
- A QIEA employs a *Q-gate* guiding the individuals toward better solutions [54] to produce the individuals at the next generation.
- A QIEA can exploit the search space for a global solution with a small number of individuals, even with one element [54].

Currently there is intensive research in this area, but there are some aspects that need to be addressed from the perspectives of theoretical research, engineering applications, comparative experiments, extensions of QIEAs and hybrid algorithms. These issues were presented in detail in [119].

## 2.4 Ant Colony Optimization

Instead of simulating the process of natural selection, some researchers introduced novel algorithms by simulating the collective behavior of decentralized, self-organized colonies. Ant colony optimization (ACO), originally proposed by Dorigo and co-workers in 1991 [33] and later explicitly defined in [32], is such a meta-heuristic approach for combinatorial optimization problem inspired by the foraging behavior of ants. In nature, to find the shortest path from the nest to a food source, ant colonies exploit a positive feedback mechanism by laying and detecting the chemical trail (pheromone) on the ground during their trips. More pheromone is left when more ants go through the trip, which improves the probability of other ants choosing this trip. Furthermore, the pheromone has a decreasing action over time because of evaporation of trail. In the ACO metaphor, a generic combinatorial optimization problem is transformed into a shortest path problem which is encoded as a graph; a number of paths are constructed by artificial ants walking on the graph based on a probabilistic model using pheromone; the cost of the generated path is utilized to modify the pheromone, and hence to bias the generation of further paths.

ACO was initially applied to solve traveling salesman problem (TSP) [32], one of the well-known NP-complete problems and most intensively studied combinatorial optimization problems in the areas of optimization, operational research, theoretical computer science, and computational mathematics. The TSP can be described as follows [121]. Given a set $C$ of $N$ cities, i.e., $C = \{c_1, c_2, \cdots, c_N\}$, and a set $D$ of the pairwise travel costs, $D = \{d_{ij} | i, j \in \{1, 2, \cdots, N\}, i \neq j\}$, it is requested to find the minimal cost of the path taken by a salesman visiting each of the cities just once and returning to the starting point. More generally, the task is to find a Hamiltonian tour with a minimal length in a connected, directed graph with a positive weight associated to each edge. If $d_{ij} = d_{ji}$, the TSP is symmetric in the sense that traveling from city $c_i$ to city $c_j$ costs just as much as traveling in the opposite direction, otherwise, it is asymmetric. This section uses symmetric TSP as an example to describe ACO.

ACO is an iterative metaheuristic. At each iteration, a number of paths are constructed based on stochastic decisions which are biased by pheromone and heuristic information. These paths are used for updating the pheromone in order to bias further solutions towards promising regions of the search space. Algorithm 5 gives the pseudocode of a generic ACO algorithm. In the pseudocode, a local search procedure may be applied for further improving the solutions constructed by ants. The use of such a procedure is optional; however, it has been observed that its use improves the algorithms's overall performance. The most used and well-known tour improvement local searches are 2-opt and 3-opt [69], in which two and three edges of a tour are exchanged, respectively.

---

**Algorithm 5** Pseudocode of a generic ACO

---
**Require:** $t = 0$
1: Pheromone trail initialization
2: **while** (not termination condition) **do**
3:     Construct tours
4:     Apply local search (*optional*)
5:     Update pheromone
6:     $t \leftarrow t + 1$
7: **end while**

---

The most well-known ACO algorithms in literature include the earliest ant system (AS) [33, 34], MAX-MIN ant system [114], hyper-cube ant system [9], and ant colony system (ACS) [29], and they differ in the way to construct tours and/or update pheromone. According to the studies in [8, 28, 31], the ACS is one of the most powerful ACO algorithms. Therefore, we take it as an example to describe the ACO algorithm. Algorithm 6 shows the pseudocode of an ACS algorithm, where each step is described below.

---

**Algorithm 6** Pseudocode algorithm of the ACS in [29]

---
**Require:** $t = 0$
1: Pheromone trail initialization
2: **while** (not termination condition) **do**
3:     Randomly place $M$ ants in the $N$ nodes
4:     **for** $k = 1, 2, \ldots, M$ **do**
5:         **for** $n = 1, 2, \ldots, N$ **do**
6:             Ants moving
7:         **end for**
8:         Evaluate the length of the path construct by ant $k$
9:         Local pheromone updating
10:    **end for**
11:    Global pheromone updating
12:    $t \leftarrow t + 1$
13: **end while**

---

Step 1: At the beginning of a run, the initial pheromone value $\tau_0$ is set to be $1/ND_a$, where $N$ is the number of cities in a TSP and $D_a$ is the length of a feasible tour generated randomly or by the nearest-neighbor heuristic.

Step 2: The termination criterion may be the prescribed maximal number of generations or the preset difference between the best path searched and the optimal/desired path of the problem.

Step 3: The $M$ ants are randomly positioned on the $N$ nodes of the TSP graph as the initial state of tour construction.

Step 4: The ants construct paths one by one.

Steps 5–7: Each ant constructs a whole path step by step using a pseudorandom proportional rule. Specifically, the $k$th ant in the $i$th city chooses the next city $j$ by using the following formula

$$j = \begin{cases} \arg\max_{l \in \mathcal{N}_i^k}\{[\tau_{il}]^\alpha [\eta_{il}]^\beta\}, & \text{if } q \leq q_0 \\ J, & \text{otherwise} \end{cases} \tag{2.7}$$

where $\arg\max\{\cdot\}$ stands for the argument of the maximum, that is to say, the set of points of the given argument for which the value of the given expression attains its maximum value; $\tau_{il}$ is the pheromone value of the edge connecting the $i$th node and the $l$th node; $\eta_{il}$ is a heuristic information value, equal to the inverse of the distance between the $i$th and $l$th cities; the parameters $\alpha$ and $\beta$ ($\alpha > 0$ and $\beta > 0$) determine the relative importance of the pheromone value $\tau_{il}$ and the heuristic information $\eta_{il}$; $\mathcal{N}_i^k$ ($\mathcal{N}_i^k \subseteq \mathcal{N}$) is the set of all nodes of the TSP graph that the $k$th ant in the $i$th city can visit; $q_0$ ($0 \leq q_0 \leq 1$) is a user-defined parameter specifying the distribution ratio of the two choices; $q$ is a random number generated by using a uniform distribution function in the interval $[0, 1]$; $J$ means that the next city $j$ is chosen by using a random proportional rule, i.e., the $k$th ant in the $i$th city visits the city $j$ at the next step according to the probability

$$p_{ij}^k = \begin{cases} \dfrac{[\tau_{ij}]^\alpha [\eta_{ij}]^\beta}{\sum_{l \in \mathcal{N}_i^k} [\tau_{il}]^\alpha [\eta_{il}]^\beta}, & j \in \mathcal{N}_i^k \\ 0, & \text{otherwise} \end{cases} \tag{2.8}$$

Step 8: At each time an ant construct a whole path, the length of this path is evaluated and compared with the best path stored. If the new path is better than the stored best path, the best path is updated.

Step 9: Ant releases a mount of pheromone on edges at its every traveling when it completes a path construction procedure. In ACS, an ant updates the pheromone value $\tau_{ij}$ of the tour by applying a local pheromone update rule, defined as follows

$$\tau_{ij} = (1 - \upsilon)\tau_{ij} + \upsilon\tau_0 \tag{2.9}$$

where $\upsilon$ ($0 < \upsilon < 1$) is a local pheromone decay coefficient. The local pheromone update is used to encourage subsequent ants to choose other edges and, hence, to produce different solutions, by decreasing the pheromone value on the traversed edges.

Step 11: In this step, the globally best ant, i.e., the ant which constructs the shortest tour form the beginning of the trial, is allowed to deposit additional pheromone via a global pheromone update rule. To be specific, the pheromone value $\tau_{ij}$ of the edge connecting the $i$th node and the $j$th node is modified by

$$\tau_{ij} = (1 - \rho)\tau_{ij} + \rho\Delta\tau_{ij} \tag{2.10}$$

where $\rho$ ($0 < \rho \leq 1$) is a global pheromone decay coefficient which is also called pheromone evaporation rate, and $\Delta\tau_{ij}$ is

$$\Delta\tau_{ij} = \begin{cases} 1/D_b, & \text{if } (i, j) \in T_b \\ 0, & \text{otherwise} \end{cases} \tag{2.11}$$

where $D_b$ is the length of the shorted path searched so far, and $T_b$ is the path corresponding to $D_b$.

Step 12: The iteration counter $t$ increases by 1.

At present, the research of ACO focuses on three main aspects, i.e., improvement of different ACO algorithms, applications, and theoretic analysis. Regarding the performance improvement, researchers proposed a large variety of ACO variants by designing new path construct schemes, pheromone update schemes, mixing with various local search operators, or even incorporating novel mechanism like chaos [75]. As for applications, although ACO was originally introduced in connection to TSP, it is now recognized as one of the state-of-the-art methods for solving other kinds of discrete optimization problems, such as assignment problems, scheduling problems, graph coloring, vehicle routing problems, design of communication networks. Furthermore, in recent years, some researchers have extended its use for continuous optimization problems, multi-objective discrete problems and dynamic problems. Since experimental results show better performance of ACO over other meta-heuristics, researchers have paid much attention to the ACO theory to explain why and how it works. The first convergence proof of ACO was given in [51]. Since then various convergence proofs for various ACO variants have been published, e.g. [19, 30, 52]. For more details of the progress of ACO, the readers can refer to the comprehensive survey papers [6, 28].

## 2.5 Particle Swarm Optimization

Particle swarm optimization (PSO) is another well-known population-based meta-heuristic approach proposed by Kennedy and Eberhart in 1995 for continuous optimization problems [65]. This technique was motivated by social behavior of bird flock. In PSO, each individual is called a "particle" with properties being described by the current position vector, its velocity vector and its personal best position vector, which represents a potential solution to a problem. Instead of using genetic operators (e.g., crossover, mutation) to evolve individuals, the trajectory of each particle is adjusted by dynamically altering its velocity according to its own flying experience and its companion's experience.

Suppose there are $N$ particles in a PSO, and each particle is treated as a point in a $D$-dimensional space, representing a candidate solution to the problem. Each particle is characterized by the current position vector $\mathbf{x}_i = (x_{i,1}, x_{i,2}, \ldots, x_{i,D})$, velocity vector $\mathbf{v}_i = (v_{i,1}, v_{i,2}, \ldots, v_{i,D})$ and its personal best position vector $\mathbf{p}_i = (p_{i,1}, p_{i,2}, \ldots, p_{i,D})$, $i = 1, 2, \ldots, N$. The particle with its personal best position which returns the best fitness value among the population is called the global best particle and its position is recorded as $\mathbf{p}_g = (p_{g,1}, p_{g,2}, \ldots, p_{g,D})$, where $g$ is the index of the global best particle. Algorithm 7 shows the pseudocode PSO algorithm, where each step is described below [25, 77, 90, 116, 123].

---

**Algorithm 7** Pseudocode algorithm of the PSO

---

**Require:** An initial population of $N$ particles with positions $P(t)$ and velocities $V(t), t = 0$
1: Evaluate the particles
2: Initialize personal best and global best
3: **while** (not termination condition) **do**
4:     **for** $i = 1, 2, \ldots, N$ **do**
5:         Change the velocity and position
6:         Evaluate the particle
7:         Update personal and global best positions
8:     **end for**
9:     $t \leftarrow t + 1$
10: **end while**

---

Step 0: In this step, uniform distribution on $[x_j^{min}, x_j^{max}]$ ($j = 1, 2, \ldots, D$) in the $j$th dimension is used to generate the initial current position vector $\mathbf{x}_i$ for the $i$th particle, where $x_j^{min}$ and $x_j^{max}$ are lower limit and upper limit of particle positions in the $j$th dimension. Similarly, the initial velocity vector $\mathbf{v}_i$ is initialized by choosing its $j$th component randomly in $[-v_j^{max}, v_j^{max}]$ ($j = 1, 2, \ldots, D$), where $v_j^{max}$ is the upper limit of velocities in the $j$th dimension. $v_j^{max}$ is an important parameter that determines the search behavior of the algorithm. If $v_j^{max}$ is too small, particles may become trapped in local optima, unable to move far away to a better position. On the other hand, if $v_j^{max}$ is too large, particles might fly past good solutions.

Step 1: The performance of each particle is measured according to a pre-defined fitness function.

Step 2: For each particle, set its personal best position as the current position, i.e., $\mathbf{p}_i = \mathbf{x}_i, i = 1, 2, \ldots, N$. Also, identify the global best position $\mathbf{p}_g$ based on the fitness value of the particles.

Step 3: The termination criterion may be the prescribed maximal number of generations or the preset difference between the best solution searched and the optimal/desired solution of the problem.

Step 4: The particle flies one by one.

Step 5: The velocity and position of the $i$th particle are updated according to the following equation,

$$\mathbf{v}_i = \mathbf{v}_i + c_1 r_1 (\mathbf{p}_i - \mathbf{x}_i) + c_2 r_2 (\mathbf{p}_g - \mathbf{x}_i) \tag{2.12}$$

$$\mathbf{x}_i = \mathbf{x}_i + \mathbf{v}_i \tag{2.13}$$

where $c_1$ and $c_2$ are acceleration coefficients, $r_1$ and $r_2$ are two different sequences of random numbers uniformly distributed over $(0, 1)$. In (2.12), the first part represents the previous velocity, which provides the necessary momentum for particles to roam across the search space; the second part is the "cognition" part, which represents the private thinking of the particle itself; the third part is the "social" part, which represents the collaboration among the particles in finding the global optimal solution. Equation (2.12) is used to calculate the particle's new velocity and the particle flies toward a new position according to (2.13). In this step, if the particle's velocity on $j$th dimension exceeds the maximum value $v_j^{max}$, then it is clamped to $v_j^{max}$.

Step 6: The performance of the particle is measured according to a pre-defined fitness function.

Step 7: Comparing particle's fitness with its personal best performance. If current value is better than its personal best fitness, then update its personal best fitness as the current fitness and set $\mathbf{p}_i = \mathbf{x}_i$. Also, comparing particle's fitness with the population's overall previous best. If current value is better that the previous best value, then update the global best fitness as the current value and set $\mathbf{p}_g = \mathbf{x}_i$.

Step 9: The iteration counter $t$ increases by 1.

The original PSO has been found performing well in solving some simple problems, however, its performance is not satisfactory when solving complex problems. Therefore, a considerable amount of work has been done in developing the original PSO. For example, in [108], to reduce the importance of $v_j^{max}$, Shi and Eberhart introduced the concept of inertia weight in the calculation of velocities to balance the local and global search, and later they further improved the algorithm performance with a linearly varying inertia weigh over the iterations. In [101], time-varying acceleration coefficients are introduced to control the local search and the convergence to the global optimum solution. In fully informed particle swarm algorithm [80], the particle is affected by all its neighbors, sometimes with no influence from its own previous success. In [77], a novel learning strategy whereby all other particles' historical best information is proposed to update a particle's velocity, which enables the diversity of the swarm to be preserved to discourage premature convergence. Instead of moving toward a kind of stochastic average of personal best position and global best position, particles moving toward points defined by personal best position and local best position is also widely investigated, where the best position is the location of the particle's neighborhood defined by a certain topology. Currently, various topologies have been studied, such as simple ring lattice, small-world modifications [64, 117], or von Neumann structure [67]. Some theoretical analysis for PSO approaches has been developed. For example, in [25], Clerc and Kennedy analyzed a particles trajectory as it moves in discrete time from the algebraic view and in continuous time from the analytical view. In [24], Clerc analyzed the distribution of velocities of a particle in order to observe algorithm behavior in stagnation phases. As for applications, PSO has been applied across various areas, such as classification, pattern recognition, planning, signal processing, power system, controller design. For more information of the important work in PSO, the readers can refer to the survey paper [96].

## 2.6 Differential Evolution

Differential Evolution (DE) is a meta-heuristic approach originally proposed by Storn and Price in 1996 for handling continuous optimization problems [112, 113]. Rather than using natural selection or colony collective behavior, DE relies on engineering aspects. In 1994, Price published a genetic annealing algorithm [97], which is a population-based, combinatorial optimization algorithm that implements an annealing criterion via thresholds. Later, Genetic Annealing has been used to solve Chebyshev polynomial fitting problem. As the performance of genetic annealing was not very satisfactory because of its slow convergence and difficulties to set effective control parameters, Price introduced floating-point encoding, arithmetic operations and differential mutation operator in genetic annealing algorithm. As a result, these alterations transformed the combinatorial algorithm genetic annealing into a numerical optimizer, which becomes the first generation of DE. Due to its distinguished characteristics, such as few control parameters, simple and straightforward implementation, remarkable performance and low complexity, DE [27, 86] has been recognized as a competitive continuous optimization technique.

Similar to GA or PSO, DE also maintains a population during the evolution. Let $P(t) = \{\mathbf{x}_1^t, \mathbf{x}_2^t, \ldots, \mathbf{x}_N^t\}$ be the population at the $t$th iteration, and $\mathbf{x}_i^t = (x_{i,1}^t, x_{i,2}^t, \ldots, x_{i,D}^t)$ $(i = 1, 2, \ldots, N)$ be the $i$th individual in $P(t)$ that represents a potential solution to the problem, where $N$ is the population size and $D$ is the number of decision variables of the problem. Starting with an initial population $P(t)(t = 0)$, the optimization process involves three basic steps, i.e., mutation, crossover and selection. Algorithm 8 shows the pseudocode of a basic DE, where each step is described below [21–23, 86].

---

**Algorithm 8** Pseudocode algorithm of the basic DE

---

**Require:** An initial population $P(t), t = 0$
1: Evaluate the population $P(t)$
2: **while** (not termination condition) **do**
3:     Mutate to form $V(t)$
4:     Crossover to form $U(t)$
5:     Evaluate $U(t)$
6:     Selection to form $P(t + 1)$
7:     $t \leftarrow t + 1$
8: **end while**

---

Step 0: The initial population $P(0) = \{\mathbf{x}_1^0, \mathbf{x}_2^0, \ldots, \mathbf{x}_N^0\}$ is produced, where each component of an individual is uniformly and randomly sampled in the feasible space, that is,

$$x_{i,j}^0 = x_j^{min} + rand(0, 1) \cdot (x_j^{max} - x_j^{min}), \qquad (2.14)$$

where $i = 1, 2, \ldots, N$; $j = 1, 2, \ldots, D$; $rand(0, 1)$ is a uniformly distributed random variable within the interval $[0,1]$, $x_j^{min}$ and $x_j^{max}$ are the lower and upper bound of the $j$th decision variable.

Step 1: The performance of each individual is evaluated according to a pre-defined fitness function.

Step 2: The termination criterion may be the prescribed maximal number of iterations or the preset difference between the best solution searched and the optimal/desired solution of the problem.

Step 3: The mutation operator is performed on $\mathbf{x}_i^t$ (called target vector) ($i = 1, 2, \ldots, N$) to create a mutant vector $\mathbf{v}_i^t$ (called donor vector) by perturbing a randomly selected vector $\mathbf{x}_{r_1}^t$ with the difference of two other randomly selected vector $\mathbf{x}_{r_2}^t$ and $\mathbf{x}_{r_3}^t$. This operation is formulated as

$$\mathbf{v}_i^t = \mathbf{x}_{r_1}^t + F \cdot (\mathbf{x}_{r_2}^t - \mathbf{x}_{r_3}^t) \tag{2.15}$$

where $\mathbf{x}_{r_1}^t$, $\mathbf{x}_{r_2}^t$ and $\mathbf{x}_{r_3}^t$ are distinct vectors randomly selected from the current population $P(t)$, and they are selected a new for each mutation operation. $F \in (0, 1)$ is a constant called differential factor, which scales the differential vector $(\mathbf{x}_{r_2}^t - \mathbf{x}_{r_3}^t)$ added to the base vector $\mathbf{x}_{r_1}^t$.

Step 4: Following mutation, the donor individual $\mathbf{v}_i^t$ ($i = 1, 2, \ldots, N$) is recombined with the target individual $\mathbf{x}_i^t$ to produce an offspring $\mathbf{u}_i^t$ (called trial vector) by using a binomial crossover operator, which is a typical case of genes' exchange, formulated as

$$u_{i,j}^t = \begin{cases} v_{i,j}^t, & \text{if } rand_j(0, 1) \leq Cr \text{ or } j = j_{rand} \\ x_{i,j}^t, & \text{otherwise} \end{cases}, \tag{2.16}$$

where $j = 1, 2, \ldots, D$; $Cr \in (0, 1)$ is a crossover rate which is used to control the diversity of the population; and $j_{rand} \in \{1, 2, \ldots, D\}$ is a random integer generated once for each individual $\mathbf{x}_i^t$. The condition $j = j_{rand}$ makes sure that at least one component of $\mathbf{u}_i^t$ inherits from $\mathbf{v}_i^t$ so that $\mathbf{u}_i^t$ will not be identical with $\mathbf{x}_i^t$.

Step 5: The trial vector $\mathbf{u}_i^t$ ($i = 1, 2, \ldots, N$) is evaluated according to a pre-defined fitness function.

Step 6: The selection operator is performed on $P(t)$ and $U(t)$ to construct the population $P(t+1)$ by choosing vectors between the trial vectors and their corresponding target vectors following the formula

$$\mathbf{x}_i^{t+1} = \begin{cases} \mathbf{u}_i^t, & \text{if } f(\mathbf{u}_i^t) \leq f(\mathbf{x}_i^t) \\ \mathbf{x}_i^t, & \text{otherwise,} \end{cases}, \tag{2.17}$$

where $f(\cdot)$ is a fitness function.

Step 7: The iteration counter $t$ increases by 1.

In spite of several advantages, DE still suffers from prematurity and/or stagnation. Hence a good volume of work in the literature has been devoted to overcome its drawbacks, mainly from the perspectives of parameter control, operator design, population structure, and hybridization with other meta-heuristics. Many attempts have been made to improve DE performance by setting appropriate parameter values [45, 105, 112] or using parameter adaptation techniques [17, 78, 99] for scale factor $F$ and crossover rate $Cr$. Also, lots of work focused on designing of new mutation operators, such as trigonometric mutation [39], "DE/current-to-$p$best" mutation [120], GPBX-$\alpha$ mutation [35], or mixing mutation operators [78, 99]. The population structure determines the way individuals share information with each other and many researchers investigate the population structure in DE, see [21, 35, 38]. Hybridization has become an attractive route in algorithm design due to its capability for handling quite complex problems. Therefore, significant work has been done on hybridizing DE with other meta-heuristics, see [7, 85, 91]. In addition to improving DE performance, DE has also been applied to various areas, like signal processing, controller design, planning, power systems, clustering, etc. For more details of DE, two most recently survey papers [27, 86], are recommended.

## 2.7 Conclusions

This chapter introduced the fundamental concepts and principles of several EA variants including GAs, QIEAs, ACO, PSO and DE. For each variant, we reviewed its history, detailed its algorithm and addressed its future research issues. The five variants will be used to design different types of membrane algorithms in the next chapter.

## References

1. Bäck, T. 1996. *Evolutionary algorithms in theory and practice: evolution strategies, evolutionary programming, genetic algorithms.* Oxford: Oxford University Press.
2. Bäck, T., U. Hammel, and H. Schwefel. 1997. Evolutionary computation: comments on the history and current state. *IEEE Transactions on Evolutionary Computation* 1 (2): 3–17.
3. Bae, S.H., and B.R. Moon. 2004. Mutation rates in the context of hybrid genetic algorithms. In *Genetic and Evolutionary Computation (GECCO 2004)*. Lecture Notes in Artificial Intelligence, vol. 3103, ed. K. Deb, R. Poli, W. Banzhaf, H.-G. Beyer, E. Burke, P. Darwen, D. Dasgupta, D. Floreano, J. Foster, M. Harman, O. Holland, P.L. Lanzi, L. Spector, A.G.B. Tettamanzi, D. Thierens, and A. Tyrrell, 381–382. Berlin: Springer.
4. Bagchi, P., and S. Pal. 2011. Controlling crossover probability in case of a genetic algorithm. In *Information Technology and Mobile Communication (AIM 2011)*, *Communications in Computer and Information Science*, vol. 147, ed. V.V. Das, G. Thomas, and F.L. Gaol, 287–290. Berlin: Springer.
5. Bennett, C.H., and D.P. DiVincenzo. 2000. Quantum information and computation. *Nature* 404: 247–255.
6. Birattari, M., P. Pellegrini, and M. Dorigo. 2007. On the invariance of ant colony optimization. *IEEE Transactions on Evolutionary Computation* 11 (6): 732–742.

7. Biswas, A., S. Dasgupta, S. Das, and A. Abraham. 2006. A synergy of differential evolution and bacterial foraging algorithm for global optimization. *Neural Network World* 17 (6): 607–626.

8. Blum, C. 2005. Ant colony optimization introduction and recent trends. *Physics of Life Reviews* 2: 353–373.

9. Blum, C., and M. Dorigo. 2004. The hyper-cube framework for ant colony optimization. *IEEE Transactions on Systems, Man, and Cybernetics, Part B: Cybernetics* 34 (2): 1161–1172.

10. Blum, C., and A. Roli. 2008. Hybrid metaheuristics: an introduction. In *Hybrid Metaheuristics: An Emerging Approach to Optimization, Studies in Computational Intelligence*, vol. 114, ed. C. Blum, M.J.B. Aguilera, A. Roli, and M. Sampels, 1–30. Berlin: Springer.

11. Blum, C., J. Puchinger, G.R. Raidl, and A. Roli. 2011. Hybrid metaheuristics in combinatorial optimization: a survey. *Applied Soft Computing* 11 (6): 4135–4151.

12. Boussa, I., J. Lepagnot, and P. Siarry. 2013. A survey on optimization metaheuristics. *Information Sciences* 237: 82–17.

13. Boozarjomehry, R.B., and M. Masoori. 2007. Which method is better for the kinetic modeling: decimal encoded or binary genetic algorithm? *Chemical Engineering Journal* 130 (1): 29–37.

14. Box, G.E.P. 1957. Evolutionary operation: a method for increasing industrial productivity. *Journal of the Royal Statistical Society. Series C (Applied Statistics)* 6 (2): 81–101.

15. Braune, R., S. Wagner, and M. Affenzeller. 2005. On the analysis of crossover schemes for genetic algorithms applied to the job shop scheduling problem. In *Proceedings of 9th World Multi-Conference on Systemics, Cybernetics and Informatics*, vol. 6, 236–241.

16. Bremermann, H.J. 1962. Optimization through evolution and recombination. In *Self-Organizing Systems*, ed. M.C. Yovits, G.T. Jacobi, and G.D. Goldstein. Washington DC: Spartan.

17. Brest, J., S. Greiner, B. Boskovic, M. Mernik, and V. Zumer. 2006. Self-adapting control parameters in differential evolution: a comparative study on numerical benchmark problems. *IEEE Transactions on Evolutionary Computation* 10 (6): 646–657.

18. Burian, R. 1996. Underappreciated pathways toward molecular genetics as illustrated by Jean Brachet's cytochemical embryology. In *The Philosophy and History of Molecular Biology: New Perspectives*, ed. S. Sarkar, 67–85. Netherlands: Kluwer Academic Publishers.

19. Carvelli, L., and G. Sebastiani. 2011. Some issues of ACO algorithm convergence. In *Ant Colony Optimization: Methods and Applications*, ed. A. Ostfeld, 39–52. Croatia: InTech Press.

20. Černý, V. 1985. Thermodynamical approach to the traveling salesman problem: an efficient simulation algorithm. *Journal of Optimization Theory and Applications* 45 (1): 41–51.

21. Cheng, J., G. Zhang, and F. Neri. 2013. Enhancing distributed differential evolution with multicultural migration for global numerical optimization. *Information Sciences* 247: 72–93.

22. Cheng, J., G.G. Yen, and G. Zhang. 2015. A many-objective evolutionary algorithm with enhanced mating and environmental selections. *IEEE Transactions on Evolutionary Computation* 19 (4): 592–605.

23. Cheng, J., G. Zhang, F. Caraffini, and F. Neri. 2015. Multicriteria adaptive differential evolution for global numerical optimization. *Integrated Computer-Aided Engineering* 22 (2): 103–117.

24. Clerc, M. 2006. Stagnation analysis in particle swarm optimization or what happens when nothing happens, Technical Report CSM-460, Department of Computer Science, University of Essex.

25. Clerc, M., and J. Kennedy. 2002. The Particle swarm-explosion, stability, and convergence in a multidimensional complex space. *IEEE Transactions on Evolutionary Computation* 6 (1): 58–73.

26. Darwin, C. 1859. *On the origin of species by means of natural selection, or the preservation of favoured races in the struggle for life*. London: Murray.

27. Das, S., and P.N. Suganthan. 2011. Differential evolution: a survey of the state-of-the-art. *IEEE Transactions on Evolutionary Computation* 15 (1): 4–31.

28. Dorigo, M., and C. Blum. 2005. Ant colony optimization theory: a survey. *Theoretical Computer Science* 344: 243–278.
29. Dorigo, M., and L.M. Gambardella. 1997. Ant Colony System: a cooperative learning approach to the traveling salesman problem. *IEEE Transactions on Evolutionary Computation* 1 (1): 53–66.
30. Dorigo, M., and T. Stutzle. 2004. *Ant Colony Optimization*. Scituate: Bradford Company.
31. Dorigo, M., M. Birattari, and T. Stützle. 2006. Ant colony optimization: artificial ants as a computational intelligence technique. *IEEE Computational Intelligence Magazine* 1: 28–39.
32. Dorigo, M., G. Caro, and L.M. Gambardella. 1999. Ant algorithms for distributed discrete optimization. *Artificial Life* 5 (2): 137–172.
33. Dorigo, M., V. Maniezzo, and A. Colorni. 1991. Positive feedback as a search strategy, Technical Report 01–016, Dipartimento di Elettronica, Politecnico di Milano, Milan, Italy.
34. Dorigo, M., V. Maniezzo, and A. Colorni. 1996. Ant System: optimization by a colony of cooperating agents. *IEEE Transactions on Systems, Man, and Cybernetics, Part B: Cybernetics* 26 (1): 29–41.
35. Dorronsoro, B., and P. Bouvry. 2011. Improving classical and decentralized differential evolution with new mutation operator and population topologies. *IEEE Transactions on Evolutionary Computation* 15 (1): 67–98.
36. Eiben, A.E., and J. Smith. 2003. *Introduction to Evolutionary Computing*. Berlin: Springer.
37. Eusuff, M.M., and K.E. Lansey. 2003. Optimization of water distribution network design using the shuffled frog leaping algorithm. *Journal of Water Resources Planning and Management* 129 (2): 210–225.
38. Falco, I.D., A.D. Cioppa, D. Maisto, U. Scafuri, and E. Tarantino. 2012. Improving classical and decentralized differential evolution. *Information Sciences* 207: 50–65.
39. Fan, H.Y., and J. Lampinen. 2003. A trigonometric mutation operator to differential evolution. *Journal of Global Optimization* 27 (1): 105–129.
40. Fogel, L., A. Owens, and M. Walsh. 1966. *Artificial intelligence through simulated evolution*. Chichester: Wiley.
41. Fraser, A.S. 1957. Simulation of genetic systems by automatic digital computers. *Australian Journal of Biological Sciences* 10 (4): 484–491.
42. Friedberg, R.M. 1958. A learning machine: Part I. *IBM Journal of Research and Development* 2 (1): 2–13.
43. Friedberg, R.M., B. Dunham, and J. North. 1959. A learning machine: Part II. *IBM Journal of Research and Development* 3 (3): 282–287.
44. Galaviz-Casas, J. 1998. Selection analysis in genetic algorithms. In *Progress in Artificial Intelligence (IBERAMIA 98)*, Lecture Notes in Artificial Intelligence, vol. 1484, ed. H. Coelho, 283–292. Berlin: Springer.
45. Gämperle, R., S.D. Müller, and P. Koumoutsakos. 2002. A parameter study for differential evolution. In *Proceedings of the Advances in Intelligent Systems, Fuzzy Systems, Evolutionary Computation*, 293–298.
46. Glassner, A. 2001. Quantum computing, Part 2. *IEEE Computer Graphics and Applications* 21 (6): 86–95.
47. Glassner, A. 2001. Quantum computing, Part 3. *IEEE Computer Graphics and Applications* 21 (6): 72–82.
48. Glover, F. 1989. Tabu search-part I. *INFORMS Journal on Computing* 1 (3): 190–206.
49. Goldberg, D.E. 1989. *Genetic algorithms in search, optimization and machine learning*. Boston: Addison-Wesley Longman Publishing Co. Inc.
50. Grover, L.K. 1999. Quantum computation. In *Proceedings of the 12th International Conference on VLSI Design*, 548–553.
51. Gutjahr, W. 2000. A graph-based ant system and its convergence. *Future Generation Computer Systems* 16 (9): 873–888.
52. Gutjahr, W. 2008. First steps to the runtime complexity analysis of ant colony optimization. *Computers and Operations Research* 35 (9): 2711–2727.

53. Han, K.H., and J.H. Kim. 2000. Genetic quantum algorithm and its application to combinatorial optimization problem. In *Proceedings of IEEE Congress on Evolutionary Computation*, 1354–1360.

54. Han, K.H., and J.H. Kim. 2002. Quantum-inspired evolutionary algorithm for a class of combinatorial optimization. *IEEE Transactions on Evolutionary Computation* 6 (6): 580–593.

55. Herrera, F., M. Lozano, and A.M. Sanchez. 2003. A taxonomy for the crossover operator for real-coded genetic algorithms: An experimental study. *International Journal of Intelligent Systems* 18 (3): 309–338.

56. Hey, T. 1999. Quantum computing: an introduction. *Computing and Control Engineering Journal* 10 (3): 105–112.

57. Hinterding, R. 1999. Representation, constraint satisfaction and the knapsack problem. In *Proceedings of IEEE Congress on Evolutionary Computation*, 1286–1292.

58. Holland, J.H. 1975. *Adaptation in natural and artificial systems*. Ann Arbor: University of Michigan Press.

59. Iba, H., and N. Noman. 2011. *New frontier in evolutionary algorithms: theory and applications*. London: Imperial College Press.

60. Jin, Y. 2005. A comprehensive survey of fitness approximation in evolutionary computation. *Soft Computing* 9: 3–12.

61. Karaboga, D. 2005. An idea based on honey bee swarm for numerical optimization, Technical Report-TR06, Erciyes University, Engineering Faculty, Computer Engineering Department.

62. Katayama, K., H. Hirabayashi, and H. Narihisa. 2003. Analysis of crossovers and selections in a coarse-grained parallel genetic algorithm. *Mathematical and Computer Modelling* 38 (11–13): 1275–1282.

63. Kaya, M. 2011. The effects of two new crossover operators on genetic algorithm performance. *Applied Soft Computing* 11 (1): 881–890.

64. Kennedy, J. 1999. Small worlds and mega-minds: effects of neighborhood topology on particle swarm performance. In *Proceedings of the IEEE International Conference on Evolutionary Computation*, 1931–1938.

65. Kennedy, J., and R. Eberhart. 1996. Particle swarm optimization. In *Proceedings of IEEE International Conference on Neural Networks*, 69–73.

66. Kennedy, J., and R.C. Eberhart. 2001. *Swarm Intelligence*. San Francisco: Morgan Kaufmann Publishers Inc.

67. Kennedy, J., and R. Mendes. 2002. Population structure and particle swarm performance. In *Proceedings of IEEE International Conference on Evolutionary Computation*, 1671–1676.

68. Kicinger, R., T. Arciszewski, and K. De Jong. 2005. Evolutionary computation and structural design: a survey of the state-of-the-art. *Computers and Structures* 83 (23–24): 1943–1978.

69. Kin, S. 1965. Computer solutions of the traveling salesman problem. *Bell System Technical Journal* 44 (10): 2245–2269.

70. Kirkpatrick, S., C.D. Gelatt, and M.P. Vecchi. 1983. Optimization by simulated annealing. *Science* 220 (4598): 671–680.

71. Koza, J.R. 1992. *Genetic programming: on the programming of computers by means of natural selection*. Cambridge: MIT Press.

72. Koza, J.R. 1994. *Genetic programming II: automatic discovery of reusable programs*. Cambridge: MIT Press.

73. Langdon, W.B., and R. Poli. 2002. *Foundations of genetic programming*. Berlin: Springer.

74. Larrañaga, P., and J.A. Lozano (eds.). 2002. *Estimation of distribution algorithms: a new tool for evolutionary computation*. Boston: Kluwer Academic Publishers.

75. Li, L., H. Peng, J. Kurths, Y. Yang, and H.J. Schellnhuber. 2014. Chaos-order transition in foraging behavior of ants. *Proceedings of the National Academy of Sciences of the United States of America* 111 (23): 8392–8397.

76. Li, L.X., Z.J. Shao, and J.X. Qian. 2002. An optimizing method based on autonomous animate: fish swarm algorithm. In *Proceeding of System Engineering Theory and Practice*, 32–38.

77. Liang, J.J., A.K. Qin, P.N. Suganthan, and S. Baskar. 2006. Comprehensive learning particle swarm optimizer for global optimization of multimodal functions. *IEEE Transactions on Evolutionary Computation* 10 (3): 281–285.

78. Mallipeddi, R., P.N. Suganthan, Q.K. Pan, and M.F. Tasgetiren. 2011. Differential evolution algorithm with ensemble of parameters and mutation strategies. *Applied Soft Computing* 11 (2): 1679–1696.

79. Martin, J.L.F.V., and M.S. Sanchez. 2002. Does crossover probability depend on fitness and hamming differences in genetic algorithms? In *Artificial Neural Networks (ICANN 2002)*, Lecture Notes in Computer Science, vol. 2415, ed. J.R. Dorronsoro, 389–394. Berlin: Springer.

80. Mendes, R., J. Kennedy, and J. Neves. 2004. The fully informed particle swarm: simpler, maybe better. *IEEE Transactions on Evolutionary Computation* 8 (3): 204–210.

81. Milton, J., P. Kennedy, and H. Mitchell. 2005. The effect of mutation on the accumulation of information in a genetic algorithm. In *AI 2005: Advances in Artificial Intelligence*, Lecture Notes in Artificial Intelligence, vol. 3809, ed. S. Zhang, and R. Jarvis, 360–368. Berlin: Springer.

82. Mitchell, M., and C.E. Taylor. 1999. Evolutionary computation: an overview. *Annual Review of Ecology and Systematics* 30: 593–616.

83. Moore, M., and A. Narayanan. 1995. Quantum-inspired computing, Department of Computer Science, University Exeter, Exeter, U.K.

84. Narayanan, A., and M. Moore. 1996. Quantum-inspired genetic algorithms. In *Proceedings of IEEE International Conference on Evolutionary Computation*, 61–66.

85. Neri, F., and V. Tirronen. 2008. On memetic differential evolution frameworks: a study of advantages and limitations in hybridization. In *Proceedings of the IEEE Congress on Evolutionary Computation*, 2135–2142.

86. Neri, F., and V. Tirronen. 2010. Recent advances in differential evolution: a survey and experimental analysis. *Artificial Intelligence Review* 33 (1): 61–106.

87. Neshat, M., G. Sepidnam, M. Sargolzaei, and A.N. Toosi. 2014. Artificial fish swarm algorithm: a survey of the state-of-the-art, hybridization, combinatorial and indicative applications. *Artificial Intelligence Review* 42 (4): 965–997.

88. Nielsen, A.M., and I.L. Chuang. 2000. *Quantum computation and quantum information*. Cambridge: Cambridge University Press.

89. Okabe, T. 2007. Theoretical analysis of selection operator in genetic algorithms. In *Proceedings of the IEEE Congress on Evolutionary Computation*, 4676–4683.

90. Olsson, A. 2011. *Particle swarm optimization: theory, techniques and applications, engineering tools, techniques and tables*. Nova Science Publishers, Incorporated.

91. Omran, M.G.H., A.P. Engelbrecht, and A. Salman. 2009. Bare bones differential evolution. *European Journal of Operational Research* 196 (1): 128–139.

92. Osaba, E., R. Carballedo, F. Diaz, E. Onieva, I. de la Iglesia, and A. Perallos. 2014. Crossover versus mutation: a comparative analysis of the evolutionary strategy of genetic algorithms applied to combinatorial optimization problems. *The Scientific World Journal* 2014. Article ID 154676, 22 p.

93. Passino, K.M. 2002. Biomimicry of bacterial foraging for distributed optimization and control. *IEEE Control Systems Magazine* 22 (3): 52–67.

94. Pelikan, M., D.E. Goldberg, and F.G. Lobo. 2002. A survey of optimization by building and using probabilistic models. *Computational Optimization and Applications* 21 (1): 5–20.

95. Pilato, C., D. Loiacono, F. Ferrandi, P.L. Lanzi, and D. Sciuto. 2008. High-level synthesis with multi-objective genetic algorithm: a comparative encoding analysis. In *Proceedings of the IEEE Congress on Evolutionary Computation*, 3334–3341.

96. Poli, R., J. Kennedy, and T. Blackwell. 2007. Particle swarm optimization-an overview. *Swarm Intelligence* 1 (1): 33–57.

97. Price, K. 1994. Genetic annealing. *Dr. Dobb's Journal* 127–132.

98. Price, K., R.M. Storn, and J.A. Lampinen. 2005. *Differential evolution: a practical approach to global optimization (Natural Computing Series)*. New York: Springer.

99. Qin, A.K., V.L. Huang, and P.N. Suganthan. 2009. Differential evolution algorithm with strategy adaptation for global numerical optimization. *IEEE Transactions on Evolutionary Computation* 13 (2): 398–417.
100. Rao, R.V., V.J. Savsani, and D.P. Vakharia. 2011. Teaching learning-based optimization: a novel method for constrained mechanical design optimization problems. *Computer Aided Design* 43 (3): 303–315.
101. Ratnaweera, A., S.K. Halgamuge, and H.C. Watson. 2004. Self-organizing hierarchical particle swarm optimizer with time-varying acceleration coefficients. *IEEE Transactions on Evolutionary Computation* 8 (3): 240–255.
102. Rechenberg, I. 1973. *Evolutionsstrategie: optimierung technischer systemenach prinzipien der biologischen evolution.* Stuttgart: Frommann-Holzboog.
103. Reynolds, R.G. 1994. An Introduction to cultural algorithms. *Proceedings of the 3rd Annual Conference on Evolutionary Programming*, 131–139. World Scientific Publishing.
104. Ronald, S. 1997. Robust encodings in genetic algorithms: a survey of encoding issues. In *Proceedings of the IEEE International Conference on Evolutionary Computation*, 43–48.
105. Rönkkönen, J., S. Kukkonen, and K.V. Price. 2005. Real-parameter optimization with differential evolution. In *Proceedings of the IEEE Congress on Evolutionary Computation*, 506–513.
106. H. Schwefel, H. 1975. Evolutionsstrategie und numerische optimierung. Ph.D. dissertation, Technische Berlin, Germany.
107. Schwefel, H. (ed.). 1995. *Evolution and optimum seeking.* New York: A Wiley-Interscience publication.
108. Shi, Y., and R.C. Eberhart. 1998. A modified particle swarm optimizer. In *Proceedings of the IEEE International Conference on Evolutionary Computation*, 69–73.
109. Shi, Y., and R.C. Eberhart. 1999. Empirical study of particle swarm optimization. In *Proceedings of the IEEE International Conference on Evolutionary Computation*, 101–106.
110. Simon, D. 2008. Biogeography-based optimization. *IEEE Transactions on Evolutionary Computation* 12 (6): 702–713.
111. Simon, D. 2013. *Evolutionary optimization algorithms: biologically-inspired and population-based approaches to computer intelligence.* New York: Wiley.
112. Storn, R., K. Price. 1995. Differential evolution-a simple and efficient adaptive scheme for global optimization over continuous spaces, Technical Report TR-95-012, Berkeley, CA.
113. Storn, R., and K. Price. 1997. Differential evolution-a simple and efficient heuristic for global numerical optimization. *Journal of Global Optimization* 11 (4): 341–359.
114. Stützle, T., and H.H. Hoos. 2000. MAX-MIN ant system. *Future Generation Computer Systems* 16 (8): 889–914.
115. Volná, E. 2013. *Introduction to soft computing.* Bookboon.com.
116. Wang, X., G. Zhang, J. Zhao, H. Rong, F. Ipate, and R. Lefticaru. 2015. A modified membrane-inspired algorithm based on particle swarm optimization for mobile robot path planning. *International Journal of Computers, Communications and Control* 10 (5): 732–745.
117. Watts, D.J., and S.H. Strogatz. 1998. Collective dynamics of 'small-world' networks. *Nature* 393: 440–442.
118. Yang, X.S. 2008. *Nature-inspired metaheuristic algorithms.* Frome: Luniver Press.
119. Zhang, G. 2011. Quantum-inspired evolutionary algorithms: a survey and empirical study. *Journal of Heuristics* 17: 303–351.
120. Zhang, J., and A. Sanderson. 2009. JADE: adaptive differential evolution with optional external archive. *IEEE Transactions on Evolutionary Computation* 13 (5): 945–958.
121. Zhang, G., J. Cheng, and M. Gheorghe. 2011. A membrane-inspired approximate algorithm for traveling salesman problems. *Romanian Journal of Information Science and Technology* 14 (1): 3–19.
122. Zhang, G., M. Gheorghe, L. Pan, and M.J. Pérez-Jiménez. 2014. Evolutionary membrane computing: a comprehensive survey and new results. *Information Sciences* 279: 528–551.
123. Zhang, G., F. Zhou, X. Huang, J. Cheng, M. Gheorghe, F. Ipate, and R. Lefticaru. 2012. A novel membrane algorithm based on particle swarm optimization for solving broadcasting problems. *Journal of Universal Computer Science* 18 (13): 1821–1841.

# Chapter 3
# Membrane Algorithms

**Abstract** Membrane Algorithms (MAs) area is focusing on developing new variants
of meta-heuristic algorithms for solving complex optimization problems by using
either the hierarchical or network membrane structures, evolution rules and computational capabilities of membrane systems and the methods and well-established
techniques employed in Evolutionary Computation. MAs studied in this volume, and
described in this Chapter, refer to four variants of meta-heuristics using the hierarchical structure of the membrane systems - nested membrane structure, one-level
membrane structure, hybrid membrane structure and dynamic membrane structure;
whereas those using the network structure consist of two subcategories - statical
network structure and dynamical network structure.

## 3.1 Introduction

Natural computing, a fast growing interdisciplinary field, aims to develop concepts,
computational paradigms and theories inspired from various natural processes and
phenomena. Membrane computing (MC) and evolutionary computation (EC) are
main branches of natural computing. MC with rigor and sound theoretical development for all variants of membrane systems provides a parallel-distributed framework and flexible evolution rules. While EC has outstanding characteristics, such as
easy-understanding, robust performance, flexibility, and good results in solving real-
world problems. These features suggest the exploration of the interactions between
membrane computing and evolutionary computation, leading to the research of evolutionary membrane computing (EMC). At present, the possible interplay of MC and
EC has produced two research topics [88]: membrane-inspired evolutionary algorithms (MIEAs), also called membrane algorithms (MAs), and automated design
of membrane computing models (ADMCM). On the one hand, MIEAs represent a
research direction in MC with great success in approaching real-world applications
[53]. On the other hand, ADMCM aims to circumvent the programmability issue of
membrane system models [13, 32] by using EC techniques. The difference between
MIEA and ADMCM is illustrated by Fig. 3.1, where they have different inputs and
outputs. This chapter only focuses on the MIEAs part.

© Springer International Publishing AG 2017

G. Zhang et al., *Real-life Applications with Membrane Computing*,
Emergence, Complexity and Computation 25, DOI 10.1007/978-3-319-55989-6_3

**Fig. 3.1** Difference between
MIEA and ADMCM

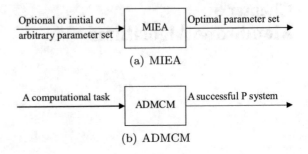

(a)  MIEA

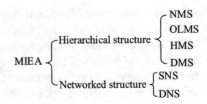

(b)  ADMCM

**Fig. 3.2**  Classification of
MIEAs with respect to the
membrane structure

$$
\text{MIEA}
\begin{cases}
\text{Hierarchical structure}
\begin{cases}
\text{NMS}\\
\text{OLMS}\\
\text{HMS}\\
\text{DMS}
\end{cases}\\
\text{Networked structure}
\begin{cases}
\text{SNS}\\
\text{DNS}
\end{cases}
\end{cases}
$$

An MIEA or MA concentrates on developing new variants of meta-heuristic algorithms for solving complex optimization problems by using the hierarchical or network membrane structures, evolution rules and computational procedures utilized by P systems and the methods and well-established techniques employed by EC [88]. In different variants of MIEAs, membrane structures have significant effects on their design. According to the membrane structure used, MIEAs can be classified into two categories: hierarchical structure based MIEAs and network structure based MIEAs. On the one hand, the hierarchical structure is subclassified into four categories: nested membrane structure (NMS: a rooted tree with only one branch), one-level membrane structure (OLMS), hybrid membrane structure (HMS) and dynamic membrane structure (DMS). The network structure is divided into two subcategories: statical network structure (SNS) and dynamical network structure (DNS). Figure 3.2 gives the classification of MIEAs from the point of view of membrane structure.

As a hybrid approximate optimization algorithm integrating P systems with meta-heuristic approaches, MIEA has been a novel and promising interdisciplinary research direction and attracted the attention of researchers from various areas. In what follows, we describe each of the above mentioned classes of MIEAs, presenting the principle behind its design, current development and some instances of it. Meanwhile, experimental results on some benchmark problems will be given to verify the algorithm performance. Finally, P systems roles in MAs are investigated in an empirical manner.

## 3.2   Membrane Algorithms with Nested Membrane Structure

A nested membrane structure (NMS, for short) is the first kind of membrane structure considered for designing an MIEA, initiating the research of such problems. This section first describes the research progress and principle of NMS-based MIEAs. Then an instance, called genetic algorithm based on P systems (GAPS, for short), is presented and its performance is verified on Knapsack problems.

### 3.2.1   Principle

The NMS-based MIEAs were designed by integrating NMS and transport rules of a cell-like P system with several meta-heuristic approaches. Algorithms in a membrane (AIM) (in fact, within the region delimited by the membrane), is a concept unifying various subalgorithms, algorithm components and an independent algorithm; communications are performed only between adjacent regions. The first version of MIEAs with NMS was introduced in [47]. For this kind of MIEAs, several meta-heuristic approaches were used as AIMs. For example, AIM in [47, 49] was designed by using a tabu search. Brownian and genetic algorithm (GA) were used to design AIM in [51, 52]. In [65], a GA was used as AIM to design a MIEA to solve a DNA sequence design problem. In [36], a GA and a local search were considered as AIM to design MIEAs for min storage problems. In [61, 69], a GA was regarded as AIM to design MIEAs for function optimization problems and proton exchange membrane fuel cell model parameter estimation problems. In [91], a sequential quadratic programming (SQP) algorithm was used as AIM to solve complex constrained problems and the gasoline blending scheduling problem.

Generally, the NMS-based MIEAs of order $m$ are composed of three components:

(i)  $m$ membranes, which is shown in Fig. 3.3, where the outermost one is the skin membrane and the innermost one is an elementary membrane;

(ii)  Inside the $i$th region delimited by the $i$th membrane, $i = 1, 2, \ldots, m$, a subalgorithm is running and a new population is obtained;

(iii)  Solution transporting mechanisms between adjoining regions.

After the initial setting, an NMS-based MIEA works as follows:

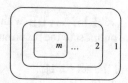

**Fig. 3.3**   A nested membrane structure [47]

(i) The tentative solutions in the $i$th region are updated by the subalgorithm with respect to the $i$th region, $i = 1, 2, \ldots, m$.

(ii) The copies of the better and worse solutions from the $i$th membrane ($1 \leq i \leq m - 1$) are sent into its adjoining inner and outer regions, respectively. The copy of the better solution from the $(m - 1)$th region is collected in the innermost membrane labeled $m$.

(iii) Several best individuals from the solutions in the region and received from the adjacent regions are selected to construct the next population in each region.

(iv) Goes to Step (i) until the termination criterion holds.

(v) The innermost membrane stores the computing result.

### 3.2.2  Genetic Algorithm Based on P System

#### 3.2.2.1  Algorithm

An approximate optimization algorithm integrating genetic algorithm and NMS, called GAPS, is presented to solve knapsack problems. In GAPS, the objects consist of binary strings; the rules are composed of selection, crossover, mutation of GA, and communication rules of P systems in each membrane. Binary strings, organized as individuals, are dealt with as multisets of objects, each of which corresponds to a candidate solution of the problem. The set of rules are responsible for evolving the system and transporting the best or worst individuals between adjacent regions. The best individual in the innermost membrane is the final solution of the problem.

More precisely, the P system used in GAPS consists of:

---
**Algorithm 1** Pseudocode algorithm of the GAPS
---
1: $t = 1$
2: Initialize the membrane structure
3: **while** (not termination condition) **do**
4:     **for** $i = 1, 2, \ldots, m$ **do**
5:         Perform GA inside the $i$th region
6:     **end for**
7:     Execute the communication rules
8:     $t \leftarrow t + 1$
9: **end while**

---

(i) a membrane structure $[[, \ldots, [\ ]_m, \ldots, ]_2]_1$ with $m$ regions, where the membranes or regions from the outmost to the innermost are labeled by $1, 2, \ldots, m$, respectively;

(ii) an alphabet that consists of the set $\{0, 1\}$;

(iii) a set of terminal symbols, $T$, $\{0, 1\}$;

(iv) initial multisets associated with each membrane

$$w_1 = \mathbf{x}_1\mathbf{x}_2 \ldots \mathbf{x}_{n_1},$$
$$w_2 = \mathbf{x}_{n_1+1}\mathbf{x}_{n_1+2} \ldots \mathbf{x}_{n_2},$$
$$\ldots \ldots$$
$$w_m = \mathbf{x}_{n_{(m-1)}+1}\mathbf{x}_{n_{(m-1)}+2} \ldots \mathbf{x}_{n_m},$$

where $\mathbf{x}_j$, $1 \leq j \leq N$, is a binary coded individual; $n_i$, $1 \leq i \leq m$, is the number of individuals in $w_i$; $\sum_{i=1}^{m} n_i = N$, where $N$ is the total number of individuals in the system;

(v) rules which are classified as

(a) evolution rules in each membrane; these are transformation-like rules including selection, crossover, and mutation of GA;

(b) communication rules which send the best or worst individual to the adjacent region.

The pseudocode algorithm for GAPS is shown in Algorithm 1. In what follows each step is described in detail.

Step 1: Initially, the iteration counter $t$ is set to 1.

Step 2: In the initialization, an NMS shown in Fig. 3.3 with $m$ membranes is constructed. $N$ individuals (e.g., let $N = Am$, where $A$ is an integer), each of which consists of $D$ binary numbers, denoted as $\mathbf{x}_1, \mathbf{x}_2, \ldots, \mathbf{x}_N$, are scattered across the $m$ regions with equal size. The population in the $i$th region is denoted as $P(i)$, and its size is $n_i = A$, $i = 1, 2, \ldots, m$.

Step 3: The termination criterion may be the prescribed maximal number of generations or the preset difference between the best solution searched and the optimal/desired solution of the problem.

Steps 4–6: In each region, the individuals are made to evolve by using GA. Specifically, the population with size $n_i$ in the $i$th region first undergoes a roulette-wheel selection to create a set of $n_i$ parents. Then each pair of parents undergoes a single point crossover with probability $p_c$ to create two new individuals. Afterwards, each new individual undergoes a mutation with a probability $p_m$. The pseudocode of the GA algorithm with one iteration is shown in Algorithm 2.

Step 7: The $i$th membrane sends the copies of the better and worse solutions into its adjoining inner and outer regions, respectively. The innermost membrane, membrane $m$, only collects the copy of the better solution coming from the $(m-1)$th region.

Step 8: The iteration counter increases by 1.

---

**Algorithm 2** Pseudocode algorithm of the GA

---

**Require:** A population $P$ with size $N$
1: $Q_1 \leftarrow \emptyset$
2: $f_i \leftarrow fitness(\mathbf{x}_j), j \in [1, N]$
3: $f_{sum} = \sum_{i=1}^{N} f_i$
4: **for** $j = 1, 2, \ldots, N$ **do**
5:      Generate a uniformly distributed random number $r \in [0, f_{sum}]$
6:      $F \leftarrow f_1$
7:      $k \leftarrow 1$
8:      **while** $F < r$ **do**
9:          $k \leftarrow k + 1$
10:          $F \leftarrow F + f_k$
11:     **end while**
12:     $Q_1 \leftarrow Q_1 \bigcup \{\mathbf{x}_k\}$
13: **end for**
14: $Q_2 \leftarrow \emptyset$
15: **for** $j = 1, 2, \ldots, N/2$ **do**
16:     Randomly select two individuals from $Q_1$, denoted as $\mathbf{x}_1, \mathbf{x}_2$
17:     **if** $rand(0, 1) < p_c$ **then**
18:         Generate a random crossover point $l \in [1, D]$
19:         Swap the last $D - l$ bits of $\mathbf{x}_1$ and $\mathbf{x}_2$ to create children $\mathbf{c}_1$ and $\mathbf{c}_2$
20:         $Q_2 \leftarrow Q_2 \bigcup \{\mathbf{c}_1, \mathbf{c}_2\}$
21:     **else**
22:         $Q_2 \leftarrow Q_2 \bigcup \{\mathbf{x}_1, \mathbf{x}_2\}$
23:     **end if**
24: **end for**
25: $Q_3 \leftarrow \emptyset$
26: **for** $j = 1, 2, \ldots, N$ **do**
27:     **for** $k = 1, 2, \ldots, D$ **do**
28:         Generate a random number $r \in [0, 1]$
29:         **if** $r < p_m$ **then**
30:             Change the $k$th bit of the $j$th individual from 0 to 1 or 1 to 0
31:         **end if**
32:     **end for**
33:     $Q_3 \leftarrow Q_3 \bigcup \{\mathbf{c}_j\}$
34: **end for**
35: $Q \leftarrow Q_3$

---

### 3.2.2.2  Examples

To show the effectiveness of GAPS algorithm, a knapsack problem is used as an application example. The knapsack problem is a combinatorial optimization problem which can be described as the selection of the most profitable items within a group given that the knapsack has a limited capacity [18]. Formally, the knapsack problem requires the selection of a subset of a given set of items so as to maximize a profit function

$$f(x_1, \ldots, x_K) = \sum_{i=1}^{K} p_i x_i, \tag{3.1}$$

**Table 3.1** Comparisons of GAPS and GA on three instances of the knapsack problem

| Item | GAPS | | | | GA | | | |
|------|------|------|---------|------|------|------|---------|------|
| | Best | Worst | Average | SD | Best | Worst | Average | SD |
| 50 | 296.54 | 281.98 | 287.29 | 3.06 | 293.91 | 278.52 | 283.64 | 3.05 |
| 200 | 1047.98 | 1017.15 | 1027.13 | 7.34 | 1044.91 | 1011.55 | 1024.59 | 8.99 |
| 400 | 2120.54 | 2086.29 | 2100.86 | 9.01 | 2115.16 | 2084.17 | 2097.83 | 8.55 |

subject to

$$\sum_{i=1}^{K} r_i x_i \leq C, \tag{3.2}$$

where $K$ is the number of items, $p_i$ is the profit of the $i$th item, $r_i$ is the weight of the $i$th item, $C$ is the capacity of the given knapsack, and $x_i$ is 0 or 1.

To verify the advantage of GAPS over GA, three stochastic knapsack problems with 50, 200 and 400 items, with $r_i$ randomly generated from [1, 10], $p_i = r_i + 5$ and $C = 0.5 \sum_{i=1}^{K} r_i$ are constructed. For both GAPS and GA, the population size, crossover rate $p_c$, mutation probability $p_m$ and the termination conditions are set to 20, 0.8, 0.05 and 20000 function evaluations, respectively. In addition, the number $m$ of membranes in GAPS is set to 4. Each algorithm independently solves each of three problems for 30 runs. Table 3.1 lists the best, worst, mean and standard deviation (SD) of the solutions.

From Table 3.1, it can be seen that GAPS outperforms GA, which demonstrates the effectiveness of combining GA with NMS. Although the superiority of NMS-based algorithm is not quite significant; however, it initiates the research of MAs. In the next few sections, some advanced MAs will be introduced.

## 3.3 Membrane Algorithms with One-Level Membrane Structure

One-level membrane structure (OLMS, for short), first proposed in [78], is currently the most widely used membrane structure in MIEAs. This section first describes the basic principle of OLMS-based MIEAs. Then three instances by using different AIMs are presented in detail, and their performances are tested on three kinds of optimization problems.

### 3.3.1 Principle

OLMS, which is a special hierarchical membrane structure coming from a cell-like P system, has a certain number of elementary membranes inside the skin membrane. The communication in OLMS is usually a global process and can be executed between any two or more elementary membranes. In the first OLMS-based MIEA [78], a quantum-inspired evolutionary algorithm (QIEA) was considered as an AIM. After that, several variants of MIEAs with OLMS were presented by using different types of AIM designed by using GA [68, 91, 92], particle swarm optimization (PSO) [55, 66, 67, 94], ant colony optimization (ACO) [83], quantum particle swarm optimization [16], quantum shuffled frog leaping algorithm [15] and differential evolution (DE) [5]. The MIEA performance was tested by applying knapsack problems, satisfiability problems or TSPs in [78, 79, 82–84]. A multi-objective MIEA with OLMS was discussed in [81]. The algorithm performance was verified in [4, 5, 12, 39, 66–68, 94] by using various benchmark functions. In addition, a wide range of engineering problems, such as digital filter design [39], broadcasting problems of P systems [85], radar emitter signal processing [5, 40, 82, 94], image decomposition [84], controller design [63], spectrum allocation problems in cognitive radio systems [15, 16] and image segmentation [55, 62], were solved by using different variants of MIEAs with OLMS.

The investigations on OLMS [87] show that OLMS-based MIEAs have better optimization performance than their counterpart approaches because of their improved capacity of balancing exploration and exploitation, which is derived from their better balance between convergence and diversity.

Generally, in an OLMS, there are $m$ elementary membranes, labeled as $1, 2, \ldots, m$, placed inside the skin membrane, denoted by 0, as shown in Fig. 3.4. The rules are of two types: evolution rules in each of the compartments 1 to $m$ which are transformation-like rules for updating an individual according to the evolutionary mechanism of a metaheuristic approach and communication rules which send the fittest individual from each of the $m$ regions delimited by $m$ elementary membranes into the skin membrane and then the overall fittest individual from the skin back to each region.

The general steps for implementing MIEAs with OLMS can be described as follows.

(i) Design the OLMS shown in Fig. 3.4.
(ii) Inside each elementary membrane, at least one individual, which is randomly chosen from the population with $N$ individuals, is put. There are totally $m$ elementary membranes ($m \leq N$). So the number of individuals in an elementary membrane varies from 1 to $N - m + 1$.
(iii) The number of iterations is decided for the subpopulation inside each elementary membrane, where a metaheuristic approach is independently performed. Specifically, the number $g_i$ ($i = 1, 2, \ldots, m$) of iterations is randomly generated between 1 and a certain integer number for the $i$th elementary membrane.

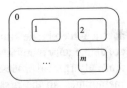

**Fig. 3.4** A one level membrane structure

(iv) Each elementary membrane independently perform its evolutionary process of the meta-heuristic approach.
(v) Specific information between regions is exchanged by using communication rules. For example, the best individual found inside each elementary membrane is sent into the skin membrane; the best individual inside the skin membrane is sent back to all the elementary membranes.

In OLMS-based MIEAs, the individuals in a population is initially scattered across the membrane structure. Inside each membrane, the number $n_i$, $1 \le i \le m$, of objects is randomly chosen. All the individuals are evaluated at each generation to select the best individual found. The communication rule is executed once at every $g_i$ ($1 \le i \le m$) generations for each compartment. If the best individual found in the skin membrane is not updated for a certain number of iterations, the algorithm terminates.

### 3.3.2 Quantum-Inspired Evolutionary Algorithm Based on P Systems

#### 3.3.2.1 Algorithm

Quantum-inspired evolutionary algorithm (QIEA) based on P systems (QEPS) [78] uses the concepts and principles of QIEAs within a P system framework. A Q-bit representation and Q-gate evolutionary rules together with an OLMS and transformation/communication-like rules are employed. The objects are organized in multisets of special strings built either over the set of Q-bits or $\{0, 1\}$. The rules will be responsible to evolve the system and select the fittest Q-bit individuals.

More precisely the P system-like framework consists of:

(i) an one-level membrane structure (OLMS) $[[\ ]_1, [\ ]_2, \ldots, [\ ]_m]_0$, which has $m$ regions labeled by $1, \ldots, m$, respectively, inside the skin membrane, labeled by $0$;
(ii) an alphabet with all possible Q-bits and the set $\{0, 1\}$;
(iii) a set of terminal symbols, $T$, $\{0, 1\}$;
(iv) initial multisets associated with each membrane

$$
\begin{aligned}
w_0 &= \lambda, \\
w_1 &= q_1 q_2 \cdots q_{n_1}, \\
w_2 &= q_{n_1+1} q_{n_1+2} \cdots q_{n_2}, \\
&\cdots \cdots \cdots \\
w_m &= q_{(n_{(m-1)}+1)} q_{(n_{(m-1)}+2)} \cdots q_{n_m},
\end{aligned}
$$

where $q_j$, $1 \le j \le N$, is a Q-bit individual; $n_i$, $1 \le i \le m$, is the number of individuals in $w_i$; $\sum_{i=1}^{m} n_i = N$, where $N$ is the total number of individuals used in this computation;

(v) rules consisting of two types:

  (a) transformation-like evolution rule in each compartment for updating a Q-bit
      individual using the current Q-gate (see Step 8 of the QIEA in Sect. 2.3);
  (b) observation rules from Q-bit individuals to binary solutions, shown in
      Algorithm 3 of Chap. 2;
  (c) communication rules for sending the best binary individual from $m$ com-
      partments into the skin membrane and then the overall best binary repre-
      sentation from the skin membrane back to each compartment.

At the beginning of QEPS, the initial population with $m$ multisets $w_1, \ldots, w_m$,
each of which represents a Q-bit individual, is scattered across OLMS. Inside each
compartment, the number $n_i$, $1 \le i \le m$, of objects is randomly chosen. All the
individuals in all the compartments are evaluated by using rules of type (b) to select
the best one, which is used to adjust the Q-gates to generate the offspring according
to evolution rules. The communication rule is executed once every $g_i(1 \le i \le m)$
iterations for the $i$th compartment. The process does not halt until the best solution
found keeps unchanged for a certain number of generations.

### 3.3.2.2  Examples

Three types of QEPS namely QEPSo, QEPSm and QEPSn are implemented by
using different Q-gate update methods. The three algorithms use different methods
for deriving the rotation angle $\theta_{ij}^t$ in $G_{ij}^t(\theta)$, illustrated by (2.4), where $\theta_{ij}^t$ is defined
as (2.6). The particular values of $s(\alpha_{ij}^t, \beta_{ij}^t)$ and $\Delta\theta_{ij}^t$ used in QEPSo, QEPSm, and
QEPSn are illustrated in Tables 3.2, 3.3 and 3.4, respectively. At the same time,
four types of QIEA are considered as compared algorithms, including bQIEAo [22],

**Table 3.2** Look-up table of $\theta_{ij}^t$ for bQIEAo, bQIEAcms and QEPSo, where $f(.)$ is the fitness,
$s(\alpha_{ij}^t, \beta_{ij}^t)$ is the sign of $\theta_{ij}^t$, and $b$ and $x$ are certain bits of the current best solution $b$ and the binary
solution $x$, respectively [22]

| $x$ | $b$ | $f(x) \ge f(b)$ | $\Delta\theta_{ij}^t$ | $s(\alpha_{ij}^t, \beta_{ij}^t)$ | | | |
|-----|-----|-----|-----|-----|-----|-----|-----|
| | | | | $\alpha_{ij}^t \beta_{ij}^t > 0$ | $\alpha_{ij}^t \beta_{ij}^t < 0$ | $\alpha_{ij}^t = 0$ | $\beta_{ij}^t = 0$ |
| 0 | 0 | False | 0 | 0 | 0 | 0 | 0 |
| 0 | 0 | True | 0 | 0 | 0 | 0 | 0 |
| 0 | 1 | False | 0 | 0 | 0 | 0 | 0 |
| 0 | 1 | True | $0.05\pi$ | $-1$ | $+1$ | $\pm 1$ | 0 |
| 1 | 0 | False | $0.01\pi$ | $-1$ | $+1$ | $\pm 1$ | 0 |
| 1 | 0 | True | $0.025\pi$ | $+1$ | $-1$ | 0 | $\pm 1$ |
| 1 | 1 | False | $\pi$ | $+1$ | $-1$ | 0 | $\pm 1$ |
| 1 | 1 | True | $\pi$ | $+1$ | $-1$ | 0 | $\pm 1$ |

**Table 3.3** Look-up table of $\theta_{ij}^t$ for bQIEAm and QEPSm, where $f(.)$ is the fitness, $s(\alpha_{ij}^t, \beta_{ij}^t)$ is the sign of $\theta_{ij}^t$, and $b$ and $x$ are certain bits of the current best solution $b$ and the binary solution $x$, respectively [23]

| $x$ | $b$ | $f(x) \geq f(b)$ | $\Delta\theta_{ij}^t$ | $s(\alpha_{ij}^t, \beta_{ij}^t)$ |
|---|---|---|---|---|
| 0 | 0 | False | 0 | $\pm 1$ |
| 0 | 0 | True | 0 | $\pm 1$ |
| 0 | 1 | False | $0.01\pi$ | $+1$ |
| 0 | 1 | True | 0 | $\pm 1$ |
| 1 | 0 | False | $0.01\pi$ | $-1$ |
| 1 | 0 | True | 0 | $\pm 1$ |
| 1 | 1 | False | 0 | $\pm 1$ |
| 1 | 1 | True | 0 | $\pm 1$ |

**Table 3.4** Look-up table of $\theta_{ij}^t$ for bQIEAn and QEPSn, where $d_1 = \alpha_{ij}'^t \beta_{ij}'^t$, $\xi_1 = \arctan(\beta_{ij}'^t/\alpha_{ij}'^t)$, $\alpha_{ij}'^t$, $\beta_{ij}'^t$ are the amplitudes of the current best solution, and $d_2 = \alpha_{ij}^t \beta_{ij}^t$, $\xi_2 = \arctan(\beta_{ij}^t/\alpha_{ij}^t)$, $\alpha_{ij}^t$, $\beta_{ij}^t$ are the amplitudes of the current solution, and $e = 0.5\pi||\alpha_{ij}'^t| - |\alpha_{ij}^t||$

| $d_1 > 0$ | $d_2 > 0$ | $\theta_{ij}^t$ | $f(\alpha_{ij}^t, \beta_{ij}^t)$ | |
|---|---|---|---|---|
| | | | $|\xi_1| \geq |\xi_2|$ | $|\xi_1| < |\xi_2|$ |
| True | True | $e$ | $+1$ | $-1$ |
| True | False | $e$ | $-1$ | $+1$ |
| False | True | $e$ | $-1$ | $+1$ |
| False | False | $e$ | $+1$ | $-1$ |

bQIEAm [23], bQIEAcms [38] and bQIEAn [80]. In bQIEAo, the look-up table in Table 3.2 is used to decide the rotation angle of the Q-gate. bQIEAm [23] determines the rotation angle of the Q-gate according to Table 3.3. bQIEAcms uses the same method as bQIEAo to determine the rotation angle of each Q-gate [38]. The rotation angle in bQIEAn is modified according to the look-up table in Table 3.4.

Extensive comparisons between three QEPS variants and four QIEA variants on three knapsack problems are shown in Table 3.5. According to the results, the three types of QEPS obtain much higher profits than four QIEA variants. QEPSm is the best among the seven algorithms with respect to profits. QEPSm and QEPSo consume less time than bQIEAm and bQIEAo, respectively.

More experiments on the knapsack problems with 200, 400 and 600 items are carried out to study the effect of parameter $m$ on the QEPSm performances. QEPSm uses 20 individuals as a population, i.e., $N = 20$. The termination criterion is that the best individual found cannot be further improved in 20 successive iterations. Let $m$ vary from 2 to $N$. $m = N$ indicates that there is only one individual inside each elementary membrane. Parameter $g_i$, $1 \leq i \leq m$, is assigned as a uniformly random integer ranged from 1 to 10. Figures 3.5, 3.6 and 3.7 show the mean best solutions

**Table 3.5** Experimental results of the knapsack problem: the number of items is 200, 400 and 600, the number of runs is 30. BS, MBS, WS, STD and ET represent best solution, mean best solution, worst solution, standard deviation and elapsed time (in seconds), respectively. IT and CRI are abbreviations for items and criteria, respectively

| IT | CRI | QEPSm | QEPSo | QEPSn | bQIEAm | bQIEAo | bQIEAn | bQIEAcms |
|----|-----|-------|-------|-------|--------|--------|--------|----------|
| 200 | BS | **1188.31** | 1089.90 | 1099.96 | 1178.33 | 1078.01 | 1088.27 | 1078.14 |
| | MBS | **1179.65** | 1056.24 | 1080.21 | 1159.27 | 1050.47 | 1064.90 | 1056.69 |
| | WS | **1168.33** | 1041.38 | 1057.85 | 1138.16 | 1032.56 | 1046.28 | 1032.90 |
| | STD | **5.07** | 10.82 | 9.81 | 9.26 | 10.91 | 11.56 | 12.43 |
| | ET | 2093.22 | **847.64** | 1076.95 | 2468.33 | 872.75 | 936.45 | 1014.00 |
| 400 | BS | **2406.43** | 2168.68 | 2215.23 | 2371.42 | 2150.47 | 2162.89 | 2170.44 |
| | MBS | **2380.60** | 2133.95 | 2177.03 | 2319.48 | 2130.82 | 2135.95 | 2132.92 |
| | WS | **2361.43** | 2101.38 | 2145.47 | 2281.34 | 2109.57 | 2110.97 | 2110.63 |
| | STD | **8.91** | 14.76 | 15.54 | 21.13 | 12.19 | 12.84 | 14.70 |
| | ET | 6988.12 | **1495.03** | 2129.05 | 7106.36 | 1574.77 | 1828.38 | 1757.16 |
| 600 | BS | **3557.69** | 3183.18 | 3262.69 | 3492.68 | 3172.15 | 3175.50 | 3177.64 |
| | MBS | **3524.35** | 3145.81 | 3202.06 | 3421.55 | 3143.61 | 3143.98 | 3177.64 |
| | WS | **3492.68** | 3116.26 | 3151.57 | 3362.53 | 3119.98 | 3115.38 | 3115.22 |
| | STD | **14.81** | 16.82 | 21.77 | 39.44 | 15.46 | 14.91 | 13.83 |
| | ET | 13231.31 | **2216.11** | 3557.66 | 13597.50 | 2525.98 | 2807.94 | 2372.56 |

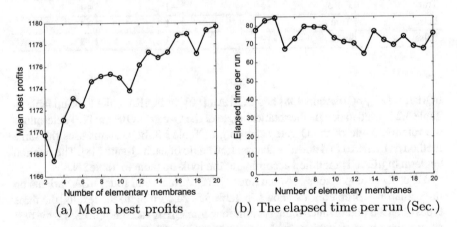

(a) Mean best profits          (b) The elapsed time per run (Sec.)

**Fig. 3.5** Experimental results of 200 items with different number of membranes

over 30 runs and the elapsed time per run for the three cases of 200, 400, and 600 items, respectively. The experimental results indicate that $m$ could be assigned as $N$ in terms of profits and the elapsed time.

To investigate the effect of the number of evolutionary generations parameter $g_i$, $1 \le i \le m$, on the QEPSm performances, experiments are conducted on knapsack problems with 200, 400 and 600 items. Let $N = 20$ and $m = 20$. Thus, $n_i = 1$

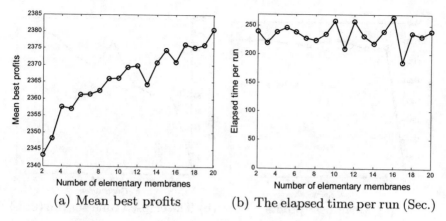

(a) Mean best profits

(b) The elapsed time per run (Sec.)

**Fig. 3.6** Experimental results of 400 items with different number of membranes

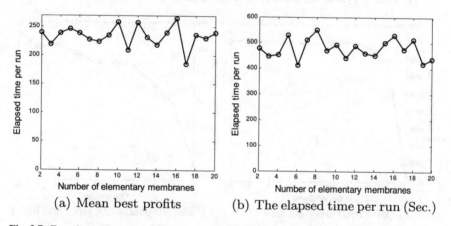

(a) Mean best profits

(b) The elapsed time per run (Sec.)

**Fig. 3.7** Experimental results of 600 items with different number of membranes

$(1 \leq i \leq m)$. The halting condition is the same as the experiments above. Let $g_i$ $(1 \leq i \leq m)$ change between 1 and 10. Figures 3.8, 3.9 and 3.10 show the mean best profits over 30 runs and the elapsed time per run. The results indicate that the profits are not affected by the change of parameter $g_i$.

### 3.3.3 Ant Colony Optimization Based on P Systems

#### 3.3.3.1 Algorithm

In this section, an approximate optimization algorithm integrating ant colony optimization techniques and P systems, called ACOPS [83], is presented to solve traveling

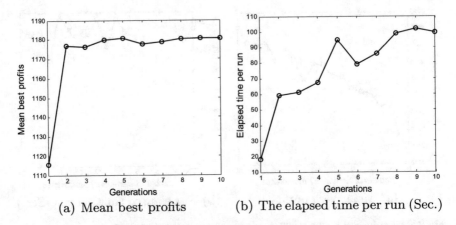

(a) Mean best profits          (b) The elapsed time per run (Sec.)

**Fig. 3.8**  Mean best profits and elapsed time of 200 items

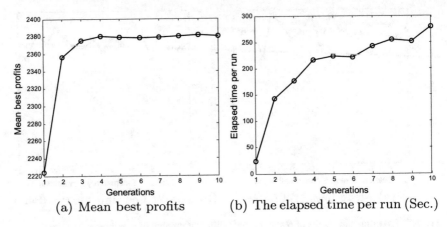

(a) Mean best profits          (b) The elapsed time per run (Sec.)

**Fig. 3.9**  Mean best profits and elapsed time of 400 items

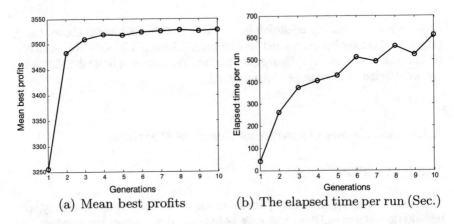

(a) Mean best profits          (b) The elapsed time per run (Sec.)

**Fig. 3.10**  Mean best profits and elapsed time of 600 items

salesman problems (TSPs). In ACOPS, the pheromone model and pheromone update rules of the Ant Colony System (ACS, described in Sect. 2.4 of Chap. 2), and the hierarchical membrane structure and transformation/communication rules of a cell-like P systems are used. More specifically, ACOPS applies OLMS to organize objects and evolution rules. These objects consist of ants or TSP construction graphs. Evolution rules, which are responsible to evolve the system and select the best ant, include a tour construction, and transformation/communication rules implemented by using local and global pheromone update rules. The P system-like framework has the following ingredients:

(i) an OLMS membrane structure $\mu = [[ \ ]_1[ \ ]_2 \ldots [ \ ]_m]_0$ with $m + 1$ compartments, which are delimited by $m$ elementary membranes labeled by $1, \ldots, m$, respectively, and the skin membrane labeled by 0;
(ii) a vocabulary with all the ants;
(iii) a set of terminal symbols, $T$ and TSP construction graphs;
(iv) initial multisets associated with each membrane

$$
\begin{aligned}
w_0 &= \lambda, \\
w_1 &= A_1 A_2 \ldots A_{n_1}, \\
w_2 &= A_{(n_1+1)} A_{(n_1+2)} \ldots A_{n_2}, \\
&\cdots\cdots\cdots \\
w_m &= A_{(n_{(m-1)}+1)} A_{(n_{(m-1)}+2)} \ldots A_{n_m},
\end{aligned}
$$

where where $A_j$ $(1 \leq j \leq N)$ is an ant; $n_i$ $(1 \leq i \leq m)$ is the number of ants in $w_i$; $\sum_{i=1}^{m} n_i = N$, where $N$ is the total number of ants;
(v) rules with two types:

(a) transformation-like tour construction rules in each compartment for constructing tours of the ants (i.e., the ACS algorithm);
(b) communication rules for updating the edges of the TSP graphs by using pheromone values, including the use of the local pheromone update strategy of the ACS to exchange information between the current ant and its subsequent ant, the use of the global pheromone update strategy in the ACS to implement communications between the best ant and the rest within a certain membrane and between the ants in the elementary membranes and those in the skin membrane.

At the beginning of ACOPS, the multisets $w_1, \ldots, w_m$, each of which is an ant in the colony, are scattered across OLMS. Each ant uses the rules of type (a) to sequentially construct its tours in its elementary membrane. Thus, an ant can sketch a whole path for the $N$ cities through $N$ steps. The evaluation is performed at each generation for all the ants to select the best one for adjusting the pheromone values in the TSP graph to communicate with the other ants in the same elementary membrane. The best ant in the $i$th compartment is sent out to the skin membrane every $g_i$ $(i = 1, 2, \ldots, m)$ generations. The process does not halt until the termination criterion, a certain number of iterations, is arrived. Algorithm 3 shows the pseudocode algorithm for ACOPS. Details are described as follows:

---

**Algorithm 3** Pseudocode algorithm of the ACOPS

---
1: $t = 1$
2: Initialize the membrane structure
3: **while** (not termination condition) **do**
4:      Scatter ants into elementary membranes
5:      Determine iterations for each of elementary membranes
6:      **for** $i = 1, 2, \ldots, m$ **do**
7:          Perform ACS inside the $i$th elementary membrane
8:      **end for**
9:      Form a colony of ants in the skin membrane
10:     Perform ACS in the skin membrane
11:     Execute global communication
12:     $t \leftarrow t + 1$
13: **end while**

---

Step 1: Set the iteration counter $t$ to 1.

Step 2: Construct the OLMS with $m$ elementary membranes shown in Fig. 3.4.

Step 3: The termination criterion may be the prescribed maximal number of generations or the preset difference between the best solution searched and the optimal/desired solution of the problem.

Step 4: $N$ ants in the colony are randomly scattered across OLMS and each elementary membrane contains at least two ants.

Step 5: The number $g_i^t$ ($i = 1, 2, \ldots, m$) of iterations is randomly generated between $g_{min}$ and $g_{max}$ for the $i$th elementary membrane, where $g_{min}$ and $g_{max}$ are lower and upper limits of iterations for elementary membranes, respectively.

Steps 6–8: The ACS algorithm is independently performed in each of the $m$ elementary membranes. This process consists of the tour construction, local pheromone update and global pheromone update.

Step 9: Each compartment sends its best ant out into the skin membrane.

Step 10: In the skin membrane, the ACS algorithm is independently executed for $g_0^t$ iterations.

Step 11: Communication rules are used to exchange some information between the ants in the skin membrane and those in the elementary membranes.

Step 12: The iteration counter $t$ increases by 1.

### 3.3.3.2   Examples

Several TSP instances are used to show the ACOPS performance. First, how to set the number of elementary membranes is empirically discussed. Then the ACOPS is compared with its counterpart ACS algorithm and the first MIEAs [47]. All the TSP benchmarks are available in [59]. To discuss how to choose the number $m$ of elementary membranes and the number $g_i$ $(i = 1, 2, \ldots, m)$ of generations for each elementary membrane, four TSP benchmarks including Eil76, Eil101, Ch130 and Ch150, are tested. Parameter setting is as follows: the number $N = 40$ of ants, the parameters $\alpha = 1$ and $\beta = 3$ for determining the relative importance of the pheromone values, the global pheromone decay coefficient $\rho = 0.6$, the local pheromone decay coefficient $\upsilon = 0.1$ and the user-defined parameter $q_0 = 0.9$, and 10000 function evaluations as the termination condition.

The effect of the number of elementary membranes on the ACOPS performance is first investigated. Let $m$ vary from 2 to 20, $g_{min} = 10$ and $g_{max} = 30$. The mean of best solutions and their corresponding mean of elapsed time per run, out of 20 independent runs, are used to evaluate the performances of ACOPS for each of the 19 cases. Experimental results are shown in Figs. 3.11, 3.12, 3.13 and 3.14. It can be seen from the figures that there are some general trends. The fluctuant behavior can be observed from the best solutions and the mean of best solutions over 20 runs. A general increase as $m$ goes up from 2 to 20 can be seen from the elapsed time per run. The experimental results indicate that $m$ could be assigned as about 4 by trading off the quality of solutions and the elapsed time.

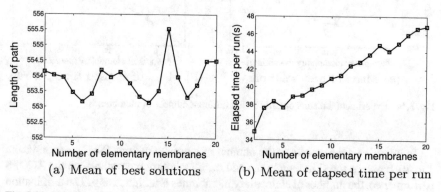

(a) Mean of best solutions      (b) Mean of elapsed time per run

**Fig. 3.11** Experimental results of Eil76 with different number of membranes

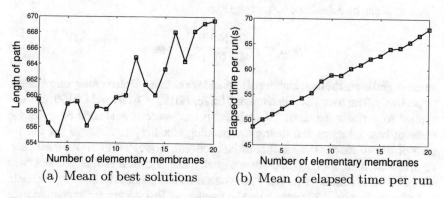

(a) Mean of best solutions      (b) Mean of elapsed time per run

**Fig. 3.12** Experimental results of Eil101 with different number of membranes

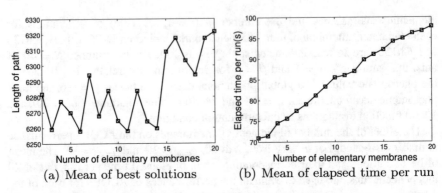

(a) Mean of best solutions          (b) Mean of elapsed time per run

**Fig. 3.13** Experimental results of Ch130 with different number of membranes

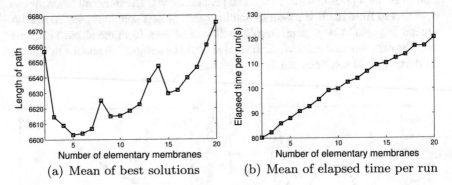

(a) Mean of best solutions          (b) Mean of elapsed time per run

**Fig. 3.14** Experimental results of Ch150 with different number of membranes

To investigate the effects of the number of communications (NoC), varying from 1 to 40, between the skin membrane and the elementary membranes on the ACOPS performance, the number of elementary membranes is assigned as 4. The termination criterion is the maximal number 10000 of function evaluations (NoFE), $g_{min} = 10$. Thus, $g_{max}$ can be calculated by using (3.3).

$$g_{max} = \frac{2 \cdot NoFE}{NoC \cdot (N + m)} - g_{min} \qquad (3.3)$$

where $N$ and $m$ are the total number of ants and the number of elementary membranes, respectively. The four TSP benchmarks, Eil76, Eil101, Ch130 and Ch150, are also applied to perform the tests. The ACOPS performance is evaluated by using the mean of best solutions and their corresponding mean of elapsed time per run for each of the 40 cases. The number of independent runs is set to 20 for each case. Figures 3.15, 3.16, 3.17 and 3.18 show experimental results. It is observed that there is a behavior oscillating between various maxima and minima underlying the mean of best solutions over 20 runs. A drastic fluctuation first occurs in the elapsed time per run and a relatively steady level is the kept. So it is recommended that NoC could

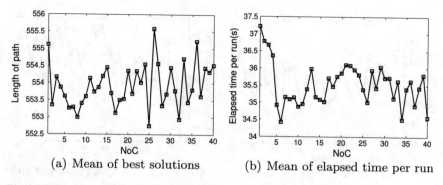

**Fig. 3.15**  Experimental results of Eil76 with different NoC

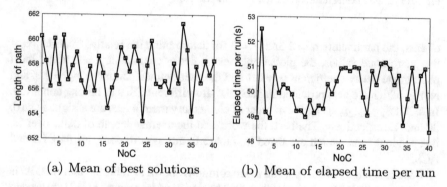

**Fig. 3.16**  Experimental results of Eil101 with different NoC

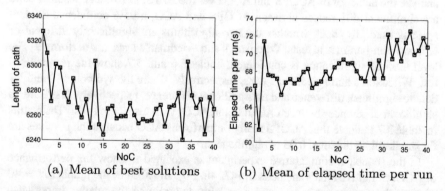

**Fig. 3.17**  Experimental results of Ch130 with different NoC

be chosen between 15 and 35 by trading off the quality of solutions and the elapsed time.

In what follows, an experimental comparison is drawn between ACOPS and its counterpart ACO. Twenty symmetric TSP benchmark problems are shown in Table 3.6. ACOPS and ACS use the same parameter setting, i.e., the number $N = 40$

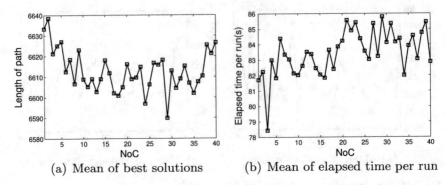

**Fig. 3.18** Experimental results of Ch150 with different NoC

of ants, the parameters $\alpha = 1$ and $\beta = 3$ for determining the relative importance of the pheromone values, the global pheromone decay coefficient $\rho = 0.6$, the local pheromone decay coefficient $\upsilon = 0.1$ and the user-defined parameter $q_0 = 0.9$, and NoFE, which are also provided in Table 3.6, for different TSPs as halting condition. In ACOPS, $g_{min}$, $g_{max}$ and the number of elementary membranes are assigned as 10, 30 and 4, respectively. The best, the worst and the average length of paths over 20 independent runs, which are listed in Table 3.6, are used to compare ACOPS and ACS.

It is observed from the best and mean solutions shown in Table 3.6 that ACOPS is superior to ACS in 19 out of 20 instances. The statistical techniques in [17] are used to analyze the behavior of ACOPS and ACO over the 20 TSPs. To check whether there are significant differences between ACOPS and ACO, a 95% confidence Student $t$-test is used. To check whether the two algorithms are significantly different or not, two non-parametric tests, Wilcoxon's and Friedman's tests, are performed. The level 0.05 of significance is considered. Tables 3.6 and 3.7 show the results of $t$-test, Wilcoxon's and Friedman's tests, respectively, where the symbols '+' and '−' denote significant difference and no significant difference, respectively. There are 16 significant differences between ACOPS and ACO, with respect to $t$-test. The results in Table 3.7 indicate that ACOPS really outperforms ACO because the $p$-values are far smaller than the level 0.05 of significance.

In the sequel, a comparative experiment is executed to show the performance of ACOPS and Nishida's methods in [47, 48, 50]. In ACOPS, the number $N = 40$ of ants, the parameters $\alpha = 1$ and $\beta = 3$ for determining the relative importance of the pheromone values, the global pheromone decay coefficient $\rho = 0.6$, the local pheromone decay coefficient $\upsilon = 0.1$ and the user-defined parameter $q_0 = 0.9, m = 4$, $g_{min} = 10$ and $g_{max} = 30$. The number 20 of trials and a prescribed NoFE as halting condition are used as the same as Nishida's methods. Tables 3.8, 3.9, 3.10 and 3.11 show the results, where NoM represents the number of membranes and the results of Nishida's algorithm are obtained from [48, 50]. Table 3.11 shows the results of Wilcoxon's and Friedman's tests. The results indicate that ACOPS is superior to

**Table 3.6** A comparison between ACOPS and ACS ('+' and '−' represent significant difference and no significant difference, respectively. '−' means no optimum available)

| TSP | NoFE | ACO | | | ACOPS | | | t-test | Optimum |
|---|---|---|---|---|---|---|---|---|---|
| | | Best | Average | Worst | Best | Average | Worst | | |
| ulysses16 | 1e+4 | 74.11 | 74.11 | 74.11 | 73.99 | 74.02 | 74.23 | 3.47e-6(+) | 74 |
| att48 | 2e+4 | 33588.34 | 33654.16 | 33740.35 | 33523.71 | 33644.97 | 34060.49 | 7.05e-1(−) | 33524 |
| eil76 | 3e+4 | 545.39 | 546.22 | 551.93 | 544.37 | 551.62 | 555.55 | 3.48e-7(+) | 538 |
| kroA100 | 4e+4 | 21577.69 | 21776.91 | 22320.91 | 21285.44 | 21365.64 | 21552.00 | 4.00e-8(+) | 21282 |
| eil101 | 4e+4 | 642.66 | 652.30 | 684.19 | 640.98 | 648.48 | 664.24 | 2.03e-1(−) | 629 |
| lin105 | 4e+4 | 14383.00 | 14472.38 | 14482.31 | 14383.00 | 14444.77 | 14612.43 | 8.61e-2(+) | 14379 |
| ch130 | 4.5e+4 | 6204.09 | 6268.43 | 6333.16 | 6148.99 | 6205.54 | 6353.69 | 1.02e-3(+) | 6110 |
| gr137 | 4.5e+4 | 718.92 | 725.02 | 749.93 | 709.91 | 718.85 | 738.35 | 6.63e-3(+) | — |
| pr144 | 5e+4 | 58587.14 | 58612.82 | 58687.80 | 58535.22 | 58596.00 | 58761.43 | 2.24e-1(−) | 58537 |
| ch150 | 5e+4 | 6595.00 | 6630.59 | 6689.79 | 6548.89 | 6570.86 | 6612.46 | 1.05e-7(+) | 6528 |
| rat195 | 6e+4 | 2370.24 | 2392.69 | 2434.39 | 2348.32 | 2355.23 | 2373.79 | 1.61e-7(+) | 2323 |
| d198 | 6e+4 | 16172.77 | 16266.93 | 16530.79 | 16073.13 | 16192.89 | 16381.91 | 3.75e-2(+) | 15780 |
| kroa200 | 6e+4 | 29597.01 | 29988.74 | 30466.71 | 29453.10 | 29552.92 | 29688.13 | 1.92e-6(+) | 29437 |
| gr202 | 6e+4 | 496.48 | 496.96 | 499.53 | 488.41 | 494.21 | 499.44 | 6.82e-4(+) | — |
| tsp225 | 7e+4 | 4067.96 | 4146.32 | 4262.76 | 3904.46 | 3971.68 | 4044.32 | 2.11e-9(+) | 3916 |
| gr229 | 7e+4 | 1739.77 | 1763.80 | 1802.44 | 1725.84 | 1756.28 | 1792.91 | 2.11e-1(−) | — |
| gil262 | 8e+4 | 2452.82 | 2487.58 | 2512.85 | 2407.68 | 2431.58 | 2450.65 | 4.86e-10(+) | 2378 |
| a280 | 9e+4 | 2626.44 | 2683.21 | 2787.61 | 2595.31 | 2636.49 | 2728.06 | 8.46e-4(+) | 2579 |
| pr299 | 10e+4 | 51050.78 | 52103.27 | 53698.23 | 49370.69 | 51021.74 | 52251.21 | 7.69e-4(+) | 48191 |
| lin318 | 10e+4 | 44058.08 | 45297.99 | 46410.50 | 42772.12 | 43433.54 | 45194.62 | 5.95e-9(+) | 42029 |

**Table 3.7**  Results of non-parametric statistical tests for ACOPS and ACS in Table 3.6. '+' represents significant difference

| Tests | ACOPS versus ACO |
| --- | --- |
| Wilcoxon test($p$-value) | 1.6286e–004(+) |
| Friedman test($p$-value) | 5.6994e–005(+) |

Nishida's algorithm with respect to the quality of solutions, the number of NoFE and statistical tests.

### 3.3.4  Differential Evolution Based on P Systems

#### 3.3.4.1  Algorithm

To have a good balance between exploration and exploitation with a limited computational budget [34], it is advisable to consider the combination of a global optimization algorithm and a local search operator. This section discusses a membrane-inspired optimization algorithm called DEPS [5] by using a P system with a distributed parallel framework to properly organize differential evolution (described in Sect. 2.6) and the simplex method, which was introduced by Nelder and Mead [46].

To be specific, DEPS uses the OLMS with $m$ elementary membranes, shown in Fig. 3.4, to arrange the objects consisting of real-valued strings, the rules composed of mutation, crossover and selection in elementary membranes and evolution rules in P systems, and simplex search in the skin membrane. A solution of the problem is represented by a real-valued string. Real-valued strings, which are organized as individuals, are processed as multisets of objects. The P system in DEPS is composed of the following elements:

(i) a membrane structure, OLMS, $[[\ ]_1[\ ]_2, \ldots, [\ ]_m]_0$ with $m + 1$ regions delimited by $m$ elementary membranes labeled by $1, \ldots, m$, respectively, and the skin membrane labeled by 0;

(ii) an alphabet with all real-valued strings used;

(iii) the set of terminal symbols formed by real-valued strings;

(iv) initial multisets associated with each membrane

$$
\begin{aligned}
w_0 &= \lambda, \\
w_1 &= x_1 x_2 \ldots x_{n_1}, \\
w_2 &= x_{(n_1+1)} x_{(n_1+2)} \ldots x_{n_2}, \\
&\ldots \ldots \ldots \\
w_m &= x_{(n_{(m-1)}+1)} x_{(n_{(m-1)}+2)} \ldots x_{n_m},
\end{aligned}
$$

**Table 3.8**  Comparisons of Nishida's algorithm and ACOPS on Eil51

| NoM | Nishida's algorithm | | | | | | ACOPS | | | | | |
|---|---|---|---|---|---|---|---|---|---|---|---|---|
| | 2 | 10 | 30 | 50 | 70 | | 4 | | | | | |
| | | | | | | | 1e+4 | 2e+4 | 3e+4 | 4e+4 | 5e+4 | |
| NoFE | 1.2e+5 | 7.6e+5 | 2.36e+6 | 3.96e+6 | 5.56e+6 | | | | | | | |
| Best | 440 | 437 | 432 | 429 | 429 | | 429.4841 | 429.4841 | 428.9816 | 428.9816 | 428.9816 | |
| Worst | 786 | 466 | 451 | 444 | 443 | | 435.5985 | 436.3928 | 434.9739 | 433.6050 | 433.8558 | |
| Average | 522 | 449 | 441 | 435 | 434 | | 432.3858 | 431.8023 | 431.3146 | 430.5506 | 430.4495 | |

**Table 3.9** Comparisons of Nishida's algorithm and ACOPS on KroA100

| NoM | Nishida's algorithm | | | | | | | ACOPS | | | | | | |
|---|---|---|---|---|---|---|---|---|---|---|---|---|---|---|
| | 2 | 10 | 30 | 50 | 70 | 100 | 4 | 1e+4 | 2e+4 | 4e+4 | 6e+4 | 8e+4 | 1e+5 |
| NoFE | 3e+5 | 1.9e+6 | 5.9e+6 | 9.9e+6 | 1.39e+7 | 1.99e+7 | 1e+4 | 21331 | 21285 | 21285 | 21285 | 21285 | 21285 |
| Best | 23564 | 21776 | 21770 | 21651 | 21544 | 21299 | 21331 | 22332 | 21665 | 21552 | 21475 | 21427 | 21575 |
| Worst | 82756 | 24862 | 23940 | 24531 | 23569 | 22954 | 22332 | 21593 | 21407 | 21367 | 21337 | 21320 | 21362 |
| Average | 34601 | 23195 | 22878 | 22590 | 22275 | 21941 | 21593 | | | | | | |

**Table 3.10** Comparisons of Nishida's algorithm and ACOPS on 8 TSPs

| TSP | Nishida's algorithm | | | | ACOPS | | | |
|---|---|---|---|---|---|---|---|---|
| | NoFE | Best | Average | Worst | NoFE | Best | Average | Worst |
| ulysses22 | 9.9e+7 | 75.31 | 75.31 | 75.31 | 2e+4 | 75.31 | 75.32 | 75.53 |
| eil51 | 9.9e+7 | 429 | 434 | 444 | 4e+4 | 429 | 431 | 434 |
| eil76 | 9.9e+7 | 556 | 564 | 575 | 6e+4 | 546 | 551 | 558 |
| eil101 | 9.9e+7 | 669 | 684 | 693 | 8e+4 | 641 | 647 | 655 |
| kroA100 | 9.9e+7 | 21651 | 22590 | 24531 | 1e+5 | 21285 | 21320 | 21427 |
| ch150 | 9.9e+7 | 7073 | 7320 | 7633 | 1.2e+5 | 6534 | 6560 | 6584 |
| gr202 | 9.9e+7 | 509.7 | 520.1 | 528.4 | 1.4e+5 | 489.2 | 492.7 | 497.1 |
| tsp225 | 9.9e+7 | 4073.1 | 4153.6 | 4238.9 | 7e+4 | 3899.6 | 3938.2 | 4048.2 |

**Table 3.11** Results of non-parametric statistical tests for Nishida's algorithm and ACOPS in Table 3.10. '+' represents significant difference

| Tests | ACOPS versus Nishida's algorithm |
|---|---|
| Wilcoxon test($p$-value) | 0.0156 (+) |
| Friedman test($p$-value) | 0.0339 (+) |

where $x_j$ $(1 \leq j \leq N)$ is an ant; $n_i$ $(1 \leq i \leq m)$ is the number of individuals in $w_i$; $\sum_{i=1}^{m} n_i = N$, where $N$ is the total number of individuals in the computation;
(v) rules with the following types:

    (a) evolution rules consisting of mutation, crossover and selection in each of the elementary membranes 1 to $m$;

    (b) evolution rules composed of reflection, expansion, contraction and shrinking in the skin membrane;

    (c) communication rule responsible for sending the best $k_i$ individuals from the $i$th elementary membrane into the skin membrane;

    (d) communication rule responsible for sending the $k_i$ individuals from the skin membrane back to the $i$th elementary membrane.

Algorithm 4 shows the DEPS pseudocode algorithm, where each step is elaborated as follows:

Step 1: Initially, the iteration counter $t$ is set to 1.

Step 2: In the initialization, an OLMS with $m$ elementary membranes shown in Fig. 3.4 is constructed. A population with $N$ individuals is randomly scattered across the $m$ elementary membranes. Each individual is formed by $D$ real-valued numbers. Inside each elementary membrane, there are at least $n_0$ individuals. Thus, the subpopulation in the $i$th elementary membrane is represented as $P_i(t)$ with a size of $n_i$ $(i = 1, 2, \ldots, m)$.

Step 3: Determine the numbers $\mathbf{k} = (k_1, k_2, \ldots, k_m)$ of the objects for executing the communication rules in the elementary membranes with the criterion that $k_i$ is

---

**Algorithm 4** Pseudocode algorithm of the DEPS

---

1: $t = 1$
2: Initialize the membrane structure
3: Generate **k**
4: **while** (not termination condition) **do**
5:     **for** $i = 1, 2, \ldots, m$ **do**
6:         Perform DE inside the $i$th elementary membrane
7:     **end for**
8:     Execute the communication rules (b)
9:     Perform simplex search in skin membrane
10:    Execute the communication rules (c)
11:    $t \leftarrow t + 1$
12: **end while**

---

proportional to $n_i$ and the sum of $k_i$ $(i = 1, 2, \ldots, m)$ is $D + 1$, where $D$ is the number of variables of the problem.

Step 4: Determine the halting condition, which could be the prescribed maximal number of generations or the preset difference between the best solution searched and the optimal/desired solution of the problem.

Steps 5–7: The evolutionary procedures of DE shown in Algorithm 8 in Sect. 2.6 are independently executed in each elementary membrane to evolve the objects.

Step 8: This step, together with Step 10, is used to fulfill the information exchange among individuals in the elementary membranes and the skin membrane. The best $k_i$ individuals from $P_i(t)$ with respect to fitness values in the $i$th elementary membrane form a subpopulation $Q_i(t)$ $(i = 1, 2, \ldots, m)$. A copy of the individual in $Q_i(t)$ is sent out into the skin membrane.

Step 9: The simplex local search is performed inside the skin membrane on the population $Q_0(t)$, $Q_0(t) = (Q_1(t), Q_2(t), \ldots, Q_m(t))$. The pseudocode simplex method is shown in Algorithm 5, where $\alpha, \beta, \gamma$ are the coefficients for the reflection, contraction, expansion and shrinking operations, respectively. This process will stop when a predetermined maximal number of iterations or the number of function evaluations is satisfied. The original population $Q_0(t)$ evolves to $\bar{Q}_0(t)$, and the sub-population $Q_i(t)$ in $Q_0(t)$ is made to evolve to $\bar{Q}_i(t)$ $(i = 1, 2, \ldots, m)$.

Step 10: Store the overall best individual among $\bar{Q}_0(t)$ into $X^*(t)$. The type (d) communication rule is used to send each individual in $\bar{Q}_i(t)$ back to the $i$th elementary membrane to replace the worst $k_i$ individuals in $P_i(t)$ $(i = 1, 2, \ldots, m)$.

Step 11: The iteration counter $t$ increases by 1.

The computation of DEPS starts from the initial configuration with initial multisets $w_i$ $(i = 1, 2, \ldots, m)$ and evolution rules inside each membrane. According to the rules in each region available in a non-deterministic and maximally parallel manner, the system goes from one configuration to a new one. If no rules can be applied to the existing objects in any region, the computation halts and the output is collected from the skin membrane.

---

**Algorithm 5** Pseudocode algorithm of simplex in [46]

---

**Require:** A population with size $D + 1$
1: **while** (not termination condition) **do**
2:    Determine the worst and the best individuals as $\mathbf{x}_h$ and $\mathbf{x}_l$
3:    Calculate the centroid of $\mathbf{x}_i$ $(i \neq h)$ as $\bar{\mathbf{x}}$
4:    $\mathbf{x}_r = (1 + \alpha)\bar{\mathbf{x}} - \alpha\mathbf{x}_h$ and $y_r = f(\mathbf{x}_r)$
5:    **if** $y_l < y_r < y_h$ **then**
6:       Replace $\mathbf{x}_h$ with $\mathbf{x}_r$
7:    **else if** $y_r < y_l$ **then**
8:       $\mathbf{x}_e = \gamma\mathbf{x}_r + (1 - \gamma)\bar{\mathbf{x}}$ and $y_e = f(\mathbf{x}_e)$
9:       **if** $y_e < y_l$ **then**
10:          Replace $\mathbf{x}_h$ with $\mathbf{x}_e$
11:       **else**
12:          Replace $\mathbf{x}_h$ with $\mathbf{x}_r$
13:       **end if**
14:    **else if** $y_r > y_i$ $(i \neq h)$ **then**
15:       **if** $y_r \leq y_h$ **then**
16:          Replace $\mathbf{x}_h$ with $\mathbf{x}_r$
17:       **end if**
18:       $\mathbf{x}_c = \beta\mathbf{x}_h + (1 - \beta)\bar{\mathbf{x}}$ and $y_c = f(\mathbf{x}_c)$
19:       **if** $y_c \leq y_h$ **then**
20:          Replace $\mathbf{x}_h$ with $(\mathbf{x}_i + \mathbf{x}_l)/2$
21:       **else**
22:          Replace $\mathbf{x}_h$ with $\mathbf{x}_c$
23:       **end if**
24:    **end if**
25: **end while**

---

### 3.3.4.2 Examples

To show the effectiveness of DEPS, the experiments are carried out on several benchmark functions with various dimensions in continuous spaces. In what follows, the description on the benchmark functions is first presented and then the number $m$ of elementary membranes is empirically investigated. Experimental comparisons between DE and DEPS are finally drawn.

The experiments use the test suit [70] shown in Table 3.12, which consists of nine difficultly unconstrained benchmark functions: four unimodal functions, $f_1, f_2,$ $f_3$ and $f_4$, and five multi-modal functions, $f_5, f_6, f_7, f_8$ and $f_9$. DEPS has the following presetting parameters: population size $N$, the differential factor $F$, the crossover rate $C$, the reflection coefficient $\alpha$, expansion coefficient $\gamma$ and contraction coefficient $\beta$ in simplex search, the minimum size $n_0$ of the population in elementary membranes, the number $m$ of the elementary membranes, and 3 sub-termination criterions which include the outmost termination condition $TC_0$ as shown in the *while* loop of Algorithm 4, the termination condition $TC_1$ of DE in elementary membranes, and the termination condition $TC_2$ of simplex search in the skin membrane.

To investigate the setting of the number $m$ of elementary membranes, the experiments are performed on six benchmark functions, $f_1, f_3, f_5, f_6, f_7$ and $f_8$ with 30 dimensions. Parameter setting is as follows: $N = 3 \times D, F = 0.6, C = 0.2, \alpha = 1, \gamma = 2,$

**Table 3.12**  Nine benchmark functions [70], where $D$ is the dimension of functions

| Function | Name | Range | Optimum |
|---|---|---|---|
| $f_1$ | Sphere model | $[-100, 100]^D$ | 0 |
| $f_2$ | Schwefel problem 1.2 | $[-100, 100]^D$ | 0 |
| $f_3$ | Schwefel problem 2.21 | $[-100, 100]^D$ | 0 |
| $f_4$ | Generalized Rosenbrock's function | $[-30, 30]^D$ | 0 |
| $f_5$ | Generalized Schwefel's problem 2.26 | $[-500, 500]^D$ | 0 |
| $f_6$ | Generalized Rastrigin's function | $[-5.12, 5.12]^D$ | 0 |
| $f_7$ | Ackley's function | $[-32, 32]^D$ | 0 |
| $f_8$ | Generalized Griewark function | $[-600, 600]^D$ | 0 |
| $f_9$ | Generalized Penalized function | $[-50, 50]^D$ | 0 |

**Table 3.13**  Mean best solutions over 30 runs. The number $m$ of elementary membranes varies from 1 to 10, and the dimensions $D$ of all functions are 30

| $m$ | $f_1$ | $f_3$ | $f_5$ | $f_6$ | $f_7$ | $f_8$ |
|---|---|---|---|---|---|---|
| 1 | 1.65e−08 | 4.57e−07 | 2.52e+01 | 6.20e+00 | 3.66e−07 | 7.33e−09 |
| 2 | 1.08e−18 | 1.14e−07 | 2.49e+01 | 4.45e+00 | 1.83e−10 | 3.02e−12 |
| 3 | 4.41e−26 | 5.13e−08 | 2.49e+01 | 5.76e+00 | 5.03e−14 | 3.35e−20 |
| 4 | 2.49e−25 | 2.63e−08 | 2.67e+01 | 6.69e+00 | 2.72e−14 | 1.32e−30 |
| 5 | 1.37e−37 | 5.67e−08 | 3.10e+01 | 9.01e+00 | 2.58e−14 | 6.88e−29 |
| 6 | 3.07e−39 | 1.60e−07 | 2.83e+01 | 8.84e+01 | 2.07e−14 | 3.70e−46 |
| 7 | 1.94e−27 | 9.63e−07 | 5.03e+01 | 7.77e+00 | 2.01e−14 | 2.84e−30 |
| 8 | 3.60e−26 | 1.90e−06 | 3.85e+01 | 9.22e+00 | 2.13e−14 | 1.66e−29 |
| 9 | 1.25e−22 | 3.11e−06 | 1.37e+02 | 1.27e+01 | 2.32e−14 | 1.65e−22 |
| 10 | 1.10e−22 | 3.74e−06 | 1.18e+02 | 1.39e+01 | 3.17e−14 | 2.32e−17 |

$\beta = 0.5$, $n_0 = 6$. $m$ varies from 1 to 10. $TC_0$ uses the maximal number $1.5e + 05$ of function evaluations. $TC_1$ and $TC_2$ refer to the halting criterions that the best solutions found are not changed at 20 and 50 successive generations, respectively. The independent runs is 30. Experimental results are shown in Table 3.13 with respect to the mean of the best solutions. To clearly show the difference, Fig. 3.19 gives the logarithm values of mean best values against the number $m$ of elementary membranes. The results indicate that the DEPS performance varies with the values of $m$ and the value of $m$ could be assigned as 4 by considering both unimodal and multimodal functions.

The following experiments are used to compare DEPS with DE. The tests use the nine benchmark functions with dimensions $D$ varying from 10 to 100 at intervals of 10. DEPS with DE have the following common parameter setting: $N = 3 \times D$, $F = 0.6$, $C = 0.2$ and the same $TC_0$ which are $1e + 05$, $1.25e + 05$, $1.5e + 05$, $1.75e + 05$, $2e + 05$, $2.25e + 05$, $2.5e + 05$, $2.75e + 05$, $3e + 05$, $3.25e + 05$ for ten different dimensions $D$, respectively. DEPS has several additional parameters: $m = 4$; $\alpha =$

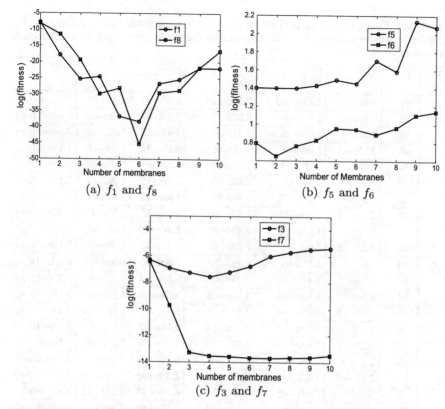

**Fig. 3.19**  The mean best solutions over 30 runs as the number $m$ of elementary membranes

$1, \gamma = 2, \beta = 0.5, n_0 = 6$, the same $TC_1$ and $TC_2$ as the above tests on the discussion of $m$. The independent runs is 30.

Experimental results are shown in Tables 3.14 and 3.15 with respect to the mean and standard deviation of the best solutions. The statistical tests [17], "$t$-test", Wilcoxon's and Friedman's tests, are used to check whether there is a significant difference between DEPS and DE or not. The "$t$-test" is a single-problem analysis technique. The confidence level is set to 0.95. The results are listed in last column in Tables 3.14 and 3.15, where the symbols "+" and "−" indicate that there is a significant difference or not between DEPS and DE with respect to each test problem. Wilcoxon's and Friedman's tests are two multi-problem analysis techniques. The significance level is set to 0.05. Table 3.16 show the results, where the symbol "+" means that there is a significant difference between the two algorithms.

As indicated in Tables 3.14 and 3.15, DEPS is superior to DE due to more better results in the test problems and 78 symbols "+" out of 90 test problems. The results in Table 3.16 of Wilcoxon's and Friedman's tests indicate that DEPS really outperforms DE due to far smaller values than the level 0.05 of significance.

**Table 3.14** Comparisons of DE and DEPS on $f_1$–$f_5$. $D$, FEs, Mean and Std represent dimension, the number of function evaluations, mean of best solutions and standard deviation over 30 independent runs, respectively

| $f$ | $D$ | FEs | DE | | DEPS | | $t$-test |
|---|---|---|---|---|---|---|---|
| | | | Mean | Std | Mean | Std | |
| $f_1$ | 10 | 1e+5 | **2.19e–120** | **7.38e–120** | 3.55e–59 | 1.93e–58 | 3.19e–01(–) |
| | 20 | 1.25e+5 | 1.30e–35 | 8.53e–36 | **2.13e–67** | **1.16e–66** | 1.69e–11(+) |
| | 30 | 1.5e+5 | 1.68e–16 | 5.41e–17 | **2.49e–25** | **7.62e–25** | 3.68e–24(+) |
| | 40 | 1.75e+5 | 1.04e–08 | 2.11e–09 | **4.59e–16** | **2.49e–15** | 1.93e–34(+) |
| | 50 | 2e+5 | 1.10e–04 | 1.73e–05 | **1.33e–09** | **4.47e–09** | 1.26e–40(+) |
| | 60 | 2.25e+5 | 3.24e–02 | 4.80e–03 | **2.13e–07** | **4.96e–07** | 3.19e–42(+) |
| | 70 | 2.5e+5 | 1.25e+00 | 1.54e–01 | **1.89e–04** | **9.84e–04** | 1.67e–46(+) |
| | 80 | 2.75e+5 | 1.70e+01 | 1.53e+00 | **2.10e–03** | **6.70e–03** | 2.70e–54(+) |
| | 90 | 3e+5 | 1.09e+02 | 8.80e+00 | **5.30e–03** | **1.25e–02** | 6.29e–57(+) |
| | 100 | 3.25e+5 | 4.46e+02 | 2.29e+01 | **3.68e–02** | **1.68e–01** | 3.19e–42(+) |
| $f_2$ | 10 | 1e+5 | 4.31e–09 | 6.76e–09 | **0.00e–00** | **0.00e–00** | 9.25e–04(+) |
| | 20 | 1.25e+5 | 9.11e+02 | 2.25e+02 | **6.19e–02** | **2.13e–01** | 5.22e–30(+) |
| | 30 | 1.5e+5 | 1.12e+04 | 1.97e+03 | **1.72c–00** | **4.87e+00** | 7.16e–38(+) |
| | 40 | 1.75e+5 | 2.80e+04 | 3.49e+03 | **5.51e–01** | **1.16e+00** | 2.88e–46(+) |
| | 50 | 2e+5 | 5.10e+04 | 5.00e+03 | **1.21e–01** | **2.10e–01** | 3.84e–52(+) |
| | 60 | 2.25e+5 | 7.94e+04 | 5.30e+03 | **2.77e+00** | **3.67e+00** | 1.10e–61(+) |
| | 70 | 2.5e+5 | 1.06e+05 | 1.01e+04 | **8.86e+00** | **4.97e+00** | 5.32e–53(+) |
| | 80 | 2.75e+5 | 1.42e+05 | 9.57e+03 | **2.85e+01** | **1.53e+01** | 1.87e–61(+) |
| | 90 | 3e+5 | 1.84e+05 | 1.30e+04 | **6.43e+01** | **2.23e+01** | 2.46e–60(+) |
| | 100 | 3.25e+5 | 2.25e+05 | 1.33e+04 | **1.27e+02** | **3.70e+01** | 9.62e–65(+) |
| $f_3$ | 10 | 1e+5 | **2.89e–27** | **2.06e–27** | 2.60e–03 | 1.20e–02 | 2.46e–01(–) |
| | 20 | 1.25e+5 | 2.00e–05 | 4.25e–06 | **1.52e–08** | **4.37e–09** | 1.81e–33(+) |
| | 30 | 1.5e+5 | 1.52e–01 | 1.88e–01 | **2.63e–08** | **2.11e–08** | 2.04e–46(+) |
| | 40 | 1.75e+5 | 3.24e+00 | 2.36e–01 | **3.87e–06** | **3.00e–06** | 1.44e–59(+) |
| | 50 | 2e+5 | 1.23e+01 | 5.12e–01 | **2.18e–04** | **1.10e–04** | 1.47e–73(+) |
| | 60 | 2.25e+5 | 2.43e+01 | 1.32e+00 | **5.00e–03** | **2.80e–03** | 6.92e–67(+) |
| | 70 | 2.5e+5 | 3.57e+01 | 1.11e+00 | **3.41e–02** | **1.66e–02** | 6.64e–81(+) |
| | 80 | 2.75e+5 | 4.46e+01 | 1.05e+00 | **1.55e–01** | **5.62e–02** | 1.17e–87(+) |
| | 90 | 3e+5 | 5.16e+01 | 1.27e+00 | **4.62e–01** | **1.53e–01** | 2.31e–86(+) |
| | 100 | 3.25e+5 | 5.65e+01 | 1.01e+00 | **9.32e–01** | **2.88e–01** | 1.85e–93(+) |
| $f_4$ | 10 | 1e+5 | 7.81e–01 | 1.29e+00 | **4.57e–02** | **1.01e–01** | 2.90e–03(+) |
| | 20 | 1.25e+5 | 1.58e+01 | **9.20e–01** | 1.41e+01 | 2.76e+00 | 1.60e–03(+) |
| | 30 | 1.5e+5 | 2.47e+01 | **1.47e+00** | 2.67e+01 | 1.11e+01 | 4.70e–07(+) |
| | 40 | 1.75e+5 | 4.71e+01 | 7.25e+00 | **3.36e+01** | **1.89e+00** | 4.28e–14(+) |
| | 50 | 2e+5 | 1.82e+02 | 3.34e+01 | **4.39e+01** | **4.47e+00** | 2.27e–30(+) |

(continued)

**Table 3.14** (continued)

| $f$ | $D$ | FEs | DE Mean | DE Std | DEPS Mean | DEPS Std | $t$-test |
|---|---|---|---|---|---|---|---|
| | 60 | 2.25e+5 | 6.37e+02 | 6.38e+01 | **6.58e+01** | **2.47e+01** | 3.30e−47(+) |
| | 70 | 2.5e+5 | 2.44e+03 | 1.95e+02 | **7.45e+01** | **2.14e+01** | 2.84e−56(+) |
| | 80 | 2.75e+5 | 1.51e+04 | 1.84e+03 | **9.60e+01** | **3.30e+01** | 1.42e−46(+) |
| | 90 | 3e+5 | 9.37e+04 | 1.22e+04 | **9.83e+01** | **2.98e+01** | 4.89e−45(+) |
| | 100 | 3.25e+5 | 4.85e+05 | 4.96e+04 | **1.06e+02** | **3.15e+01** | 4.67e−51(+) |
| $f_5$ | 10 | 1e+5 | **1.27e−04** | **0.00e+00** | 1.27e−04 | 0.00e+00 | 1.00e+00(−) |
| | 20 | 1.25e+5 | **2.55e−04** | 5.51e−20 | 2.55e−04 | **0.00e+00** | 1.00e+00(−) |
| | 30 | 1.5e+5 | **3.82e−04** | **1.65e−19** | 3.83e−04 | 3.86e−06 | 3.22e−01(−) |
| | 40 | 1.75e+5 | 6.13e+00 | 6.04e+00 | **5.09e−04** | **1.10e−19** | 7.09e−07(+) |
| | 50 | 2e+5 | 5.75e+03 | 2.49e+02 | **1.58e+01** | **4.09e+01** | 3.88e−72(+) |
| | 60 | 2.25e+5 | 9.81e+03 | 3.69e+02 | **2.76e+01** | **5.62e+01** | 1.09e−75(+) |
| | 70 | 2.5e+5 | 1.33e+04 | **3.66e+02** | **2.65e+03** | 2.09e+03 | 7.66e−35(+) |
| | 80 | 2.75e+5 | 1.66e+04 | 2.39e+03 | **7.19e+03** | **4.58e+02** | 4.47e−29(+) |
| | 90 | 3e+5 | 2.02e+04 | **4.57e+02** | **1.09e+04** | 2.15e+03 | 4.12e−31(+) |
| | 100 | 3.25e+5 | 2.35e+04 | **4.46e+02** | **1.36e+04** | 2.44e+03 | 9.73e−30(+) |

**Table 3.15** Comparisons of DE and DEPS on $f_6$–$f_9$. $D$, FEs, Mean and Std represent dimension, the number of function evaluations, mean of best solutions and standard deviation over 30 independent runs, respectively

| $f$ | $D$ | FEs | DE Mean | DE Std | DEPS Mean | DEPS Std | $t$-test |
|---|---|---|---|---|---|---|---|
| $f_6$ | 10 | 1e+5 | **0.00e+00** | **0.00e+00** | 6.63e−01 | 1.12e+00 | 1.90e−03(−) |
| | 20 | 1.25e+5 | **2.37e−08** | **7.73e−08** | 1.93e+00 | 1.84e+00 | 3.52e−07(−) |
| | 30 | 1.5e+5 | 5.45e+01 | 4.36e+00 | **6.69e+00** | **3.57e+00** | 1.98e−47(+) |
| | 40 | 1.75e+5 | 1.21e+02 | 6.69e+00 | **1.17e+01** | **4.30e+00** | 1.95e−59(+) |
| | 50 | 2e+5 | 1.97e+02 | **9.46e+00** | **2.75e+01** | 1.29e+01 | 2.00e−53(+) |
| | 60 | 2.25e+5 | 2.86e+02 | **1.19e+01** | **7.52e+01** | 3.72e+01 | 1.11e−36(+) |
| | 70 | 2.5e+5 | 3.81e+02 | **1.10e+01** | **1.24e+02** | 4.55e+01 | 4.79e−37(+) |
| | 80 | 2.75e+5 | 4.84e+02 | **1.19e+01** | **2.06e+02** | 6.23e+01 | 7.26e−32(+) |
| | 90 | 3e+5 | 5.90e+02 | **1.11e+01** | **2.62e+02** | 8.54e+01 | 1.38e−28(+) |
| | 100 | 3.25e+5 | 7.01e+02 | **1.61e+01** | **2.49e+02** | 1.22e+02 | 1.07e−27(+) |
| $f_7$ | 10 | 1e+5 | 4.20e−15 | **9.01e−16** | **3.73e−15** | 1.45e−15 | 1.33e−01(−) |
| | 20 | 1.25e+5 | 4.20e−15 | **9.01e−16** | **3.73e−15** | 1.45e−15 | 1.56e−01(−) |
| | 30 | 1.5e+5 | 3.29e−09 | 6.55e−10 | **2.72e−14** | **2.72e−14** | 5.62e−35(+) |
| | 40 | 1.75e+5 | 2.19e−05 | 5.61e−06 | **2.21e−08** | **1.13e−07** | 5.62e−35(+) |
| | 50 | 2e+5 | 2.10e−03 | 1.70e−04 | **1.83e−06** | **5.10e−06** | 4.64e−57(+) |

(continued)

**Table 3.15** (continued)

| $f$ | $D$ | FEs | DE | | DEPS | | $t$-test |
|---|---|---|---|---|---|---|---|
| | | | Mean | Std | Mean | Std | |
| | 60 | 2.25e+5 | 3.57e–02 | 2.50e–03 | **6.17e–05** | **1.26e–04** | 5.74e–60(+) |
| | 70 | 2.5e+5 | 3.10e–01 | 2.07e–02 | **6.85e–04** | **1.90e–03** | 1.79e–61(+) |
| | 80 | 2.75e+5 | 1.77e+00 | 8.21e–02 | **3.00e–03** | **6.10e–03** | 1.03e–70(+) |
| | 90 | 3e+5 | 3.29e+00 | 7.30e–02 | **6.00e–03** | **1.33e–02** | 6.53e–89(+) |
| | 100 | 3.25e+5 | 4.52e+00 | 7.92e–02 | **6.00e–03** | **1.32e–02** | 6.27e–95(+) |
| $f_8$ | 10 | 1e+5 | **7.43e–102** | **2.20e–101** | 2.07e–02 | 2.33e–02 | 9.17e–06(–) |
| | 20 | 1.25e+5 | 3.93e–34 | 1.31e–33 | **7.56e–66** | **3.04e–65** | 1.06e–01(+) |
| | 30 | 1.5e+5 | 3.85e–15 | 4.64e–15 | **1.32e–30** | **5.78e–30** | 2.92e–05(+) |
| | 40 | 1.75e+5 | 4.43e–08 | 2.76e–08 | **5.14e–15** | **2.81e–14** | 2.75e–12(+) |
| | 50 | 2e+5 | 2.46e–04 | 6.78e–05 | **1.33e–10** | **4.01e–10** | 1.36e–27(+) |
| | 60 | 2.25e+5 | 4.52e–02 | 1.21e–02 | **2.16e–07** | **5.47e–07** | 3.49e–28(+) |
| | 70 | 2.5e+5 | 7.13e–01 | 4.15e–02 | **4.11e–05** | **1.38e–04** | 4.01e–65(+) |
| | 80 | 2.75e+5 | 1.15e+00 | 1.08e–02 | **5.45e–04** | **1.40e–03** | 9.66e–111(+) |
| | 90 | 3e+5 | 1.96e+00 | 6.36e–02 | **1.01e–02** | **3.32e–02** | 1.27e–76(+) |
| | 100 | 3.25e+5 | 5.02e+00 | 3.03e–01 | **9.60e–04** | **3.30e–03** | 2.58e–64(+) |
| $f_9$ | 10 | 1e+5 | **4.71e–32** | **0.00e+00** | **4.71e–32** | **0.00e+00** | 1.00e+00(–) |
| | 20 | 1.25e+5 | **2.36e–32** | **0.00e+00** | **2.36e–32** | **0.00e+00** | 1.00e+00(–) |
| | 30 | 1.5e+5 | 3.62e–17 | 1.77e–17 | **7.78e–24** | **4.26e–23** | 3.76e–16(+) |
| | 40 | 1.75e+5 | 4.80e–09 | 1.19e–09 | **1.89e–15** | **1.03e–14** | 7.29e–30(+) |
| | 50 | 2e+5 | 1.89e–04 | 4.76e–05 | **1.80e–14** | **8.05e–14** | 1.60e–29(+) |
| | 60 | 2.25e+5 | 2.83e–01 | 6.86e–02 | **6.61e–11** | **3.37e–10** | 2.07e–30(+) |
| | 70 | 2.5e+5 | 4.69e+00 | 4.32e–01 | **2.52e–10** | **6.35e–10** | 1.17e–53(+) |
| | 80 | 2.75e+5 | 1.84e+01 | 2.10e+00 | **4.72e–09** | **1.47e–08** | 2.49e–48(+) |
| | 90 | 3e+5 | 1.65e+03 | 1.19e+03 | **2.28e–07** | **8.07e–07** | 3.06e–10(+) |
| | 100 | 3.25e+5 | 1.29e+05 | 4.21e+04 | **2.49e–07** | **5.01e–07** | 6.95e–24(+) |

**Table 3.16** Results of non-parametric statistical tests for DE and DEPS

| Tests | DE versus DEPS |
|---|---|
| Wilcoxon test($p$-value) | 1.8239e–013(+) |
| Friedman test($p$-value) | 8.2157e–015(+) |

This subsection further investigates the DEPS scalability by showing the changing trends of the mean best solutions as the dimension $D$. The experimental results are show in Fig. 3.20. It is observed that DE has advantage over DEPS with respect to small $D$ and DEPS outperforms DE as $D$ increases.

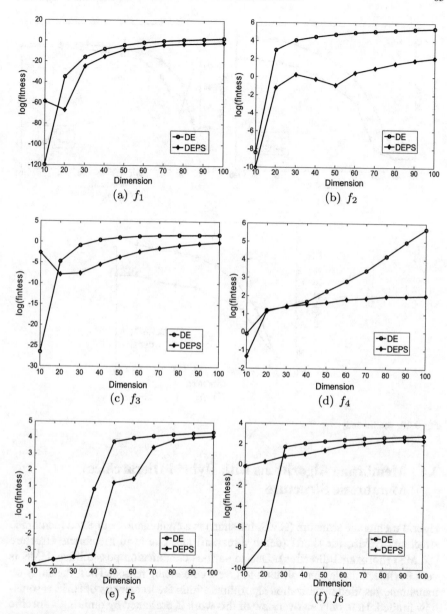

**Fig. 3.20** Logarithm values of mean best solutions over 30 runs for functions $f_1$–$f_9$ with various dimensions

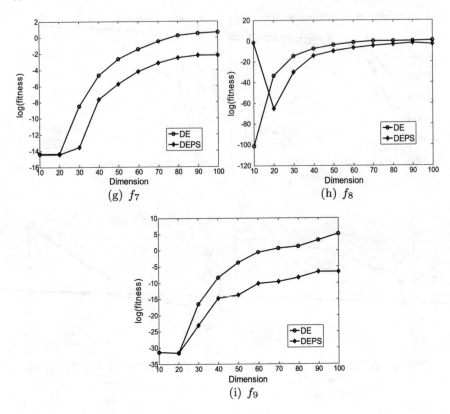

**Fig. 3.20** (continued)

## 3.4 Membrane Algorithms with Hybrid Hierarchical Membrane Structure

Hybrid membrane structure (HMS, for short) is a combination of nested membrane structures (NMS, for short) (depth direction) and one-level membrane structure (OLMS) (horizontal direction). Due to various combination possibilities, HMS is very flexible, complex and difficult to give a convincing reason. Thus, the study on membrane-inspired optimization algorithms within the framework of HMS is somehow limited. Here only an overview of this work is presented by omitting a specific example.

In [73], the HMS-based MIEAs was designed by using an NMS and a star topology to organize differential evolution. In [26, 31, 58, 76], two NMS were put inside the skin membrane to arrange GA or PSO and rules including transformation, communication, rewriting, splicing and/or uniport to design an HMS-based algorithm. Benchmark functions were used to test the algorithm performance. In [54], several NMS, each of which has two membranes, were placed inside the skin membrane to organize differential evolution and transformation and communication rules to build an HMS-based algorithm. An image segmentation problem was solved. HMS was also constructed by using two layers of OLMS [93]. In [26, 31], optimal controllers for a marine diesel engine and a time-varying unstable plant were discussed by using a single objective and a dynamic multi-objective MIEA, respectively.

## 3.5 Membrane Algorithms with Dynamic Hierarchical Membrane Structure

Differing from the static membrane structure with a fixed P system framework through the whole evolution like NMS, OLMS and HMS, dynamic membrane structure (DMS, for short) refers to a membrane structure which can be changed during the evolution. In what follows, the basic principle of DMS-based MIEAs is first discussed and then an approximate algorithm with instances of the P systems with active membranes and QIEA follows.

### 3.5.1 Brief Introduction

As a special hierarchical membrane structure coming from a cell-like P system with active membranes, DMS can produce an exponential workspace (in terms of the number of membranes) in linear time and consequently are able to efficiently solve NP-complete problems, from a mathematical perspective. Thus, the dynamic structure underlying the P systems might have a certain degree of influence on the population diversity which is a direct factor affecting the algorithm performance. Therefore, much attention has been paid to DMS.

In [27, 42–44], the DMS can be described as follows: at the beginning, a certain number of membranes are placed inside the skin membrane; as the system goes forward from a configuration to another, all the membranes inside the skin membrane merge into one membrane by using the merge rule of a P system; then the membrane inside the skin membrane is divided into a certain number of membranes by applying the division rule of a P system; the two processes of merging and division are repeated until the computation halts. In [42–44], DMS can be regarded as a changing OLMS. To be specific, the division and merging rules are performed on elementary membranes inside the skin membrane. In [72], a membrane structure with a star topology was considered as the initial membrane structure; then the dissolution of a membrane and the division of a membrane were performed at the same time, so the total number of regions is fixed. In [27], a certain number of NMS contained in the skin membrane were used to form the initial membrane structure and then recurrent division operations were applied to produce a new membrane structure. In [90], an MIEA was designed by using the separation and merging rules of the P system with active membranes. In this DMS, the number of elementary membranes inside the skin membrane varies with the number of iterations in a non-deterministic way. In [64], in response to the requirements of the mobile robot path planning problem, a dynamic double OLMS was introduced to organize the particles with various dimensions and fulfill the communications between particles in different membranes. DMS-based MIEAs have been used to various applications, such as benchmark functions [27, 43, 44], satisfiability problems [90], S-boxes in cryptographic algorithms [72], radar emitter signal analysis [42], and robot path planning [64].

### 3.5.2 Approximate Algorithm Using P Systems with Active Membranes

#### 3.5.2.1 Algorithm

The approach combining a QIEA and a P system with active membranes is called QEAM [90]. It is based on the hierarchical arrangement of the compartments and developmental rules (e.g., membrane separation, merging, transformation/communication-like rules) of a P system with active membranes model, and the objects consisting of Q-bit individuals, a probabilistic observation and the evolutionary rules designed with Q-gates to specify the membrane algorithms.

The dynamic P system in QEAM is composed of the following elements:

(i) a dynamic membrane structure $[[ \ ]_1[ \ ]_2, \ldots, [ \ ]_m]_0$ with $m$ regions labeled by $1, \ldots, m$, respectively, contained in the skin membrane, labeled by 0, where $m$ is a changing number with the evolution process;

(ii) an alphabet with the binary set $\{0, 1\}$ and all possible Q-bits;

(iii) a set of terminal symbols, $T = \{0, 1\}$;

(iv) initial multisets associated with each membrane

$$
\begin{aligned}
w_0 &= \lambda, \\
w_1 &= q_1 q_2 \cdots q_{n_1}, \\
w_2 &= q_{(n_1+1)} q_{(n_1+2)} \cdots q_{n_2}, \\
&\cdots \cdots \cdots \\
w_m &= q_{(n_{(m-1)}+1)} q_{(n_{(m-1)}+2)} \cdots q_{n_m},
\end{aligned}
$$

where $q_j$ $(1 \leq j \leq N)$ is a Q-bit individual; $n_i$ $(1 \leq i \leq m)$ is the number of individuals in $w_i$; $\sum_{i=1}^{m} n_i = N$, where $N$ is the total number of individuals in this computation;

(v) rules include the following types:

(a) $[a \to v]_h$, for $h \in H, a \in V, v \in V^*$; (*object evolution rules*, associated with membranes and depending on the label, but not directly involving the membranes, in the sense that the membranes are neither taking part in the application of these rules nor are they modified by them);

---

**Algorithm 6** Pseudocode algorithm of the QEAM

---

1: $t = 1$
2: Initialize the membrane structure
3: **while** (not termination condition) **do**
4:     Produce $\mathbf{k}_t = (k_{t_1}, k_{t_2}, \ldots, k_{t_m})$
5:     Perform object evolution rule (a) in elementary membranes
6:     Perform communication rule (c)
7:     Perform object evolution rule (a) in the skin membranes
8:     $t \leftarrow t + 1$
9:     Determine the number $m_{t+1}$ of elementary membranes
10:     **if** $m_{t+1} < m_t$ **then**
11:         Perform membrane merging rule (d)
12:     **else if** $m_{t+1} > m_t$ **then**
13:         Perform membrane separation rule (e)
14:     **end if**
15:     Perform communication rule (b)
16: **end while**

---

(b) $a[\ ]_h \rightarrow [b]_h$, for $h \in H, a, b \in V$; (*communication rules*; an object is introduced in membrane $h$, possibly modified during this process);

(c) $[a]_h \rightarrow [\ ]_h b$, for $h \in H, a, b \in V$; (*communication rules*; an object is sent out of membrane $h$, possibly modified during this process);

(d) $[\ ]_h [\ ]_h \rightarrow [\ ]_h$, for $h \in H$; (*merging rules for elementary membranes*; two elementary membranes with the same label are merged into a single membrane; the objects of the former membranes are put together in the new membrane);

(e) $[W]_h \rightarrow [U]_h [W - U]_h$, for $h \in H, U \subset W$; (*separation rules for elementary membranes*; the membrane is separated into two membranes with the same labels; the objects from $U$ are placed in the first membrane, those from $(W - U)$ are placed in the other membrane);

To easily understand QEAM, the steps are listed with pseudocodes in Algorithm 6. More details are as follows:

Step 1: Set the iteration counter $t$ to 1.

Step 2: Construct the initial membrane structure, a one-level membrane structure $[[\ ]_1, [\ ]_2, \ldots, [\ ]_m]_0$ with $m$ elementary membranes $(1, 2, \ldots, m)$ inside the skin membrane denoted by 0. The parameter $m$ is a random number ranged from 1 to the number $N$ of Q-bit individuals. At the initial configuration, $N$ objects formed by $N$ Q-bit individuals are scattered across the $m$ elementary membranes in a random and non-deterministic way, which satisfies that each elementary membrane contains at least one object. Thus, the number of objects in each elementary membrane varies between 1 to $N - m + 1$.

Step 3: Choose a halting criterion, which may be the difference between the best solution found and the optimal or close-to-optimal solution, and a prescribed number of maximal iterations.

---

**Algorithm 7** Pseudocode algorithm of the TABU search

---
1: $t = 1$
2: Initialize Tabu search
3: **while** $(t < t_{max})$ **do**
4:     $t \leftarrow t + 1$
5:     Search the neighborhood
6:     Evaluate candidate solutions
7:     Update tabu list
8: **end while**

---

Step 4: Determine the numbers $\mathbf{k}_t = (k_{t_1}, k_{t_2}, \ldots, k_{t_m})$, where $k_{t_i}$ $(i = 1, 2, \ldots, m)$ is the number of evolutionary generations for independently performing object evolution rule (a) in the $i$th elementary membranes. $k_{t_i}$ is randomly produced between 1 and a certain integer number $t_{max}$.

Step 5: Perform the QIEA evolutionary operations in each elementary membrane. The halting criterion for the $i$th elementary membrane is the maximal number $k_{t_i}$ $(i = 1, 2, \ldots, m)$ of evolutionary generations. To transform a current Q-bit $[\alpha \ \beta]^T$ into the corresponding Q-bit $[\alpha' \ \beta']^T$ at the next generation, the Q-gate update procedure defined in (2.5) and (2.6) is used, where the rotation angle $\theta$ in the Q-gate $G(\theta)$ in (2.6) can be obtained from the Table 2.1.

Step 6: The communication rule (b) is used to exchange information among the objects in the elementary membranes and the skin membrane, which is implemented by applying the best individual searched to operate on the Q-gate update procedure to generate the offspring. This step is to send the best binary solution in each elementary membrane out into the skin membrane.

Step 7: The local search, tabu search [19], is executed on the best binary solution in the skin membrane. The pseudocode algorithm of tabu search is shown in Algorithm 7. In the "Initialize tabu search" step, an empty tabu list is created and tabu length is set to a value. Tabu search is to explore the neighborhood of the best binary solution in the skin membrane to generate a set of candidate solutions. Then a fitness function is used to evaluate each candidate solution. Finally, the tabu list is updated by using the best one among the candidate solutions.

Step 8: The iteration counter $t$ increases by 1.

Step 9: Generate the number $m_{t+1}$ of elementary membranes at iteration $t + 1$, where $m_{t+1}$ is a random number between 1 and $N$. This step decides the membrane structure at the next iteration.

Steps 10–11: Perform the merging rules for elementary membranes. If $m_{t+1} < m_t$, the $(m_t - m_{t+1})$ elementary membranes are merged into the $m_{t+1}$ elementary membranes. The merging process is shown in Algorithm 8. The merging process is as follows: any two arbitrary elementary membranes $i$ and $j$ are chosen from $M$ elementary membranes $(1 \le i, j \le M$ and $i \ne j)$; then the elementary membranes $i$ and $j$ are merged into a single membrane; accordingly, all the objects in the elementary membranes $i$ and $j$ are placed into the merged membrane. The initial value of $M$ is $m_t$.

---

**Algorithm 8** Membrane merging process

---

1: $M \leftarrow m_t$
2: **while** $(M > m_{t+1})$ **do**
3:     Choose any two arbitrary elementary membranes
4:     Perform the merging rule (d)
5:     $M \leftarrow M - 1$
6: **end while**

---

**Algorithm 9** Membrane separation process

---

1: $M \leftarrow m_t$
2: **while** $(M < m_{t+1})$ **do**
3:     Choose any one elementary membrane whose content is $W$
4:     **if** $|W| \geq 2$ **then**
5:         Perform the separation rule (e)
6:     **end if**
7:     $M \leftarrow M + 1$
8: **end while**

---

Steps 12–14: Perform the separation rule (e) for elementary membranes. If $m_{t+1} > m_t$, each of the $(m_{t+1} - m_t)$ elementary membranes are separated into two membranes. The separation process is shown in Algorithm 9, where $|W|$ is the number of objects in the pre-separation membrane. To be specific, the separation process is described as follows: any one elementary membrane $i$ ($1 \leq i \leq M$) with at least two objects is chosen from $M$ elementary membranes; $|U|$ ($|U| < |W|$) objects are put in the first membrane and the $|W| - |U|$ objects are placed in the other membrane. The initial value of $M$ is $m_t$.

Step 15: Perform the communication rule (c) between the skin membrane and the elementary membranes. In this step, the best binary solution in the skin membrane is sent into each elementary membrane.

### 3.5.2.2 Examples

To show the QEAM performance, a well-known NP-complete problem, the satisfiability problem (SAT), is used to carried out experiments. As SAT is a fundamentally paradigmatic problem in artificial intelligence applications, automated reasoning, mathematical logic, and related research areas [14], much attention has been paid for many years. SAT can be briefly described as follows [20]: given a Boolean formula in conjunctive normal form (CNF), determine whether or not it is satisfiable, that is, whether there exists an assignment to its variables on which it evaluates to true. To be specific, a SAT instance is the following: given a Boolean formula $f(\mathbf{x})$, where $\mathbf{x}$ is a set of Boolean variables $x_1, x_2, \ldots, x_n$, i.e., $x_i \in \{0, 1\}$, $1 \leq i \leq n$, to search a variable assignment $a_1, a_2, \ldots, a_n$, with $a_i \in \{0, 1\}$, $1 \leq i \leq n$, to variables $x_1, x_2, \ldots, x_n$ such that $f(a_1, a_2, \ldots, a_n)$ becomes true. Here the experiments consider only 3-SAT problems, where each clause has exactly three literals.

**Table 3.17** Comparisons of QIEA, QEPS and QEAM on 55 instances of the SAT problem

| SAT | QIEA | | QEPS | | QEAM | | QEAM versus QIEA | | QEAM versus QEPS | |
|-----|------|-----|------|-----|------|-----|--------|--------|--------|--------|
| | Mean | Std | Mean | Std | Mean | Std | t-test | Imp.(%) | t-test | Imp.(%) |
| 1 | 7.67 | 0.82 | 6.60 | 1.18 | 0.93 | 0.26 | 5.23e–23(+) | +87.87 | 5.30e–17(+) | +85.91 |
| 2 | 8.27 | 2.12 | 7.40 | 1.12 | 1.53 | 0.64 | 2.32e–12(+) | +81.50 | 1.13e–16(+) | +79.32 |
| 3 | 7.40 | 1.40 | 6.27 | 0.88 | 1.00 | 0.53 | 5.90e–16(+) | +86.49 | 5.64e–18(+) | +84.05 |
| 4 | 7.87 | 1.19 | 6.33 | 0.98 | 1.00 | 0.38 | 7.31e–19(+) | +87.29 | 5.74e–18(+) | +84.20 |
| 5 | 7.13 | 1.41 | 6.20 | 0.86 | 0.07 | 0.26 | 1.30e–17(+) | +99.02 | 2.50e–21(+) | +98.87 |
| 6 | 7.27 | 0.80 | 5.93 | 0.80 | 0.20 | 0.41 | 5.39e–23(+) | +97.25 | 1.53e–20(+) | +96.63 |
| 7 | 7.73 | 1.16 | 6.40 | 1.18 | 1.07 | 0.46 | 1.73e–18(+) | +86.16 | 8.26e–16(+) | +83.28 |
| 8 | 8.73 | 0.70 | 7.67 | 1.18 | 1.53 | 0.83 | 5.99e–21(+) | +82.47 | 6.01e–16(+) | +80.05 |
| 9 | 7.87 | 1.13 | 6.87 | 1.19 | 1.47 | 0.64 | 1.27e–17(+) | +81.32 | 2.84e–15(+) | +78.60 |
| 10 | 8.33 | 0.82 | 6.93 | 0.88 | 1.07 | 0.59 | 5.74e–22(+) | +87.15 | 7.33e–19(+) | +84.56 |
| 11 | 16.67 | 1.11 | 15.40 | 0.99 | 2.00 | 0.93 | 5.05e–26(+) | +88.00 | 9.32e–26(+) | +87.01 |
| 12 | 15.07 | 1.49 | 14.60 | 0.74 | 1.47 | 0.74 | 1.76e–23(+) | +90.25 | 1.38e–28(+) | +89.93 |
| 13 | 16.33 | 0.82 | 14.80 | 2.01 | 2.07 | 0.88 | 6.61e–28(+) | +87.32 | 1.84e–19(+) | +86.01 |
| 14 | 14.00 | 1.20 | 12.87 | 1.60 | 1.87 | 0.64 | 1.53e–24(+) | +86.64 | 1.41e–20(+) | +85.47 |
| 15 | 15.07 | 1.03 | 14.87 | 0.92 | 1.60 | 1.06 | 9.13e–25(+) | +89.38 | 3.01e–25(+) | +89.24 |
| 16 | 16.13 | 1.06 | 14.87 | 1.19 | 2.20 | 0.68 | 4.29e–27(+) | +86.36 | 5.80e–25(+) | +85.21 |
| 17 | 15.33 | 1.40 | 14.53 | 1.64 | 1.67 | 0.90 | 1.54e–23(+) | +89.11 | 2.00e–21(+) | +88.51 |
| 18 | 15.73 | 1.44 | 14.53 | 1.51 | 2.20 | 0.86 | 2.54e–23(+) | +86.01 | 8.03e–22(+) | +84.86 |
| 19 | 14.93 | 1.49 | 13.80 | 1.97 | 1.80 | 0.41 | 6.02e–24(+) | +87.94 | 9.25e–20(+) | +86.96 |
| 20 | 14.40 | 1.45 | 13.93 | 1.22 | 2.00 | 0.76 | 1.48e–22(+) | +86.11 | 1.19e–23(+) | +85.64 |
| 21 | 24.53 | 1.81 | 22.93 | 1.39 | 3.0 | 0.93 | 1.44e–26(+) | +87.77 | 5.29e–28(+) | +86.92 |
| 22 | 24.20 | 1.21 | 22.80 | 2.14 | 3.33 | 0.62 | 4.78e–31(+) | +86.24 | 3.08e–24(+) | +85.39 |
| 23 | 24.27 | 1.28 | 22.93 | 1.79 | 4.20 | 0.94 | 1.15e–28(+) | +82.69 | 6.05e–25(+) | +81.68 |
| 24 | 23.40 | 1.68 | 22.80 | 1.08 | 3.53 | 0.92 | 2.63e–26(+) | +84.91 | 1.51e–29(+) | +84.52 |
| 25 | 24.27 | 1.22 | 22.80 | 1.47 | 3.73 | 0.88 | 1.45e–29(+) | +84.63 | 4.13e–27(+) | +83.64 |
| 26 | 24.00 | 1.51 | 22.47 | 1.60 | 3.60 | 0.74 | 3.53e–28(+) | +85.00 | 1.06e–26(+) | +83.98 |
| 27 | 24.13 | 1.06 | 23.40 | 1.35 | 3.87 | 0.74 | 2.99e–31(+) | +83.96 | 1.08e–28(+) | +83.46 |
| 28 | 24.00 | 1.85 | 23.40 | 1.30 | 3.73 | 0.80 | 6.33e–26(+) | +84.46 | 6.40e–29(+) | +84.06 |
| 29 | 25.13 | 1.68 | 24.13 | 2.00 | 4.27 | 1.28 | 1.06e–25(+) | +83.01 | 9.18e–24(+) | +82.30 |
| 30 | 22.93 | 2.15 | 22.27 | 1.98 | 3.40 | 0.74 | 4.81e–24(+) | +85.17 | 1.64e–24(+) | +84.73 |
| 31 | 33.53 | 1.96 | 32.87 | 1.51 | 5.67 | 0.90 | 6.07e–29(+) | +83.09 | 3.87e–31(+) | +82.75 |
| 32 | 33.80 | 1.90 | 32.07 | 2.12 | 5.40 | 1.06 | 6.07e–29(+) | +84.02 | 2.76e–27(+) | +83.16 |
| 33 | 34.47 | 1.81 | 33.73 | 1.33 | 6.07 | 1.33 | 4.37e–28(+) | +82.39 | 1.85e–30(+) | +82.00 |
| 34 | 34.06 | 1.84 | 33.27 | 2.34 | 5.13 | 0.74 | 1.13e–30(+) | +84.94 | 1.78e–27(+) | +84.58 |
| 35 | 34.67 | 1.76 | 33.60 | 2.20 | 5.20 | 1.32 | 2.25e–29(+) | +85.00 | 4.32e–27(+) | +84.52 |
| 36 | 34.80 | 2.21 | 33.93 | 1.44 | 6.20 | 1.47 | 9.51e–27(+) | +82.18 | 1.93e–29(+) | +81.73 |
| 37 | 32.80 | 2.86 | 32.93 | 1.33 | 5.27 | 1.28 | 2.49e–24(+) | +83.93 | 1.05e–30(+) | +84.00 |
| 38 | 32.80 | 1.97 | 32.47 | 1.19 | 5.27 | 1.10 | 3.02e–28(+) | +83.93 | 4.14e–32(+) | +83.77 |
| 39 | 34.20 | 2.08 | 33.40 | 1.76 | 5.67 | 0.98 | 1.78e–28(+) | +83.42 | 1.09e–29(+) | +83.02 |
| 40 | 33.93 | 1.67 | 33.20 | 2.34 | 5.67 | 0.98 | 1.96e–30(+) | +83.29 | 7.19e–27(+) | +82.92 |

(continued)

**Table 3.17** (continued)

| SAT | QIEA | | QEPS | | QEAM | | QEAM versus QIEA | | QEAM versus QEPS | |
|-----|------|-----|------|-----|------|-----|----------------|---------|----------------|---------|
| | Mean | Std | Mean | Std | Mean | Std | $t$-test | Imp.(%) | $t$-test | Imp.(%) |
| 41 | 42.60 | 1.50 | 41.20 | 2.01 | 6.87 | 1.13 | 1.29e–33(+) | +83.87 | 1.13e–30(+) | +83.33 |
| 42 | 43.07 | 1.67 | 40.00 | 2.67 | 5.40 | 0.91 | 4.16e–34(+) | +87.46 | 2.66e–28(+) | +86.50 |
| 43 | 41.73 | 1.71 | 41.53 | 1.06 | 6.27 | 1.75 | 2.55e–30(+) | +84.97 | 2.08e–32(+) | +84.90 |
| 44 | 42.80 | 3.28 | 41.07 | 1.94 | 6.47 | 1.25 | 2.73e–26(+) | +84.88 | 1.01e–30(+) | +84.25 |
| 45 | 43.93 | 1.67 | 42.60 | 2.64 | 7.13 | 1.06 | 2.38e–33(+) | +83.77 | 1.66e–28(+) | +83.26 |
| 46 | 43.00 | 3.23 | 42.07 | 1.62 | 7.60 | 1.30 | 4.55e–26(+) | +82.33 | 6.09e–32(+) | +81.93 |
| 47 | 43.20 | 2.04 | 42.73 | 1.87 | 7.87 | 1.30 | 2.12e–30(+) | +81.78 | 5.60e–31(+) | +81.58 |
| 48 | 44.27 | 2.19 | 43.60 | 2.32 | 7.47 | 1.06 | 7.50e–31(+) | +83.13 | 4.97e–30(+) | +82.87 |
| 49 | 44.67 | 2.26 | 43.53 | 2.10 | 8.93 | 1.67 | 9.21e–29(+) | +80.01 | 6.37e–29(+) | +79.49 |
| 50 | 43.13 | 1.55 | 41.47 | 2.50 | 6.53 | 0.99 | 3.87e–34(+) | +84.86 | 5.45e–29(+) | +84.25 |
| 51 | 83.07 | 3.45 | 81.27 | 2.91 | 12.93 | 1.33 | 1.48e–33(+) | +84.43 | 5.51e–35(+) | +84.09 |
| 52 | 83.40 | 3.44 | 82.73 | 2.96 | 15.27 | 1.67 | 8.05e–33(+) | +81.69 | 4.07e–34(+) | +81.54 |
| 53 | 83.00 | 2.73 | 81.60 | 2.50 | 12.67 | 1.50 | 1.05e–35(+) | +84.73 | 3.04e–36(+) | +84.47 |
| 54 | 85.13 | 3.23 | 83.60 | 3.14 | 15.80 | 1.66 | 1.15e–33(+) | +81.44 | 1.14e–33(+) | +81.10 |
| 55 | 84.87 | 2.83 | 80.80 | 2.31 | 14.20 | 1.52 | 2.22e–35(+) | +83.27 | 1.76e–36(+) | +82.43 |

The experiments consider QIEA [23] and QEPS [78] as benchmark algorithms, together with QEAM, to solve fifty-five 3-SAT benchmark problems, where each of the first ten, the second ten, the third ten, the fourth ten, the fifth ten and the last five problems has $50, 75, 100, 125, 150$ and $250$ Boolean variables and $218, 325, 430, 538, 645$ and $1065$ clauses, respectively. Parameter setting for QEAM, QIEA and QEPS is as follows: a population with 50 individuals, the prescribed number of $2.75 \times 10^6$ evaluations to solutions as the halting criterion, the number of clauses that are not satisfied by the variable assignment as the fitness function, and the independent runs for each of them is 15. In QEAM and QEPS, $k_{t_i}$ $(i = 1, 2, \ldots, m)$ is set to a uniformly random integer ranged from 1 to 10, and the number of elementary membranes is assigned to 15. The mean of best solutions (Mean, for short) and the standard deviation of best solutions (Std, for short) are listed in Table 3.17, where the symbol '(+)' represents the significant difference according to $t$-test with a confidence level of 95%. Table 3.17 also show the percentage of improvement (Imp., for short) in the average number of false clauses due to the QEPS algorithm over QIEA and QEPS. To check whether there are significant differences between the two pairs of algorithms, QEAM versus QIEA and QEAM versus QEPS over all tested SAT instances, two non-parametric tests, Wilcoxon's and Friedman's tests, are performed. The results are shown in Table 3.18, where symbols '+' and '−' represent significant difference and no significant difference, respectively.

It is observed in Table 3.17 that QEAM is superior to QIEA and QEPS. This is also verified by the $t$-test results with 55 significant differences between the two pairs of algorithms, QEAM versus QIEA and QEAM versus QEPS. The same conclusion can

**Table 3.18** Non-parametric statistical tests results between QEAM and QIEA or QEPS

| Tests | QEAM versus QIEA | QEAM versus QEPS |
|---|---|---|
| Wilcoxon test ($p$-value) | 1.21e–13 (+) | 1.21e–13 (+) |
| Friedman test ($p$-value) | 1.11e–10 (+) | 1.11e–10 (+) |

be also drawn from Wilcoxon's and Friedman's tests. The $p$-values far smaller than the level of significance 0.05 in Table 3.18 indicates that QEAM really outperforms QIEA and QEPS.

## 3.6  Membrane Algorithms with Static Network Structure

This section describes the static network structure-based (SNS-based, for short) MIEAs. Similar to the previous sections, a brief introduction and some main work of this kind of MIEAs is given. Then an instance called a hybrid approach based on DE and Tissue P Systems (DETPS) is presented.

### 3.6.1  Brief Introduction

A tissue membrane system has a network membrane structure consisting of several one-membrane cells in a common environment and a certain number of channels connecting the cells. The network membrane structure is flexible and adaptable to various network topologies. These features are very useful in organizing algorithms with distinct evolutionary mechanisms and characteristics. Two types of SNS were reported in the literature.

- In [28, 30], the SNS-based MIEA was designed by using a tissue P system with seven cells to organize GA evolutionary operators to solve multi-objective problems. Three cells are for three single-objective optimizations and three cells are for optimizing all the objectives. The last one cell is for collecting results. Benchmark functions, simulated moving bed and controller design were considered as application examples.
- In [86], a tissue P system with five cells in a common environment was used to arrange five representative and widely used variants of differential evolution (DE) algorithms. The five cells use fully connected channels to communicate with each other. The solution of constrained manufacturing parameter optimization problems were discussed. More details can be referred to Sect. 3.6.2.

## 3.6.2  A Hybrid Approach Based on Differential Evolution and Tissue P Systems

### 3.6.2.1  Algorithm

The hybrid MIEA based on differential evolution and tissue P systems (DETPS, for short) [86] uses a tissue-like P system framework with five fully connected cells to arrange five representative and widely used variants of DE algorithms. Specifically, the network membrane structure of a tissue-like P system with five cells was used to design a MIEA by placing each of five DE variants into each of the five cells. The fully connected channels are used for the mutual communications of the five cells. Inside each cell, the DE variant evolves independently according to its own evolutionary mechanism for a certain number of generations. The communication between the cells happens every a certain number of iterations. In DETPS, real-valued strings corresponding to DE individuals are organized as multisets; the transformation and communication rules in a tissue P system and evolutionary operators of DE are used to construct the DETPS evolution rules, which are responsible to evolve the system. Figure 3.21 shows the tissue P system of DETPS, where ovals represent the cells and arrows indicate the channels.

The tissue-like P system are described as the following tuple

$$\Pi = (O, \sigma_1, \sigma_2, \ldots, \sigma_5, syn, i_{out}),$$

where

(1) $O = \{\mathbf{x}_1, \mathbf{x}_2, \ldots, \mathbf{x}_N\}$, with $\mathbf{x}_i$ ($i = 1, 2, \ldots, N$) a real-valued string (*objects*) and $N$ is the total number of individuals involved in the system;
(2) $syn \subseteq \{1, 2, \ldots, 5\} \times \{1, 2, \ldots, 5\}$ (*channels* among cells, here is a fully-connected network);
(3) $i_{out} = 1$ designates the *output cell*;
(4) $\sigma_1, \sigma_2, \ldots, \sigma_5$ are cells of the form $\sigma_i = (Q_i, s_{i,0}, w_{i,0}, P_i)$, $1 \leq i \leq 5$,

**Fig. 3.21** The tissue-like P system in DETPS

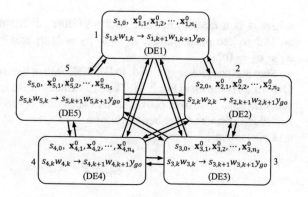

where

(a) $Q_i = \{s_{i,0}, s_{i,1}, \ldots, s_{i,t_{max}}\}$ where $s_{i,j}$ is the state of cell $\sigma_i$ at generation $j$ and $t_{max}$ is the number of generations when the system halts; $1 \le i \le 5, 1 \le j \le t_{max}$;
(b) $s_{i,0} \in Q_i$ is the *initial state*;
(c) $w_{i,0} = \{\mathbf{x}_{i,1}^0, \mathbf{x}_{i,2}^0, \ldots, \mathbf{x}_{i,n_i}^0\}$, where $n_i$ is the number of individuals in cell $\sigma_i$ and satisfies $\sum_{i=1}^5 n_i = N$;
(d) $P_i$ is a finite set of *rules* of the form $s_{i,k}w_{i,k} \to s_{i,k+1}w_{i,k+1}y_{go}$, where $s_{i,k}, s_{i,k+1} \in Q_i; w_{i,k}, w_{i,k+1} \in O^*, y_{go} \in (O \times \{go\})^*, k = 1, 2, \ldots, t_{max}$.

In DETPS, there are two types of evolution rules: transformation-like rules and communication rules. The former are put inside the cells and are responsible for guiding individuals towards better solutions generation by generation according to the evolutionary mechanism of DE. The latter are used to send the best individual in each cell into the rest four cells for exchanging the information between the five cells. In what follows, the details on each step of the DETPS algorithm are provided.

Step 1: *Initialization*. Create a network membrane structure of a tissue P system with five cells in a common environment, shown in Fig. 3.21. Each pair of the five cells are linked through channels. An initial population with $N$ individuals consisting of five group $n_i = \frac{N}{5}, 1 \le i \le 5$, is produced. Each group is placed inside each cell. Each individual is formed by a real-valued string composed of $D$ random numbers uniformly generated in the interval $[0, 1]$, where $D$ is the number of genes of an individual or the number of parameters.

Step 2: *Parameter setting*. Two parameters, the scaling factor $F$ and the crossover probability $Cr$, need to be set. There is a direct influence from the two parameters on the DETPS performance, such as search capability, convergence and robustness. In DETPS, the value of the scaling factor is assigned as a real number following a uniform distribution in the range $[0.4, 1]$; the crossover probability $Cr$ follows the distribution

$$Cr = \begin{cases} \frac{1}{\sqrt{2\pi}\sigma_1} e^{-\frac{(\alpha-\mu_1)^2}{2\sigma_1^2}}, 0 \le \alpha < 0.5 \\ \frac{1}{\sqrt{2\pi}\sigma_2} e^{-\frac{(\alpha-\mu_2)^2}{2\sigma_2^2}}, 0.5 \le \alpha \le 1. \end{cases} \tag{3.4}$$

where $\alpha$ is a random number with a uniform distribution in the interval $[0, 1]$; $\mu_1$ and $\mu_2$ are mean values, $\mu_1 = 0, \mu_2 = 1$; $\sigma_1$ and $\sigma_2$ are standard deviations, $\sigma_1 = \sigma_2 = 0.2$.

Step 3: *Evolutionary process*. In DETPS, each of the five cells contains a variant of DE, which is shown in Fig. 3.21. The five variants of DE have the identical population initialization, crossover operation and selection operation, but they have distinct mutation operators described as follows:

Cell 1: rand/1

$$\mathbf{v}_{i_1}^t = \mathbf{x}_{r_{11}}^t + F \cdot \left(\mathbf{x}_{r_{12}}^t - \mathbf{x}_{r_{13}}^t\right) \tag{3.5}$$

Cell 2: rand/2

$$\mathbf{v}_{i_2}^t = \mathbf{x}_{r_{21}}^t + F \cdot \left(\mathbf{x}_{r_{22}}^t - \mathbf{x}_{r_{23}}^t\right) + F \cdot \left(\mathbf{x}_{r_{24}}^t - \mathbf{x}_{r_{25}}^t\right) \tag{3.6}$$

Cell 3: best/1

$$\mathbf{v}_{i_3}^t = \mathbf{x}_{b_3}^t + F \cdot \left(\mathbf{x}_{r_{31}}^t - \mathbf{x}_{r_{32}}^t\right) \tag{3.7}$$

Cell 4: best/2

$$\mathbf{v}_{i_4}^t = \mathbf{x}_{b_4}^t + F \cdot \left(\mathbf{x}_{r_{41}}^t - \mathbf{x}_{r_{42}}^t\right) + F \cdot \left(\mathbf{x}_{r_{43}}^t - \mathbf{x}_{r_{44}}^t\right) \tag{3.8}$$

Cell 5: current-to-best/1

$$\mathbf{v}_{i_5}^t = \mathbf{x}_{i_5}^t + F \cdot \left(\mathbf{x}_{b_5}^t - \mathbf{x}_{i_5}^t\right) + F \cdot \left(\mathbf{x}_{r_{51}}^t - \mathbf{x}_{r_{52}}^t\right) \tag{3.9}$$

where the scaling factor $F$ is a random value with a uniform distribution in the range [0.4, 1] for each mutant vector; $\mathbf{v}_{i_j}^t$ is the mutant vector at generation $t$ in cell $j$, $1 \leq i_j \leq n_j$ $(1 \leq j \leq 5)$; $\mathbf{x}_{r_{11}}^t, \mathbf{x}_{r_{12}}^t$ and $\mathbf{x}_{r_{13}}^t$ are any three arbitrary individuals in the subpopulation with $n_1$ individuals inside cell 1 and four individuals $\mathbf{x}_{b_2}^t, \mathbf{x}_{b_3}^t, \mathbf{x}_{b_4}^t$ and $\mathbf{x}_{b_5}^t$; $r_{11}, r_{12}$ and $r_{13}$ are different integers in the range $[1, n_1 + 4]$; $\mathbf{x}_{r_{21}}^t, \mathbf{x}_{r_{22}}^t, \mathbf{x}_{r_{23}}^t, \mathbf{x}_{r_{24}}^t$ and $\mathbf{x}_{r_{25}}^t$ are any five arbitrary individuals in the subpopulation composed of $n_2$ individuals inside cell 2 and four individuals $\mathbf{x}_{b_1}^t, \mathbf{x}_{b_3}^t, \mathbf{x}_{b_4}^t$ and $\mathbf{x}_{b_5}^t$; $r_{21}, r_{22}, r_{23}, r_{24}$ and $r_{25}$ are different integers in the range $[1, n_2 + 4]$; $\mathbf{x}_{r_{31}}^t$ and $\mathbf{x}_{r_{32}}^t$ are any two arbitrary individuals in the subpopulation consisting of $n_3$ individuals inside cell 3 and four individuals $\mathbf{x}_{b_1}^t, \mathbf{x}_{b_2}^t, \mathbf{x}_{b_4}^t$ and $\mathbf{x}_{b_5}^t$; $r_{31}$ and $r_{32}$ are different integers in the range $[1, n_3 + 4]$; $\mathbf{x}_{r_{41}}^t, \mathbf{x}_{r_{42}}^t, \mathbf{x}_{r_{43}}^t$ and $\mathbf{x}_{r_{44}}^t$ are any four arbitrary individuals in the subpopulation made up of $n_4$ individuals inside cell 4 and four individuals $\mathbf{x}_{b_1}^t, \mathbf{x}_{b_2}^t, \mathbf{x}_{b_3}^t$ and $\mathbf{x}_{b_5}^t$; $r_{41}, r_{42}, r_{43}$ and $r_{44}$ are different integers in the range $[1, n_4 + 4]$; $\mathbf{x}_{r_{51}}^t$ and $\mathbf{x}_{r_{52}}^t$ are any two arbitrary individuals in the subpopulation consisting of $n_5$ individuals inside cell 5 and four individuals $\mathbf{x}_{b_1}^t, \mathbf{x}_{b_2}^t, \mathbf{x}_{b_3}^t$ and $\mathbf{x}_{b_4}^t$; $r_{51}$ and $r_{52}$ are different integers in the range $[1, n_5 + 4]$; $\mathbf{x}_{b_1}^t, \mathbf{x}_{b_2}^t, \mathbf{x}_{b_3}^t, \mathbf{x}_{b_4}^t$ and $\mathbf{x}_{b_5}^t$ are the best individuals in cells 1–5, respectively, and each of them is sent out into the other four cells through the channels connecting them.

This step also uses the objective function to evaluate individuals. To cope with linear and nonlinear constraints, the parameterless penalty strategy in [9] is employed in DETPS. This strategy uses a pair-wise comparison method by considering the following three scenarios:

(i) Any feasible solution is preferred to any infeasible solution, to be specific, when one feasible and one infeasible solutions are compared, the feasible solution is chosen.

(ii) Among two feasible solutions, the one having better objective function value is preferred, that is, when two feasible solutions are compared, the one with better objective function value is chosen.

(iii) Among two infeasible solutions, the one with smaller constraint violation is preferred, i.e., when two infeasible solutions are compared, the one having smaller constraint violation is chosen.

A feasible solution refers to the one satisfying all linear and non-linear constraints. An infeasible solution refers to the one for which at least one constraint is violated.

Step 4: *Termination criterion*. A prescribed number of function evaluations or a difference between the optimal solution and the best one found can be regarded as the halting criterion.

Step 5: *Output*. The computation result coming from cell 1 is the best one among the five solutions corresponding to the five individuals $\mathbf{x}_{b_1}^{t_{max}}$, $\mathbf{x}_{b_2}^{t_{max}}$, $\mathbf{x}_{b_3}^{t_{max}}$, $\mathbf{x}_{b_4}^{t_{max}}$ and $\mathbf{x}_{b_5}^{t_{max}}$.

### 3.6.2.2  Examples

To show the effectiveness of DETPS, five constrained problems are used to conduct experiments and several algorithms in the literature are considered as comparisons. The five problems, with formulations shown in (3.10)–(3.14), have different characteristics of objective functions and multiple constraints, such as linear, nonlinear and quadratic. Some special features of the problems are discussed in detail in the following description.

Problem 1:

$$\min f(x) = 5 \sum_{i=1}^{4} x_i - 5 \sum_{i=1}^{4} x_i^2 - \sum_{i=5}^{13} x_i \tag{3.10}$$

subject to

$$\begin{cases} g_1(x) = 2x_1 + 2x_2 + x_{10} + x_{11} - 10 \leq 0 \\ g_2(x) = 2x_1 + 2x_3 + x_{10} + x_{12} - 10 \leq 0 \\ g_3(x) = 2x_2 + 2x_3 + x_{11} + x_{12} - 10 \leq 0 \\ g_4(x) = -8x_1 + x_{10} \leq 0 \\ g_5(x) = -8x_2 + x_{11} \leq 0 \\ g_6(x) = -8x_3 + x_{12} \leq 0 \\ g_7(x) = -2x_4 - x_5 + x_{10} \leq 0 \\ g_8(x) = -2x_6 - x_7 + x_{11} \leq 0 \\ g_9(x) = -2x_8 - x_9 + x_{12} \leq 0 \end{cases}$$

where $0 \leq x_i \leq 1$, $i = 1, 2, 3, \ldots, 9$; $0 \leq x_i \leq 100$, $i = 10, 11, 12$; $0 \leq x_{13} \leq 1$. The optimal solution is $f(x^*) = -15$ at $x^* = (1, 1, 1, 1, 1, 1, 1, 1, 1, 3, 3, 3, 1)$.

Problem 2:

$$\max f(x) = (\sqrt{D})^D \prod_{i=1}^{D} x_i \tag{3.11}$$

subject to

$$g(x) = \sum_{i=1}^{D} x_i^2 - 1 = 0$$

where $n = 10$ and $0 \le x_i \le 10, i = 1, 2, \ldots, n$. The global maximum $f(x^*) = 1$ at $x^* = (1/\sqrt{n}, 1/\sqrt{n}, 1/\sqrt{n}, \ldots)$.

Problem 3:

$$\min f(x) = (x_1 - 10)^2 + 5(x_2 - 12)^2 + x_3^4 + 3(x_4 - 11)^2 \\ + 10x_5^6 + 7x_6^2 + x_7^4 - 4x_6x_7 - 10x_6 - 8x_7 \tag{3.12}$$

subject to

$$\begin{cases} g_1(x) = -127 + 2x_1^2 + 3x_2^4 + x_3 + 4x_4^2 + 5x_5 \le 0 \\ g_2(x) = -282 + 7x_1 + 3x_2 + 10x_3^2 + x_4 - x_5 \le 0 \\ g_3(x) = -196 + 23x_1 + x_2^2 + 6x_6^2 - 8x_7 \le 0 \\ g_4(x) = 4x_1^2 + x_2^2 - 3x_1x_2 + 2x_3^2 + 5x_6 - 11x_7 \le 0 \end{cases}$$

where $-10 \le x_i \le 10$, $i = 1, 2, \ldots, 7$. The optimal solution is $f(x^*) = 680.6300573$ at $x^* = (2.330499, 1.951372, -0.4775414, 4.365726, -0.6244870, 1.1038131, 1.594227)$.

Problem 4:

$$\min f(x) = x_1 + x_2 + x_3 \tag{3.13}$$

subject to

$$\begin{cases} g_1(x) = -1 + 0.0025 (x_4 + x_6) \le 0 \\ g_2(x) = -1 + 0.0025 (x_5 + x_7 - x_4) \le 0 \\ g_3(x) = -1 + 0.01 (x_8 - x_5) \le 0 \\ g_4(x) = -x_1x_6 + 833.33252x_4 + 100x_1 - 83333.333 \le 0 \\ g_5(x) = -x_2x_7 + 1250x_5 + x_2x_4 - 1250x_4 \le 0 \\ g_6(x) = -x_3x_8 + 1250000 + x_3x_5 - 2500x_5 \le 0 \end{cases}$$

where $100 \le x_1 \le 10000, 1000 \le x_i \le 10000, i = 2, 3, 100 \le x_i \le 10000, i = 4, 5, \ldots, 8$. The optimal solution is $f(x^*) = 7049.248021$ at $x^* = (579.3066, 1359.9709, 5109.9707, 182.0177, 295.601, 217.982, 286.165, 395.6012)$.

Problem 5:

$$\max f(x) = \frac{100 - (x_1 - 5)^2 - (x_2 - 5)^2 - (x_3 - 5)^2}{100} \tag{3.14}$$

subject to

$$g(x) = (x_1 - p)^2 - (x_2 - q)^2 - (x_3 - r)^2 \leq 0$$

where $0 \leq x_i \leq 10$, $i = 1, 2, 3$, $p, q, r = 1, 2, 3, \ldots, 9$. The optimal solution is $f(x^*) = 1$ at $x^* = (5, 5, 5)$.

In <u>Problem 1</u>, there are thirteen variables and nine linear inequality constraints. The ratio of the feasible search space to the entire search space is approximately 0.0003% [41]. There are six active constraints at the optimal point. Two kinds of experiments with different halting criterions are carried out.

(1) A prescribed number 120,000 of function evaluations or a tolerance 1.0e–6 between the optimal solution and the best one found is considered as the halting criterion. A population with 80 individuals in DETPS is equally divided into 5 groups, i.e., $N = 80$, $n_1 = n_2 = n_3 = n_4 = n_5 = 16$. Benchmark algorithms include hybrid immune-hill climbing algorithm (HIHC) [71], genetic algorithm with an artificial immune system (GAIS) [7] and genetic algorithm based on immune network modeling (GAINM) [21]. Table 3.19 lists the statistical results over 30 independent runs.

(2) A prescribed number 25,000 of function evaluations or a tolerance 1.0e–2 between the optimal solution and the best one found is regarded as the halting criterion. A population with 40 individuals in DETPS is equally divided into 5 groups, i.e., $N = 40$, $n_1 = n_2 = n_3 = n_4 = n_5 = 8$. Six optimization algorithms are considered as benchmark algorithms: teaching-learning-based optimization (TLBO) [56], multimembered evolutionary strategy (M-ES) [45], particle evolutionary swarm optimization (PESO) [74], cultural differential evolution (CDE) [1], co-evolutionary differential evolution (CoDE) [29] and artificial bee colony (ABC) [33]. Table 3.20 lists the statistical results over 30 independent runs.

Tables 3.19 and 3.20 show that DETPS with a much smaller number of function evaluations obtains competitive results than the nine optimization algorithms: HIHC, GAIS, GAINM, TLBO, M-ES, PESO, CDE, CoDE and ABC. According to Table 3.19, DETPS has only 53.46% function evaluations of HIHC and only 42.77% function evaluations of GAIS and GAINM, respectively. Table 3.20 indicates that

**Table 3.19** Statistical results of DETPS, HIHC, GAIS and GAINM to test problem 1. The results of HIHC, GAIS and GAINM are referred from [7, 21, 71], respectively. Best, Mean, Worst and SD represent best solution, mean best solution, worst solution and standard deviation over independent 30 runs, respectively

| Methods | Best | Mean | Worst | SD | Function evaluations |
|---------|------|------|-------|-----|---------------------|
| DETPS | −15.0000 | −15.0000 | −15.0000 | 2.2362e–6 | 64,156 |
| HIHC | −15 | −14.8266 | −14.3417 | 0.145 | 120,000 |
| GAIS | −14.7841 | −14.5266 | −13.8417 | 0.2335 | 150,000 |
| GAINM | −5.2735 | −3.7435 | −2.4255 | 0.9696 | 150,000 |

**Table 3.20** Statistical results of seven algorithms to test problem 1. The results of TLBO, M-ES, PESO, CDE, CoDE and ABC are referred from [1, 29, 33, 45, 56, 74], respectively. Best, Mean and Worst represent best solution, mean best solution and worst solution over independent 30 runs, respectively

| Methods | Best | Mean | Worst | Function evaluations |
|---------|------|------|-------|----------------------|
| DETPS | −15.0 | −15.0 | −15.0 | 20,875 |
| TLBO | −15.0 | −15.0 | −15.0 | 25,000 |
| M–ES | −15.0 | −15.0 | −15.0 | 2,40,000 |
| PESO | −15.0 | −15.0 | −15.0 | 3,50,000 |
| CDE | −15.0 | −15.0 | −15.0 | 1,00,100 |
| CoDE | −15.0 | −15.0 | −15.0 | 2,48,000 |
| ABC | −15.0 | −15.0 | −15.0 | 2,40,000 |

**Table 3.21** Statistical results of six algorithms to test problem 2. The results of TLBO, M-ES, PESO, CDE and ABC are referred from [1, 33, 45, 56, 74], respectively. Best, Mean and Worst represent best solution, mean best solution and worst solution over independent 30 runs, respectively

| Methods | Best | Mean | Worst | Function evaluations |
|---------|------|------|-------|----------------------|
| DETPS | 1.001 | 0.992 | 0.955 | 90,790 |
| TLBO | 1 | 1 | 1 | 1,00,000 |
| M-ES | 1.000 | 1.000 | 1.000 | 2,40,000 |
| PESO | 1.005 | 1.005 | 1.005 | 3,50,000 |
| CDE | 0.995 | 0.789 | 0.640 | 1,00,100 |
| ABC | 1.000 | 1.000 | 1.000 | 2,40,000 |

the number of function evaluations of DETPS is 83.50, 8.70, 5.96, 20.85, 8.42 and 8.70% of TLBO, M-ES, PESO, CDE, CoDE and ABC, respectively.

In <u>Problem 2</u>, there are ten variables and one nonlinear equality constraint. The ratio of the feasible search space to the entire search space approaches 0 [41]. There is only one active constraint at the optimal point. A population with 50 individuals in DETPS is equally divided into 5 groups, i.e., $N = 50$, $n_1 = n_2 = n_3 = n_4 = n_5 = 10$. The halting criterion is a prescribed number 100,000 of function evaluations or a tolerance 1.0e–6 between the optimal solution and the best one found. The experiment considers five optimization algorithms, TLBO [56], M-ES [45], PESO [74], CDE [1] and ABC [33], as benchmarks. Table 3.21 lists the statistical results over 30 independent runs. It can be observed that DETPS obtains competitive results to TLBO, M-ES, PESO, CDE and ABC, by using 90.79, 37.83, 25.94, 90.70 and 37.83% function evaluations, respectively.

In <u>Problem 3</u>, there are seven variables and four nonlinear inequality constraints. The ratio of the feasible search space to the entire search space is approximately 0.5256% [41]. There are two active constraints at the optimal point. A population with 50 individuals in DETPS is equally classified into 5 groups, i.e., $N = 50$, $n_1 =$

**Table 3.22** Statistical results of seven algorithms to test problem 3. The results of TLBO, M-ES, PESO, CDE, CoDE and ABC are referred from [1, 29, 33, 45, 56, 74], respectively. Best, Mean and Worst represent best solution, mean best solution and worst solution over independent 30 runs, respectively

| Methods | Best | Mean | Worst | Function evaluations |
|---------|------|------|-------|---------------------|
| DETPS | 680.630 | 680.630 | 680.630 | 32,586 |
| TLBO | 680.630 | 680.633 | 680.638 | 1,00,000 |
| M-ES | 680.632 | 680.643 | 680.719 | 2,40,000 |
| PESO | 680.630 | 680.630 | 680.630 | 3,50,000 |
| CDE | 680.630 | 680.630 | 680.630 | 1,00,100 |
| CoDE | 680.771 | 681.503 | 685.144 | 2,40,000 |
| ABC | 680.634 | 680.640 | 680.653 | 1,00,000 |

**Table 3.23** Statistical results of six algorithms to test problem 4. The results of TLBO, M-ES, PESO, CDE and ABC are referred from [1, 33, 45, 56, 74], respectively. Best, Mean and Worst represent best solution, mean best solution and worst solution over independent 30 runs, respectively

| Methods | Best | Mean | Worst | Function evaluations |
|---------|------|------|-------|---------------------|
| DETPS | 7049.257 | 7050.834 | 7063.406 | 1,00,000 |
| TLBO | 7049.248 | 7083.673 | 7224.497 | 1,00,000 |
| M-ES | 7051.903 | 7253.047 | 7638.366 | 2,40,000 |
| PESO | 7049.459 | 7099.101 | 7251.396 | 3,50,000 |
| CDE | 7049.248 | 7049.248 | 7049.248 | 1,00,100 |
| ABC | 7053.904 | 7224.407 | 7604.132 | 2,40,000 |

$n_2 = n_3 = n_4 = n_5 = 10$. A prescribed number 1,00,000 of function evaluations or a tolerance 1.0e–6 between the optimal solution and the best one found is regarded as the halting criterion. Table 3.22 lists the statistical results of DETPS over 30 independent runs. The experiment considers six comparison algorithms: TLBO [56], M-ES [45], PESO [74], CDE [1], CoDE [29] and ABC [33]. Table 3.22 shows that DETPS requires only 32.59, 13.85, 13.85 and 32.59% function evaluations of TLBO, M-ES, CoDE and ABC, respectively, to gain better results, and requires only 9.31 and 32.55% function evaluations of PESO and CDE, respectively, to obtain the same solution.

In <u>Problem 4</u>, there are eight variables, three nonlinear inequality and three linear inequality constraints. The ratio of feasible search space to entire search space is approximately 0.0005% [41]. There are three active constraints at the optimal point. A population with 50 individuals in DETPS is equally classified into 5 groups, i.e., $N = 50$, $n_1 = n_2 = n_3 = n_4 = n_5 = 10$. A prescribed number 100,000 of function evaluations or a tolerance 1.0e-4 between the optimal solution and the best one found is considered as the halting criterion. Table 3.23 lists the statistical results of DETPS over 30 independent runs and the results of TLBO [56], M-ES [45], PESO

**Table 3.24** Statistical results of seven algorithms to test problem 5. The results of TLBO, M-ES, PESO, CDE, CoDE and ABC are referred from [1, 29, 33, 45, 56, 74], respectively. Best, Mean and Worst represent best solution, mean best solution and worst solution over independent 30 runs, respectively

| Methods | Best | Mean | Worst | Function evaluations |
|---------|------|------|-------|----------------------|
| DETPS | 1 | 1 | 1 | 6,540 |
| TLBO | 1 | 1 | 1 | 50,000 |
| M-ES | 1 | 1 | 1 | 2,40,000 |
| PESO | 1 | 1 | 1 | 3,50,000 |
| CDE | 1 | 1 | 1 | 1,00,100 |
| CoDE | 1 | 1 | 1 | 2,40,000 |
| ABC | 1 | 1 | 1 | 1,00,000 |

[74], CDE [1] and ABC [33]. It can be observed that DETPS is competitive to other five algorithms with respect to the quality of solutions and the number of function evaluations. DETPS requires 41.67, 28.57 and 41.67% function evaluations of M-ES, PESO and ABC, respectively, to obtain better results. But a little bit worse solutions than TLBO and CDE is gained by DETPS by applying a similar number of function evaluations.

In Problem 5, there are three variables and $9^3 = 729$ nonlinear inequality constraints. The ratio of feasible search space to entire search space is approximately 4.779% [41]. There is not any active constraint at the optimal point. A population with 50 individuals in DETPS is equally divided into 5 groups, i.e., $N = 50$, $n_1 = n_2 = n_3 = n_4 = n_5 = 10$. A prescribed number 50,000 of function evaluations or a tolerance 1.0e–6 between the optimal solution and the best one found is considered as the shalting criterion. Table 3.24 lists the statistical results of DETPS and the results of TLBO [56], M-ES [45], PESO [74], CDE [1], CoDE [29] and ABC [33]. It can be observed that DETPS requires a much smaller number of function evaluations, only 13.08, 2.73, 1.87, 6.53, 2.73 and 6.54% of TLBO, M-ES, PESO, CDE, CoDE and ABC, respectively, to obtain the same results.

## 3.7 Membrane Algorithms with Dynamic Network Structure

A population P system is a special kind of tissue P systems except for two important differences that the structure can be dynamically changed by using bond making rules and cells are allowed to communicate indirectly by means of the environment [2]. This feature could be utilized to design novel MIEAs. In this section, the basic principle and some main work of the dynamical network structure based (DNS-based, for short) MIEAs are given. Then two instances for knapsack and multi-objective function optimization problems are presented, respectively.

### 3.7.1   Brief Introduction

DNS-based MIEAs are representative optimization algorithms of a quite new and promising research direction, where the communication channel between cells can be dynamically built if they are necessary. This class of MIEAs have great potential to be extended to a complex structure with a number of cells for solving complex problems. In [89], a MIEA was introduced by using the DNS of a population P system with three cells to arrange three variants of QIEAs. The communication between a pair of cells is executed at the level of genes, which is different from the usual level of individuals. The channel for communication is created in response to the requirement. In [6], DNS was used to design a MIEA for multi-objective optimization problems. In the algorithm, the cells are divided into two groups with different functions. The first group focuses on evolving objects/individuals, while the second kind aims at selecting objects/individuals and re-distributing them across evolving cells for the next generation. Meanwhile, local communications and global communications, respectively performed among neighboring evolving cells and among all cells, are designed to promote the convergence and diversity.

### 3.7.2   Population Membrane-System-Inspired Evolutionary Algorithm

#### 3.7.2.1   Algorithm

PMSIEA in [89] uses the dynamic network of a population P system to organize three representative variants of QIEAs, including QIEA02 [23], QIEA04 [24] and QIEA07 [75]. The three QIEAs are placed inside three cells of the population P system in a common environment. In PMSIEA, both Q-bits and classical bits form the objects; there are several types of rules for evolving the system: transformation rules like in the population P system, observation and Q-gate update rules like in QIEAs, evaluation rules for candidate solutions, communication rules for information exchange between cells, and bond making rules for modifying the structure of the system. Each cell independently performs the processes consisting of initialization, observation, evaluation and Q-gate update for generating offspring. The individuals in different cells exchange some information at the level of genes through the communication channels. In PMSIEA, a candidate solution of a problem is represented by a binary string. Figure 3.22 shows the framework of the population P system used in PMSIEA. The ovals and dashed lines represent the cells and the links, respectively.

**Fig. 3.22** The framework of
the population P system
involved in PMSIEA

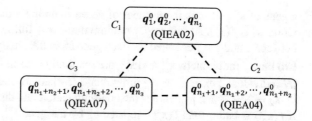

The population P system used in PMSIEA is described as the following tuple

$$\mathcal{P} = (V, \gamma, \alpha, w_e, C_1, C_2, C_3, c_e),$$

where

(i) $V$ is a finite alphabet that consists of all possible Q-bits and classical bits
   (*objects*);
(ii) $\gamma = (\{1, 2, 3\}, E)$, with $E = \{\{1, 2\}, \{1, 3\}, \{2, 3\}\}$, is a finite undirected graph;
(iii) $\alpha$ is a finite set of bond making rules $(i, \lambda; \lambda, j)$ or $\emptyset$ if no new bond can be
   added;
(iv) $w_e = \lambda$;
(v) $C_i = (w_i, S_i, R_i)$, for each $1 \le i \le 3$, with

   (a) $w_i = q_1^0 q_2^0 \ldots q_{n_i}^0$, where $q_j^0$, $1 \le j \le n_i$, is a Q-bit individual as shown in
      (2.3) of Chap. 2; $n_i$ is the number of individuals in cell $C_i$ and satisfies
      $\sum_{i=1}^{3} n_i = N$, where $N$ is the total number of individuals in this system;
   (b) $S_i$ is a finite set of communication rules; each rule has one of the following
      forms: $(\lambda; b, in)$, $(b, exit)$, for $b \in V$,
   (c) $R_i$ is a finite set of transformation rules of the form $a \to y$, for $a \in V$, and
      $y \in V^+$;

(vi) $c_e$ indicates that the result is collected in the environment.

In what follows PMSIEA is described by using algorithmic elaboration.

Step 1: *Initialization*. An initial population with $N$ individuals, each of which
consisting of a certain number of Q-bits, is randomly scattered across the membrane
structure of a population P system with three cells in a common environment. $n_i > 1$
and $\sum_{i=1}^{3} n_i = N$.

Step 2: *Observation*. PMSIEA uses a probabilistic observation process in QIEAs
to link genotypes, Q-bits, with phenotypes, classical bits. For example, as for the Q-
bit $[\alpha \ \beta]^T$, if a random number $r$ between 0 and 1 is less than $|\beta|^2$, i.e., $r < |\beta|^2$, the
observed classical bit equals 1, otherwise, it is 0. Thus, a Q-bit individual corresponds
to a binary solution. The observation process is shown in Algorithm 3 of Chap. 2.

Step 3: *Evaluation*. All the binary solutions obtained at Step 2 are evaluated by
using a specific criterion with respect to a problem.

Step 4: *Communication*. Let $P_k$ represent the binary individuals gained at Step
2 in cell $C_k$, $P_k = \mathbf{x}_1^k, \mathbf{x}_2^k, \ldots, \mathbf{x}_{n_k}^k$, where $k = 1, 2, 3$; $\mathbf{x}_i^k = b_{i1}^k b_{i2}^k \ldots b_{im}^k$, where $b_{ij}^k$ is

a gene of $x_i^k$ and $m$ is the number of genes in an individual; the fitness of the individual $x_i^k$ is $f(x_i^k)$. At this step, a random number $r_c$ following a uniform distribution between 0 and 1 is generated for each gene $b_{ij}^k$ in the binary individual $x_i^k$; if $r_c < p_c$, two binary individuals, $x_{c_1}^{k_1}$ and $x_{c_2}^{k_2}$, are randomly chosen from the whole population ($N$ individuals) except for the individual $x_i^k$, where $k_1$ and $k_2$ are the labels of cells, $k_1, k_2 = 1, 2, 3$ and $p_c$ denotes the communication rate discussed in the next section. If $f(x_{c_1}^{k_1})$ is better than $f(x_{c_2}^{k_2})$, replace $b_{ij}^k$ by the gene $b_{c_1j}^{k_1}$, otherwise, replace $b_{ij}^k$ by the gene $b_{c_2j}^{k_2}$. Thus another binary individual $\bar{x}_i^k$ corresponding to $x_i^k$ can be obtained. During the replacement operation, the values of $k_1$, $k_2$ and $k$ determine what links will be created to carry out the communication between cells $k$ and $k_1$ or $k_2$. There are three cases for the values of $k_1$, $k_2$ and $k$: (1) $k_1 = k_2 = k$ indicates no communication, i.e., the dashed lines in Fig. 3.22 do not work and the three cells are separate; (2) $k_1 = k_2 \neq k$ or $k_1 = k \neq k_2$ or $k_1 \neq k_2 = k$ indicates the communication between two cells, i.e., only one of the dashed lines in Fig. 3.22 works and the communication rule $(\lambda; b, in)$ is executed between the two cells with the channel that works; (3) $k_1 \neq k_2 \neq k$ indicates that the three dashed lines in Fig. 3.22 work and the three cells communicate with each other. Thus the communication rule $(\lambda; b, in)$ is executed between each pair of cells.

Step 5: *Q-gate update*. The objects in each of the three cells evolve according to transformation rules of the form $a \rightarrow y$ according to evolutionary mechanisms of QIEAs, instead of the semantics of P systems. The Q-gate update procedure is shown in (2.3)–(2.4), where the rotation angle $\theta$ in different cells has different definitions. Specifically, the rotation angle in cell 1 is defined as $\theta = s(\alpha, \beta) \cdot \Delta\theta$, where $\Delta\theta$ is the value of $\theta$ determining the convergence speed of the algorithm and $s(\alpha, \beta)$ is the sign of $\theta$ deciding the search direction. The approach for looking up the rotation angle $\theta$ in [23] is shown in Table 3.25, where $f(.)$ is the fitness function; $\alpha$ and $\beta$ are the probabilities of the current Q-bit.

**Table 3.25** Q-gate update approach in cell 1, where $f(.)$ is the fitness function, $\Delta\theta$ and $s(\alpha, \beta)$ are the value and the sign of $\theta$, $x$ and $b$ are the bits of the binary individuals $x_i^1$ and $\bar{x}_i^1$, respectively [23]

| $x$ | $b$ | $f(x) \geq f(b)$ | $\Delta\theta$ | $s(\alpha, \beta)$ | |
|-----|-----|------------------|----------------|--------------------|----|
|     |     |                  |                | $\alpha = 0$ | $\beta = 0$ |
| 0 | 0 | False | 0 | – | – |
| 0 | 0 | True | 0 | – | – |
| 0 | 1 | False | $0.01\pi$ | $+1$ | $-1$ |
| 0 | 1 | True | 0 | – | – |
| 1 | 0 | False | $0.01\pi$ | $+1$ | $-1$ |
| 1 | 0 | True | 0 | – | – |
| 1 | 1 | False | 0 | – | – |
| 1 | 1 | True | 0 | – | – |

**Table 3.26** Q-gate update approach in cell 3, where $f(.)$ is the fitness function, $x$ and $b$ are the bits of the binary individuals $\mathbf{x}_i^1$ and $\overline{\mathbf{x}}_i^1$, respectively [75]

| $x$ | $b$ | $f(\mathbf{x}) \geq f(\mathbf{b})$ | $s(\alpha, \beta)$ | | | $f(\gamma_\alpha, \gamma_\beta)$ |
|---|---|---|---|---|---|---|
| | | | $\alpha\beta \geq 0$ | $\alpha\beta < 0$ | $\alpha\beta = 0$ | |
| 0 | 0 | False | $-1$ | $+1$ | $\pm 1$ | $\exp(-\gamma_\beta)$ |
| 0 | 0 | True | $-1$ | $+1$ | $\pm 1$ | $\exp(-\gamma_\beta)$ |
| 0 | 1 | False | $+1$ | $-1$ | $\pm 1$ | $\exp(-\gamma_\alpha)$ |
| 0 | 1 | True | $-1$ | $+1$ | $\pm 1$ | $\exp(-\gamma_\beta)$ |
| 1 | 0 | False | $-1$ | $+1$ | $\pm 1$ | $\exp(-\gamma_\beta)$ |
| 1 | 0 | True | $+1$ | $-1$ | $\pm 1$ | $\exp(-\gamma_\alpha)$ |
| 1 | 1 | False | $+1$ | $-1$ | $\pm 1$ | $\exp(-\gamma_\alpha)$ |
| 1 | 1 | True | $+1$ | $-1$ | $\pm 1$ | $\exp(-\gamma_\alpha)$ |

In cell 2, the Q-gate update procedure in cell 1 is firstly applied. Then an additional process is used to modify the Q-bit $\left[\alpha^{t+1} \ \beta^{t+1}\right]^T$. The modification method is as follows:

(i) If $|\alpha^{t+1}|^2 \leq \epsilon$ and $|\beta^{t+1}|^2 \geq 1 - \epsilon$ then $\left[\alpha^{t+1} \ \beta^{t+1}\right]^T = \left[\sqrt{\epsilon} \ \sqrt{1-\epsilon}\right]^T$;
(ii) If $|\alpha^{t+1}|^2 \geq 1 - \epsilon$ and $|\beta^{t+1}|^2 \leq \epsilon$ then $\left[\alpha^{t+1} \ \beta^{t+1}\right]^T = \left[\sqrt{1-\epsilon} \ \sqrt{\epsilon}\right]^T$.

According to the investigation in [24], parameter $\epsilon$ is usually assigned as 0.01.

In cell 3, the approach for deciding the quantum rotation angle was defined by using the ratio of the probabilities of Q-bits [75]. The rotation angle $\theta$ is defined as

$$\theta = \theta_0 \cdot s(\alpha, \beta) \cdot f(\gamma_\alpha, \gamma_\beta) \tag{3.15}$$

where $\alpha$ and $\beta$ represent the probabilities of a Q-bit; $\theta_0$ is an initial rotation angle and is usually set to $0.05\pi$; $s(\alpha, \beta)$ is a function determining the search direction of the algorithm; $f(\gamma_\alpha, \gamma_\beta)$ is a function of $\gamma_\alpha$ or $\gamma_\beta$, where $\gamma_\alpha = |\beta|/\alpha$ and $\gamma_\beta = 1/\gamma_\alpha$. The values of $s(\alpha, \beta)$ and $f(\gamma_\alpha, \gamma_\beta)$ can be obtained in Table 3.26.

Step 6: *Halting.* A preset number of evolutionary generations is considered as the halting condition.

Step 7: *Output.* The computation result is the best solution collected in the environment.

### 3.7.2.2 Examples

To show the PMSIEA performance, a well-known NP-hard combinatorial optimization problem, knapsack problem described in Sect. 3.2.2.2, is used to conduct the experiments. The knapsack problem uses strongly correlated sets of unsorted data, i.e., $r_i$ randomly generated from $[0, 50]$, $p_i = r_i + 25$ and $C = 0.5 \sum_{i=1}^{K} r_i$.

In what follows, how to choose the parameter $p_c$ in PMSIEA is experimentally discussed by considering five instances of the knapsack problem with 600, 1200, 1600, 1800, 2400 and 3000 items, respectively. Population size $N$ is set to 20. The halting conditions for instances of the knapsack problem with 600, 1200, 1600, 1800, 2400 and 3000 items, respectively, use the numbers, 20000, 30000, 40000, 60000 and 60000, of function evaluations. Twenty-one cases for $p_c$ are considered when $p_c$ varies from 0 to 1 with interval 0.05. The best, mean and worst solutions over 30 independent runs and the elapsed time per run are used to evaluate the algorithm performance. Figure 3.23 shows experimental results. It is observed that the results of the two cases, $p_c = 0.9$ and $p_c = 0.95$, are better than other cases. So it is better to set $p_c$ to 0.9 in the subsequent experiments.

In the experiments to show the PMSIEA performance, 15 instances of the knapsack problem with the items varying from 200 to 3000 items with interval 200 are used. The four variants of QIEAs: QIEA02 in [23], QIEA04 in [24], QIEA07 in [75] and QIEA08 [60], and the QEPS in Sect. 3.3.2 are considered as comparison methods. The population consists of 20 individuals in all the algorithms. Each test for each of the 15 cases of each algorithm is independently repeated for 30 times. The maximal numbers of function evaluations as the halting condition for the six algorithms are set as follows: 20000 for the first 4 instances of the knapsack problem; 30000 for the 3 instances of the knapsack problem with 1000, 1200 and 1400 items; 40000 function evaluations for the 4 instances of the knapsack problem with 1600, 1800, 2000 and 2200 items; 60000 for the last 4 instances of the knapsack problem. Tables 3.27 and 3.28 show the best, mean and worst solutions over 30 independent runs and the elapsed time per run. To verify that PMSIEA really outperforms the other five algorithms, the statistical techniques are used to analyze the behavior of the six algorithms over the 15 instances of the knapsack problem. To check whether there are significant differences between each pair of algorithms, PMSIEA versus QIEA02, QIEA04, QIEA07, QIEA08 and QEPS, a parametric test, the $t$-test with 95% confidence, and two non-parametric tests, the Wilcoxon's and Friedman's tests with significance level 0.05, are considered. Tables 3.29 and 3.30 list the test results. The symbols "+" and "−" represent significant difference and no significant difference, respectively.

It is observed from Tables 3.27 and 3.28 that PMSIEA is better than QIEA02, QIEA04, QIEA07, QIEA08 and QEPS due to better quality of best, mean and worst solutions and the elapsed time. The test results in Tables 3.29 and 3.30 show that there are significant differences between the two pairs of algorithms, PMSIEA versus QIEA02, QIEA04, QIEA07, QIEA08 and QEPS, with respect to the $t$-test. Table 3.30 indicates that the $p$-values of the Wilcoxon's and Friedman's tests are far smaller than the level of significance 0.05, which means that PMSIEA really outperforms QIEA02, QIEA04, QIEA07, QIEA08 and QEPS.

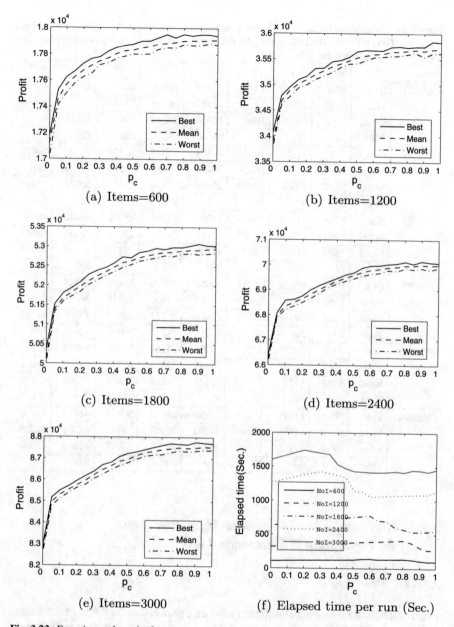

**Fig. 3.23** Experimental results for $p_c$

**Table 3.27** Experimental results of the first 8 instances of the knapsack problem. Best, Mean, Worst and Time represent the best, mean and worst solutions over 30 independent runs and the elapsed time per run, respectively

| Items | | 200 | 400 | 600 | 800 | 1000 | 1200 | 1400 | 1600 |
|---|---|---|---|---|---|---|---|---|---|
| QIEA02 | Best | 5885 | 11650 | 17403 | 22940 | 28673 | 34399 | 39560 | 45277 |
| | Mean | 5786 | 11553 | 17173 | 22659 | 28333 | 33984 | 39149 | 44864 |
| | Worst | 5359 | 11396 | 16851 | 22010 | 27954 | 33424 | 38488 | 44423 |
| | Time | 24 | 48 | 72 | 96 | 182 | 221 | 259 | 413 |
| QIEA04 | Best | 5749 | 11272 | 16561 | 21684 | 27024 | 32429 | 37329 | 42892 |
| | Mean | 5674 | 11081 | 16327 | 21499 | 26812 | 32210 | 37134 | 42596 |
| | Worst | 5627 | 10666 | 15680 | 20809 | 26524 | 31213 | 36028 | 42096 |
| | Time | 29 | 57 | 89 | 115 | 225 | 272 | 320 | 547 |
| QIEA07 | Best | 5935 | 11850 | 17749 | 23390 | 29204 | 35099 | 40490 | 46403 |
| | Mean | 5893 | 11760 | 17627 | 23286 | 29080 | 34949 | 40267 | 46184 |
| | Worst | 5859 | 11700 | 17527 | 23139 | 28929 | 34822 | 40114 | 46002 |
| | Time | 26 | 52 | 78 | 104 | 197 | 249 | 402 | 420 |
| QIEA08 | Best | 5456 | 10699 | 15734 | 20956 | 26073 | 31419 | 36384 | 41750 |
| | Mean | 5367 | 10615 | 15659 | 20747 | 25901 | 31244 | 36081 | 41499 |
| | Worst | 5325 | 10536 | 15591 | 20634 | 25775 | 31071 | 35942 | 41387 |
| | Time | 29 | 58 | 92 | 126 | 249 | 309 | 376 | 599 |
| QEPS | Best | 5959 | 11873 | 17702 | 23403 | 29531 | 35441 | 40886 | 47242 |
| | Mean | 5945 | 11837 | 17647 | 23257 | 29373 | 35292 | 40722 | 47018 |
| | Worst | 5909 | 11778 | 17575 | 23109 | 29198 | 35061 | 40364 | 46672 |
| | Time | 22 | 44 | 70 | 92 | 178 | 215 | 246 | 398 |
| PMSIEA | Best | 5984 | 11975 | 18000 | 23859 | 29822 | 35845 | 41412 | 47470 |
| | Mean | 5963 | 11965 | 17945 | 23782 | 29750 | 35782 | 41301 | 47365 |
| | Worst | 5959 | 11946 | 17902 | 23729 | 29687 | 35727 | 41214 | 47300 |
| | Time | 32 | 64 | 97 | 129 | 240 | 288 | 337 | 512 |

### 3.7.3 Multi-objective Membrane Algorithm Based on Population P Systems and DE

#### 3.7.3.1 Algorithm

In this section, a membrane algorithm based on a population P system and differential evolution (PPSDE, for short) in [6] is discussed for multi-objective optimization problems. Some basic concepts of multi-objective optimization is first presented. Then the procedures of PPSDE are described in detail. Finally, the PPSDE performance is tested on some benchmark problems.

A multi-objective optimization problem (MOP) contains several objectives to be simultaneously optimized. Without loss of generality, a minimization MOP can be formulated as

**Table 3.28** Experimental results of the last 7 instances of the knapsack problem. Best, Mean, Worst and Time represent the best, mean and worst solutions over 30 independent runs and the elapsed time per run, respectively

| Items | | 1800 | 2000 | 2200 | 2400 | 2600 | 2800 | 3000 |
|-------|-------|------|------|------|------|------|------|------|
| QIEA02 | Best | 50784 | 56453 | 61645 | 66683 | 72546 | 77511 | 83294 |
| | Mean | 50163 | 55879 | 61175 | 65984 | 71992 | 76734 | 82608 |
| | Worst | 49506 | 55129 | 59820 | 64981 | 71497 | 75924 | 82020 |
| | Time | 475 | 538 | 619 | 1056 | 1176 | 1310 | 1454 |
| QIEA04 | Best | 47920 | 53276 | 58723 | 62952 | 68858 | 73355 | 79068 |
| | Mean | 47513 | 53018 | 58278 | 62523 | 68448 | 72938 | 78548 |
| | Worst | 46444 | 51889 | 56942 | 62250 | 66910 | 71360 | 76743 |
| | Time | 569 | 636 | 708 | 1268 | 1356 | 1448 | 1560 |
| QIEA07 | Best | 51882 | 57579 | 63199 | 68351 | 74531 | 79471 | 85343 |
| | Mean | 51669 | 57414 | 62985 | 68093 | 74237 | 79215 | 85073 |
| | Worst | 51459 | 57277 | 62768 | 67932 | 73998 | 78685 | 84753 |
| | Time | 475 | 530 | 587 | 964 | 1049 | 1133 | 1222 |
| QIEA08 | Best | 46507 | 52127 | 57221 | 61294 | 67228 | 71600 | 77142 |
| | Mean | 46293 | 51816 | 57008 | 61063 | 66950 | 71308 | 76867 |
| | Worst | 46155 | 51618 | 56811 | 60894 | 66817 | 71121 | 76709 |
| | Time | 702 | 815 | 983 | 1706 | 1950 | 2230 | 2515 |
| QEPS | Best | 52772 | 58775 | 64513 | 70402 | 76621 | 81918 | 88207 |
| | Mean | 52600 | 58543 | 64230 | 70015 | 76245 | 81486 | 87657 |
| | Worst | 52395 | 58065 | 63680 | 69726 | 75296 | 80683 | 87044 |
| | Time | 464 | 523 | 605 | 1051 | 1170 | 1289 | 1441 |
| PMSIEA | Best | 53201 | 59091 | 64955 | 70244 | 76569 | 81806 | 87899 |
| | Mean | 53071 | 58968 | 64785 | 70134 | 76442 | 81664 | 87740 |
| | Worst | 52982 | 58819 | 64669 | 70024 | 76311 | 81505 | 87565 |
| | Time | 576 | 643 | 709 | 1156 | 1253 | 1352 | 1451 |

$$\min \mathbf{f}(\mathbf{x}) = (f_1(\mathbf{x}), f_2(\mathbf{x}), \ldots, f_M(\mathbf{x})) \tag{3.16}$$

subject to

$$\mathbf{x} \in \Omega$$

where $\mathbf{x} = (x_1, x_2, \ldots, x_D)$; $\Omega$ is the feasible region in the decision space and $\mathbf{f}: \Omega \to \mathcal{R}^M$ is composed of $M$ real-valued objective functions $f_1, \ldots, f_M$, where $\mathcal{R}$ is the set of real numbers. The attainable objective space is $\{f(\mathbf{x})|\mathbf{x} \in \Omega\}$.

As usual, a MOP has several contradictory objectives. The improvement of one objective often deteriorates other objectives. So it is impossible to obtain a solution for minimizing all the objectives at the same time. Instead, the result of a MOP is a set of solutions called Pareto optimal solutions. The description on Pareto optimal

**Table 3.29** The results of $t$-test. Symbols "+" and "–" represent significant difference and no significant difference, respectively

| PMSIEA versus | QIEA02 | QIEA04 | QIEA07 | QIEA08 | QEPS |
|---|---|---|---|---|---|
| 200 items | 3.41e–14(+) | 4.57e–49(+) | 3.12e–24(+) | 1.41e–65(+) | 7.83e–08(+) |
| 400 items | 6.71e–40(+) | 2.81e–49(+) | 9.19e–38(+) | 1.93e–81(+) | 1.81e–33(+) |
| 600 items | 7.30e–36(+) | 6.41e–53(+) | 7.29e–35(+) | 1.99e–92(+) | 1.72e–43(+) |
| 800 items | 4.39e–38(+) | 1.66e–60(+) | 7.09e–47(+) | 6.87e–83(+) | 1.24e–44(+) |
| 1000 items | 5.93e–44(+) | 3.49e–77(+) | 1.71e–48(+) | 4.70e–94(+) | 1.77e–28(+) |
| 1200 items | 1.67e–43(+) | 8.17e–64(+) | 7.48e–50(+) | 5.28e–89(+) | 4.92e–38(+) |
| 1400 items | 5.58e–49(+) | 6.68e–66(+) | 3.65e–51(+) | 1.06e–92(+) | 8.74e–33(+) |
| 1600 items | 4.75e–54(+) | 6.57e–82(+) | 6.17e–53(+) | 2.12e–98(+) | 4.43e–21(+) |
| 1800 items | 1.43e–48(+) | 1.32e–71(+) | 1.61e–59(+) | 3.02e–98(+) | 1.06e–32(+) |
| 2000 items | 3.98e–50(+) | 5.76e–73(+) | 6.57e–63(+) | 1.23e–94(+) | 3.71e–19(+) |
| 2200 items | 9.07e–51(+) | 2.02e–69(+) | 6.06e–59(+) | 4.94e–97(+) | 7.45e–22(+) |
| 2400 items | 9.34e–50(+) | 8.21e–88(+) | 6.78e–63(+) | 7.97e–101(+) | 2.70e–03(+) |
| 2600 items | 6.98e–62(+) | 3.20e–72(+) | 2.24e–62(+) | 4.07e–101(+) | 6.71e–04(+) |
| 2800 items | 8.78e–59(+) | 3.91e–73(+) | 1.45e–58(+) | 7.47e–102(+) | 1.60e–03(+) |
| 3000 items | 5.86e–64(+) | 9.18e–73(+) | 1.36e–64(+) | 1.08e–102(+) | 1.16e–01(–) |

**Table 3.30** The $p$-values of Wilcoxon's and Friedman's tests. The symbol + represents significant difference

| PMSIEA versus | QIEA02 | QIEA04 | QIEA07 | QIEA08 | QEPS |
|---|---|---|---|---|---|
| Wilcoxon | 6.1035e−5(+) | 6.1035e-5(+) | 6.1035e-5(+) | 6.1035e-5(+) | 6.1035e-5(+) |
| Friedman | 0.0142(+) | 0.0142(+) | 0.0142(+) | 0.0142(+) | 0.0142(+) |

solutions is as follows: for $\mathbf{u}, \mathbf{v} \in R^M$, $\mathbf{u}$ is said to dominate $\mathbf{v}$, represented as $\mathbf{u} \prec \mathbf{v}$, if and only if $u_i \leq v_i$ for all $i = 1, 2, \ldots, M$ and $\mathbf{u} \neq \mathbf{v}$. Given a set $S$ in $R^M$, a point is called non-dominated if no other point in $S$ can dominate it. A point $\mathbf{x} \in \Omega$ is called a Pareto optimal solution if it is non-dominated in the attainable objective space. Thus $\mathbf{f}(\mathbf{x})$ is a Pareto optimal objective vector, that is, there is no $\mathbf{z} \in \Omega$ such that $f(\mathbf{z})$ dominates $f(\mathbf{x})$. The set of all the Pareto optimal solutions is called the Pareto set and the set of all the Pareto optimal objective vectors is called Pareto front. In practice, it is impossible to find all Pareto optimal solutions, especially for MOPs with continuous objective functions. Without preference on objectives, the goal of solving MOPs is to find a limited amount of approximation solutions that converge well to the true Pareto front and are distributed uniformly along the approximated Pareto front.

PPSDE uses the dynamic network membrane structure of a population P system to organize DE. The cells in population P systems are composed of two categories with

**Fig. 3.24** A population P system used in PPSDE when $NP = 8\ (H = 7), M = 2$ and $R = 0.4$

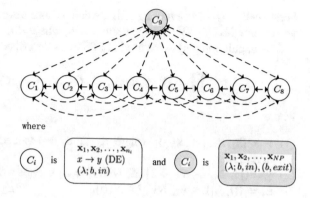

where

$C_i$ is $\begin{array}{c} x_1, x_2, \ldots, x_{n_i} \\ x \rightarrow y \text{ (DE)} \\ (\lambda; b, in) \end{array}$ and $C_i$ is $\begin{array}{c} x_1, x_2, \ldots, x_{NP} \\ (\lambda; b, in), (b, exit) \end{array}$

different functions: evolving cells $C_k, k = 1, 2, \ldots, NP$, and surviving cell $C_{NP+1}$. $NP$ is population size and it is also the size of the desired solution set. Thus, there are totally $NP + 1$ cells in the system. The evolving cells are responsible for searching individuals toward Pareto front. The surviving cell handles surviving individuals and re-scatters them across the evolving cells for the next evolution process.

In PPSDE, a certain number of real-valued strings, each of which corresponds an individual, forms the objects; The evolution rules, transformation-like rules, in each evolving cell are responsible for guiding individuals toward Pareto front by appying the DE evolutionary mechanism. The neighboring evolving cells communicate with each other through local communication controlled by E1 to enhance the convergence. The individuals in surviving cells survive by following a special selection and are re-scattered across evolving cells through global communication controlled by E2. An example is given in clearly illustrate the details of PPSDE. Figure 3.24 shows a population P system, where $NP = 8\ (H = 7), M = 2$ and $R = 0.4$. The white cell $C_k\ (k = 1, \ldots, 8)$ are evolving cells with respect to a direction vector $\mathbf{r}_k = \left(\frac{k-1}{7}, 1 - \frac{k-1}{7}\right), k = 1, 2, \ldots, 8$. The gray cell $C_9$ is the surviving cell. Each dash line with arrow $(i, j) \in E$ represents the bond made dynamically if necessary between two cells.

First of all, to obtain a uniformly distributed and well converged solution set, $M$ direction vectors following a uniform distribution are generated by using simplex-lattice design. The neighboring cells of each evolving cell are defined, which is similar to the work in [37]. That is, a direction vector $\mathbf{r} = [r_1, r_2, \ldots, r_M]$ satisfies

$$r_m \in \left\{ \frac{0}{H}, \frac{1}{H}, \ldots, \frac{H}{H} \right\}, m = 1, 2, \ldots, M, \sum_{m=1}^{M} r_m = 1 \qquad (3.17)$$

where $M$ a positive integer representing the number of objectives in a MOP. The total number of direction vectors generated by simplex-lattice design is $NV = \binom{H+M-1}{M-1}$. For instance, $NV = 100$ when $M = 2$ and $H = 99$, while $NV = 105$ when $M = 3$ and $H = 13$. Evolving cell $i$ is associated with a unique direction vector $r_i$. The neighboring cells $N(i)$ of an evolving cell $i$ are defined as the evolving cells whose direction vectors are the $\lfloor R \cdot NV \rfloor$ closest vectors to $r_i$, where $R \in (0, 1)$ is a parameter.

The population P systems used in PPSDE is described as the tuple

$$\prod = (V, \gamma, \alpha, \omega_e, C_1, C_2, \ldots, C_{NV}, C_{NV+1}, c_o)$$

where

(1) $V = \{\mathbf{x}_1, \mathbf{x}_2, \ldots, \mathbf{x}_{NV}\}$, where $\mathbf{x}_i$ is a real-valued string (objects), $i = 1, 2, \ldots, NP$;

(2) $\gamma = (\{1, 2, \ldots, NV + 1\}, E)$ with $E = E_1 \cup E_2$ is a finite directed graph, where $E_1 = \{(i, j) | 1 \le i \le NV, j \in N(i)\}$, $\qquad E_2 = \{(i, j) | i = NV + 1, 1 \le j \le NV\}$; $N(i)$ represents the direction vectors of the evolving cells;

(3) $\alpha$ is a finite set of bond making rules $(i, x_1; x_2, j)$, $(i, j) \in E$;

(4) $\omega_e = \lambda$;

(5) $C_k = (\omega_k, S_k, R_k)$, for each evolving cell $k$, $1 \le k \le NV$,

   (i) $\omega_k = \{\mathbf{x}_1, \mathbf{x}_2, \ldots, \mathbf{x}_{n_k}\}$, where $n_k$ is the number of individuals in cell $C_k$, satisfying $\sum_{k=1}^{NV} n_k = NV$;

   (ii) $S_k$ is a finite set of communication rules of the form $(\lambda; b, in)$, $b \in V$;

   (iii) $R_k$ is a finite set of transformation rules of the form $x \to y$, consisting of the mutation, crossover, and selection operators of DE;

and for the surviving cell $C_{NV+1}$,

   (i) $\omega_{NV+1} = \{\mathbf{x}_1, \mathbf{x}_2, \ldots, \mathbf{x}_{NV}\}$;

   (ii) $S_{NV+1}$ is a finite set of communication rules of the forms $(b, in)$ and $(b, exit)$ with $b \in V$;

   (iii) $R_{NV+1} = \varnothing$;

(6) $c_o = NV + 1$ is the label of the output cell.

More details on PPSDE algorithmic elaboration is presented step by step as follows:

Step 1: *Initialization.* Create a membrane structure of a population P system with $NV + 1$ cells in a common environment, which include $NV$ evolving cells labeled $k = 1, 2, \ldots, NV$ and a surviving cell labeled $NV + 1$. Evolving cell $k$ is associated with a direction vector $\mathbf{r}_k$. $NV$ individuals forming an initial population are produced and randomly scattered across evolving cells. Thus, there are $n_k$ individuals (objects) inside each evolving cell, $0 \le n_k \le NV$ and $\sum_{k=1}^{NV} n_k = NV$. The value of $n_k$ may vary between 0 and $NV$ in the process of evolution so that $\sum_{k=1}^{NV} n_k = NV$. When the computation stops, each evolving cell has only one individual, i.e., $n_k = 1$ ($k \in \{1, 2, \ldots, NV\}$).

Step 2: *Parameter setting.* PPSDE uses the adaptation scheme in [77] to dynamically adjust the scaling factor $F$ and crossover rate $Cr$. To be specific, the scaling factor $F_i$ and crossover rate $Cr_i$ at generation $g$ for the $i$th individual in the whole population are produced by

$$F_i = \texttt{randc}(\mu_F, 0.1) \tag{3.18}$$

$$Cr_i = \texttt{randn}(\mu_{Cr}, 0.1) \tag{3.19}$$

where $\texttt{randc}(\mu_F, 0.1)$ and $\texttt{randn}(\mu_{Cr}, 0.1)$ are a Cauchy distribution with location parameter $\mu_F$ and scale parameter 0.1, and a normal distribution with mean $\mu_{Cr}$ and standard deviation 0.1, respectively. If the values of $F_i$ and $Cr_i$ are out of the range of $(0,1)$, $i = 1, 2, \ldots, NP$, they will be generated again by using (3.18) and (3.19). The initial values $\mu_F$ and $\mu_{Cr}$ are set to 0.5 and they will be updated at each generation by using the following formula:

$$\mu_F = \sum_{F \in sF} F^2 / \sum_{F \in sF} F \tag{3.20}$$

$$\mu_{Cr} = \sum_{Cr \in sCr} Cr/W \tag{3.21}$$

where $W$ is a parameter; $sF$ and $sCr$ are the sets consisting of most recent $W$ values of $F$ and $Cr$.

Step 3: *Individual evolution*. The individuals in each evolving cell evolve together with the ones in neighboring cells through local communication. To be specific, for each individual $\mathbf{x}_i$ in an evolving cell $k$ ($i = 1, 2, \ldots, n_k$), we select three individuals $\mathbf{x}_{r1}$, $\mathbf{x}_{r2}$ and $\mathbf{x}_{r3}$ from cells $k_{r1}$, $k_{r2}$ and $k_{r3}$ in a random way, where $k_{r1}, k_{r2}, k_{r3} \in \{k\} \cup \{N(k)|n_k \neq 0\}$. Then the individuals $\mathbf{x}_{r1}$, $\mathbf{x}_{r2}$ and $\mathbf{x}_{r3}$ are used to produce an individual $\mathbf{v}_i$ by using mutation operator in (2.15) of Chap. 2 and parameter value $F_i$ generated by (3.18). Then, individual $\mathbf{x}_i$ is recombined with $\mathbf{v}_i$ to produce a trail vector $\mathbf{u}_i$ by using crossover operator in (2.16) of Chap. 2 and parameter value $Cr_i$ generated by (3.19). Subsequently, a selection operation is performed on the objective vector $\mathbf{f}(\mathbf{v}_i)$ calculated and vector $\mathbf{f}(\mathbf{x}_i)$. If $\mathbf{f}(\mathbf{v}_i) \prec \mathbf{f}(\mathbf{x}_i)$, $\mathbf{x}_i$ is replaced by $\mathbf{v}_i$; otherwise, nothing is done. The number of individuals in the system is resized into $NP$ through global communication at the end of each generation. The values of $k_{r1}, k_{r2}$ and $k_{r3}$ decide what membrane structure will be created to perform communication between cell $k$ and its neighboring cells during the selection process of three individuals $\mathbf{x}_{r1}$, $\mathbf{x}_{r2}$ and $\mathbf{x}_{r3}$. Thus, there are four scenarios for the values of $k_{r1}, k_{r2}$ and $k_{r3}$: (1) $k_{r1} = k_{r2} = k_{r3} = k$ indicates no communication; (2) $k_{r1} = k, k_{r2} = k_{r3} \neq k$ or $k_{r2} = k, k_{r1} = k_{r3} \neq k$ or $k_{r3} = k, k_{r1} = k_{r2} \neq k$ or $k_{r1} \neq k, k_{r2} = k_{r3} = k$ or $k_{r2} \neq k, k_{r1} = k_{r3} = k$ or $k_{r3} \neq k, k_{r1} = k_{r2} = k$ indicates the communication between cell $k$ and one cell from its neighboring cells $N(k)$; (3) $k_{r1} = k, k_{r2} \neq k_{r3} \neq k$ or $k_{r2} = k, k_{r1} \neq k_{r3} \neq k$ or $k_{r3} = k, k_{r1} \neq k_{r2} \neq k$ indicates the communication between cell $k$ and two cells from $N(k)$; (4) $k_{r1} \neq k_{r2} \neq k_{r3} \neq k$ indicate the communication between cell $k$ and three different neighboring cells.

Step 4: *Global communication*. This step uses global communication among all cells to re-scatter the individuals across $NP$ evolving cells satisfying the following three criterions: (1) the number of individuals in the system should be $NP$; (2) each individual should be re-assigned to an evolving cell that the individual approximates best to the direction vector associated with this cell; and (3) the number of evolving cells with individuals should be as large as possible. To be specific, the surviving cell $c_o$ receives all individuals (including $\mathbf{x}_i$ and temporally stored $\mathbf{u}_i$) from all evolving cells and forms a temporary population $P'$ with size $NV$ and $NV \in [NP, 2NP]$. Then, $NP$ individuals survived in $P'$ are re-scattered across $NP$ evolving cells. The rest individuals are released to environment. The temporary population $P'$ is divided into fronts $\mathcal{F}_1, \mathcal{F}_2, \ldots, \mathcal{F}_L$ using non-dominated sorting [11] and the individuals in fronts $\mathcal{F}_1, \mathcal{F}_2, \ldots, \mathcal{F}_{l-1}, l \leq L$ first survive by satisfying $\sum_{i=1}^{l-1} |\mathcal{F}_i| < NP$ and $\sum_{i=1}^{l} |\mathcal{F}_i| \geq NP$. For each individual $\mathbf{x}$ in fronts $\mathcal{F}_1, \mathcal{F}_2, \ldots, \mathcal{F}_l$, an evolving cell $k$ is determined so that $\mathbf{x}$ can enter according to the closest perpendicular distance between $\mathbf{f}(\mathbf{x})$ and $\mathbf{r}_i$, i.e.,

$$k = \underset{i \in \{1, \ldots, NP\}}{\operatorname{argmin}} \ \|\mathbf{f}(\mathbf{x}) - \mathbf{r}_i^T \mathbf{f}(\mathbf{x})\mathbf{r}_i\| \tag{3.22}$$

Next, all individuals in $\mathcal{F}_1, \mathcal{F}_2, \ldots, \mathcal{F}_{l-1}$ are sent to their corresponding cells. According to the number of individuals, all evolving cells are scanned in ascending order. If an evolving cell contains the fewest individuals and an individual in $\mathcal{F}_l$ could enter this evolving cell, the individual is sent into this cell and the number of individuals in this cell increases by 1; otherwise, goes to the next evolving cell with the fewest individuals. If all evolving cells with the fewest individuals have been checked and there are still some individuals needed to be kept alive, the evolving cells with the second fewest individuals are scanned. Repeat this scanning procedure until $NP - \sum_{i=1}^{l-1} |\mathcal{F}_i|$ individuals have been survived from $\mathcal{F}_l$. Finally, the current values of $F_i$ and $Cr_i$ generating the survived individuals are stored into $sF$ and $sCr$ for updating $\mu_F$ and $\mu_{Cr}$ at the next generation.

Step 5: *Termination condition*. A preset number of evolutionary generations or function evaluations is used as the halting criterion.

Step 6: *Output*. The final solution set is collected from cell $c_o$.

### 3.7.3.2    Examples

To show the PPSDE performance, twelve widely used test problems, including five two-objective problems (ZDT1-ZDT4, ZDT6) from ZDT test suit [95] and seven three-objective problems (DTLZ1-DTLZ7) from DTLZ test suit [10], are used in the experiment. Three widely used performance metrics [8], generational distance (GD), inverted generational distance (IGD) and hypervolume (HV), are applied to evaluate the algorithm performance. The better the algorithm is, the smaller values of GD and IGD, and larger value on HV there are. About 1000 points for ZDT problems and 5000 points for DTLZ problems in the true Pareto front are sampled to compute GD and IGD. The reference points (2, 2), (2, 2, 2), and (2, 2, 7) for all ZDT problems, DTLZ1-DTLZ6, and DTLZ7, respectively, are used to compute HV. Benchmark

**Table 3.31** Comparison of the mean and standard deviation of GD values

|        | PPSDE      | NSGA-II    | GDE3       | DEMO        | $\epsilon$-MyDE | MOEA/D-DE  |
|--------|------------|------------|------------|-------------|-----------------|------------|
| ZDT1   | 4.281E–4   | 5.910E–4   | 4.449E–4   | **4.017E-4** | 6.113E–3        | 6.926E–3   |
|        | (1.203E–4) | (6.436E–5) | (2.335E–4) | **(7.656E–5)** | (9.366E–3)   | (5.838E–3) |
| ZDT2   | **6.342E–4** | 7.969E–4 | 6.601E–4   | 6.643E–4    | 8.337E–3        | 9.860E–3   |
|        | **(2.243E–4)** | (5.820E–4) | (3.758E–4) | (3.240E–5) | (1.649E–3)   | (7.644E–3) |
| ZDT3   | **4.001E–4** | 5.953E–4 | 1.191E–3   | 1.016E–3    | 1.017E–2        | 1.219E–2   |
|        | **(1.322E–4)** | (2.509E–4) | (9.416E–5) | (1.054E–4) | (1.160E–2)   | (1.026E–2) |
| ZDT4   | **1.996E–3** | 1.376E–2 | 9.398E–2   | 1.535E–2    | 1.224E–1        | 7.324E–2   |
|        | **(3.514E–3)** | (5.445E–3) | (1.023E–1) | (4.527E–3) | (4.527E–2)   | (9.236E–3) |
| ZDT6   | 3.621E–3   | 2.604E–2   | 1.732E–3   | **1.161E–3** | 3.398E–1       | 7.324E–2   |
|        | (8.390E–4) | (8.453E–4) | (4.137E–3) | **(5.068E–5)** | (6.525E–2)   | (9.236E–3) |
| DTLZ1  | 2.825E+0   | 3.083E+0   | 3.425E+0   | **2.553E+0** | 4.294E+0       | 3.452E+0   |
|        | (2.536E–1) | (4.295E–1) | (4.533E–1) | **(1.403E–1)** | (3.582E–1)   | (5.482E–1) |
| DTLZ2  | 3.301E–3   | 3.579E–3   | 3.341E–3   | 3.387E–3    | **9.417E–4**    | 3.402E–2   |
|        | (1.096E–4) | (3.100E–4) | (9.805E–5) | (8.074E–5)  | **(1.319E–4)**  | (1.080E–2) |
| DTLZ3  | 1.372E+1   | **3.153E+0** | 1.126E+1( | 1.985E+1    | 7.677E+1        | 5.629E+1   |
|        | (5.443E+0) | **(4.484E+0)** | (1.633E+1) | (2.236E+0) | (8.366E+0)   | (5.233E+0) |
| DTLZ4  | **2.319E–4** | 4.357E–3 | 2.231E–3   | 2.262E–3    | 1.581E–3        | 4.250E–2   |
|        | **(9.564E–5)** | (3.438E–4) | (6.653E–5) | (6.478E–5) | (2.982E–4)   | (9.572E–3) |
| DTLZ5  | 8.265E–4   | 8.758E–4   | 8.963E–4   | 8.856E–4    | **1.547E–4**    | 1.438E–3   |
|        | (2.651E–5) | (1.925E–4) | (3.146E–5) | (2.656E–5)  | **(2.579E–5)**  | (1.560E–2) |
| DTLZ6  | 4.164E–1   | 6.569E–1   | **8.064E–2** | 2.360E–1  | 3.388E–1        | 7.477E–1   |
|        | (9.853E–2) | (6.826E–2) | **(3.070E–3)** | (9.825E–3) | (1.304E–2)  | (1.935E–2) |
| DTLZ7  | 3.334E–3   | 6.289E–3   | 3.648E–3   | **3.307E–3** | 5.447E–3       | 4.878E–1   |
|        | (1.691E–4) | (5.108E–4) | (2.353E–4) | **(1.782E–4)** | (1.960E–3)   | (5.615E–2) |

algorithms consist of five well-known multi-objective EAs: NSGA-II [11], GDE3 [35], DEMO [57], $\epsilon$-MyDE [10] and MOEA/D-DE [37].

The algorithm will stop when the maximal number 30,000 of function evaluations arrives. Population size $NP$ is set to 105 for PPSDE and MOEA/D-DE to three-object problems. The rest cases use $NP = 100$. The reason is that the exact number 100 of uniformly distributed direction vectors or weight vectors cannot be produced by the simplex-lattice design for PPSDE and MOEA/D-DE to three-objective problems. According to parameter analysis, $R$ and $W$ in PPSDE are set to 0.3 and 80, respectively. The parameter setting of the other algorithms is referred to their original paper. The number of independent runs for each algorithm on each test problem is 25.

Tables 3.31, 3.32 and 3.33 list the experimental results consisting of the mean and standard deviation of GD, IGD and HV values for the six algorithms over twelve test problems. The boldfaced texts highlight the best mean value (i.e., minimal value for GD and IGD, and maximal value for HV) among six algorithms on each test problem. Wilcoxon's rank sum test at a significance level of 0.05 is used to check

**Table 3.32** Comparison of the mean and standard deviation of IGD values

|        | PPSDE | NSGA-II | GDE3 | DEMO | $\epsilon$-MyDE | MOEA/D-DE |
|--------|-------|---------|------|------|--------|-----------|
| ZDT1 | **4.099E–3** | 4.776E–2 | 4.455E–3 | 4.509E–3 | 2.919E–2 | 3.685E–2 |
|        | **(1.881E–4)** | (1.591E–3) | (5.090E–4) | (1.503E–4) | (4.196E–3) | (4.685E–3) |
| ZDT2 | **1.237E–2** | 4.912E–2 | 2.877E–2 | 5.303E–2 | 4.884E–2 | 4.601E–2 |
|        | **(2.096E–3)** | (2.271E–3) | (1.210E–1) | (1.675E–1) | (6.658E–3) | (9.761E–3) |
| ZDT3 | 4.773E–2 | 4.863E–2 | **4.974E–3** | 5.107E–3 | 2.735E–2 | 5.281E–2 |
|        | (1.372E–2) | (8.053E–3) | **(9.788E–5)** | (1.364E–4) | (3.588E–3) | (2.055E–2) |
| ZDT4 | **6.971E–2** | 5.450E–1 | 1.014E–1 | 1.780E–1 | 1.573E–1 | 6.286E–1 |
|        | **(1.962E–2)** | (7.413E–2) | (2.382E–2) | (3.740E–2) | (9.225E–2) | (1.320E–1) |
| ZDT6 | 4.931E–2 | 7.124E–2 | 6.406E–2 | **1.489E–2** | 9.748E–1 | 7.415E–1 |
|        | (1.184E–3) | (5.762E–3) | (2.126E–1) | **(7.178E–4)** | (1.537E–1) | (8.095E–2) |
| DTLZ1 | 5.525E+0 | 6.472E+0 | 7.054E+0 | **5.354E+0** | 8.904E+0 | 8.240E+0 |
|        | (6.425E–1) | (1.536E+0) | (2.925E+0) | **(1.082E+0)** | (2.142E+0) | (2.435+0) |
| DTLZ2 | **5.431E–2** | 6.991E–2 | 6.444E–2 | 6.434E–2 | 5.960E–2 | 7.230E–2 |
|        | **(9.078E–3)** | (2.394E–3) | (1.603E–3) | (1.639E–3) | (1.051E–3) | (1.247E–3) |
| DTLZ3 | 1.885E+1 | 2.484E+1 | **1.607E+1** | 2.159E+1 | 2.321E+2 | 1.135E+2 |
|        | (6.801E+0) | (2.159E+1) | **(2.486E+0)** | (5.474E+0) | (2.614E+1) | (3.303E+1) |
| DTLZ4 | **6.058E–2** | 6.878E–2 | 6.544E–2 | 6.452E–2 | 8.398E–2 | 7.312E–2 |
|        | **(9.251E–3)** | (2.180E–3) | (2.305E–3) | (1.459E–3) | (3.202E–3) | (1.565E–3) |
| DTLZ5 | 8.593E–3 | 5.653E–3 | **5.612E–3** | 5.657E–3 | 1.329E–2 | 1.370E–2 |
|        | (1.095E–3) | (2.692E–4) | **(2.046E–4)** | (3.015E–4) | (2.803E–4) | (8.196E–4) |
| DTLZ6 | 9.319E–1 | 1.129E+0 | **9.214E–1** | 2.135E+0 | 4.577E+0 | 6.535E+0 |
|        | (2.199E–1) | (1.036E–1) | **(2.954E–2)** | (1.033E–1) | (1.146E–1) | (2.196E–1) |
| DTLZ7 | 7.662E–2 | 2.618E–1 | 7.863E–2 | **7.549E–2(** | 7.653E–2 | 2.568E–1 |
|        | (1.217E–2) | (1.626E–1) | (5.505E–3) | **(5.060E–3)** | (4.660E–3) | (4.345E–2) |

whether PPSDE really outperforms other algorithms with respect to three metrics in a statistical sense. Table 3.34 lists the statistical results, where the symbols "+", "=", and "−" represent the numbers of problems where PPSDE performs significantly better than, equivalent to, and worse than the corresponding algorithm, respectively.

It is observed in Table 3.31 that PPSDE and DEMO are better than the rest four algorithms with respect to the number of the best mean GD values. PPSDE, GDE3 and DEMO in Table 3.32 win the best mean IGD values on fives, four and three problems, respectively. PPSDE achieves the best HV mean values on most test problems according to Table 3.33. The statistical test results in Table 3.34 indicate that PPSDE really outperforms the other methods due to the more number of problems that PPSDE performs significantly better than five benchmark algorithms.

In PPSDE, the setting of parameter $R$ (i.e., size of $N(i)$) deciding the number of neighbors of each evolving cell and parameter $W$ (i.e., size of $sF$ and $sCr$) controlling parameter adaptation behavior has much effect on the algorithm performance. So a parameter sensitivity analysis is made by using DTLZ2 to seek an appropriate range for the two parameters. Five cases for both $R$ and $W$, i.e., $R \in \{0.2, 0.4, 0.6, 0.8, 1.0\}$

**Table 3.33** Comparison of the mean and standard deviation of HV values

|      | PPSDE | NSGA-II | GDE3 | DEMO | $\epsilon$-MyDE | MOEA/D-DE |
|------|-------|---------|------|------|--------|-----------|
| ZDT1 | **3.661E+0** | 3.650E+0 | 3.660E+0 | 3.659E+0 | 3.572E+0 | 3.568E+0 |
|      | **0(2.122E–3)** | (3.337E–4) | (3.239E–4) | (2.652E–3) | (1.417E–2) | (1.639E–2) |
| ZDT2 | **3.298E+0** | 3.286E+0 | 3.283E+0 | 3.237E+0 | 3.117E+0 | 3.144E+0 |
|      | **(5.672E–3)** | (4.240E–4) | (2.423E–2) | (3.364E–2) | (3.234E–2) | (3.627E–2) |
| ZDT3 | 4.633E+0 | 4.789E+0 | **4.815E+0** | 4.814E+0 | 4.726E+0 | 4.570E+0 |
|      | (5.638E–2) | (9.302E–2) | **(2.387E–4)** | (2.604E–4) | (1.643E–2) | (7.785E–2) |
| ZDT4 | **3.653E+0** | 3.456E+0 | 3.359E+0 | 3.275E+0 | 3.214E+0 | 3.004E+0 |
|      | **(1.048E–2)** | (2.996E–3) | (7.708E–1) | (5.570E–1) | (4.234E–1) | (2.492E–1) |
| ZDT6 | **3.033E+0** | 2.939E+0 | 2.961E+0 | 3.008E+0 | 1.382E+0 | 1.583E+0 |
|      | **(1.181E–3)** | (2.129E–2) | (2.406E–1) | (2.428E–3) | (2.473E–1) | (1.805E–1) |
| DTLZ1 | **4.224E+0** | 3.314E+0 | 3.125E+0 | 3.653E+0 | 2.235E+0 | 2.535E+0 |
|      | **(5.284E–1)** | (7.982E–1) | (5.093E–1) | (2.573E–1) | (4.583E–1) | (4.713E–1) |
| DTLZ2 | **7.369E+0** | 7.304E+0 | 7.341E+0 | 7.359E+0 | 7.143E+0 | 7.283E+0 |
|      | **(2.259E–2)** | (5.495E–2) | (1.312E–2) | (2.344E–2) | (4.217E–2) | (6.470E–2) |
| DTLZ3 | **1.846E+0** | 1.745E+0 | 2.931E–3 | 0.000E+0 | 0.000E+0 | 0.000E+0 |
|      | **(1.323E+0)** | (1.185E+0) | (1.605E–2) | (0.000E+0) | (0.000E+0) | (0.000E+0) |
| DTLZ4 | **7.383E+0** | 7.369E+0 | 7.382E+0 | 7.361E+0 | 7.260E+0 | 7.251E+0 |
|      | **(1.970E–2)** | (1.970E–2) | (2.132E–2) | (1.051E–2) | (5.056E–2) | (6.683E–2) |
| DTLZ5 | 6.074E+0 | **6.105E+0** | 6.093E+0 | 6.104E+0 | 6.070E+0 | 6.010E+0 |
|      | (5.386E–3) | **(2.601E–2)** | (3.247E–2) | (4.618E–2) | (2.864E–2) | (9.821E–2) |
| DTLZ6 | 2.472E+0 | 1.824E+0 | **2.534E+0** | 1.335E–2 | 0.000E+0 | 0.000E+0 |
|      | (3.714E–1) | (1.284E–1) | **(1.561E–1)** | (4.747E–2) | (0.000E+0) | (0.000E+0) |
| DTLZ7 | **1.336E+1** | 1.224E+1 | 1.313E+1 | 1.330E+1 | 1.294E+1 | 1.052E+1 |
|      | **(9.477E–2)** | (2.272E–1) | (5.521E–1) | (9.145E–2) | (3.224E–1) | (4.908E–1) |

**Table 3.34** Statistical test results in terms of GD, IGD and HV

|          | GD | | | IGD | | | HV | | |
|----------|----|----|----|-----|----|----|----|----|----|
|          | + | = | – | + | = | – | + | = | – |
| NSGA-II  | 11 | 0 | 1 | 10 | 1 | 1 | 9 | 1 | 2 |
| GDE3     | 8 | 2 | 2 | 7 | 2 | 3 | 8 | 2 | 2 |
| DEMO     | 6 | 2 | 4 | 7 | 1 | 4 | 8 | 2 | 2 |
| $\epsilon$-MyDE | 9 | 0 | 3 | 10 | 1 | 1 | 10 | 1 | 1 |
| MOEA/D-DE | 12 | 0 | 0 | 11 | 1 | 0 | 11 | 1 | 0 |

and $W \in \{20, 40, 60, 80, 100\}$, are considered. Thus, there are totally 25 combinations/configurations. The performance metrics, GD, IGD and HV over 25 independent runs for each configuration, are used. The mean values of all configurations on three metrics are shown in Fig. 3.25, where the results indicate that the recommended values for the two parameters are $R \in [0.2, 0.6]$ and $W \in [60, 100]$.

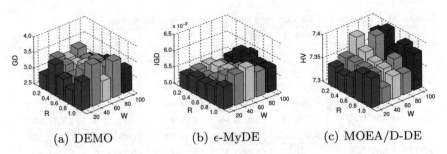

(a) DEMO                    (b) ε-MyDE                (c) MOEA/D-DE

**Fig. 3.25** Mean metric values of 25 independent runs obtained by PPSDE with 25 different combinations of R and W on DTLZ2

## 3.8  P Systems Roles in Membrane Algorithms

A membrane-inspired evolutionary algorithm (MIEA, for short), also called a membrane algorithm, is regarded as a successful instance with a practical use of an approach combining MC with EC [53]. However, going with this research direction, a question is always haunting in the researchers in the areas of MC and EC. That is, what is the role of a P system in a MIEA? Or what advantage can a P system bring to a MIEA? Obviously, this is a tricky question. This motivates the dynamic behavior analysis [87] of MIEAs in this section. In what follows, the sets of population diversity and convergence measures are introduced to experimentally perform the analysis on the MIEA, QEPS (introduced in Sect. 3.3.2) and its counterpart algorithm, QIEA (described in Sect. 2.3).

### 3.8.1  Population Diversity Analysis

Population diversity is one of the most factors that determines the performance of a population-based search method. The diversity of a population is usually measured by using an average distance between individuals. A larger distance means a better population diversity. Diversity measures are used to evaluate the levels and types of varieties of individuals in a population [3]. In what follows, six diversity measures are described to compare QEPS and QIEA:

(1) $D_{qbw}$: This is a distance in the Q-bit space for measuring the average Q-bit distance between the best and worst Q-bit individuals corresponding to the best and worst fitness values in a population, respectively. $D_{qbw}$ is defined as

$$D_{qbw} = \frac{1}{m} \sum_{j=1}^{m} \left| |a_{bj}|^2 - |a_{wj}|^2 \right| \tag{3.23}$$

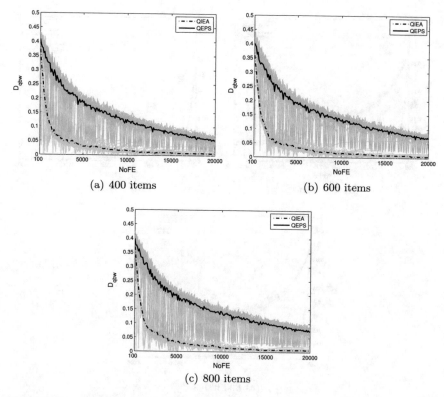

**Fig. 3.26** $D_{qbw}$ of QEPS and QIEA with items 400, 600 and 800

where $m$ is the number of Q-bits in a Q-bit individual; $|a_{bj}|^2$ and $|a_{wj}|^2$ are prob-
abilities of the $j$-th Q-bit in the best and worst Q-bit individuals, respectively.
It is verified that $0 \leq D_{qbw} \leq 1$. A larger distance between the best and worst
Q-bit individuals shows a larger value of $D_{qbw}$.

(2) $D_{qa}$: This is also a distance in the Q-bit space for measuring the average Q-bit
distance of all Q-bit individuals in a population. $D_{qa}$ is defined as

$$D_{qa} = \frac{2}{n(n-1)} \sum_{i=1}^{n} \sum_{j=i+1}^{n} \left\{ \frac{1}{m} \sum_{k=1}^{m} \left| |a_{ik}|^2 - |a_{jk}|^2 \right| \right\} \tag{3.24}$$

where $n$ and $m$ are the numbers of individuals in a population and Q-bits in
a Q-bit individual, respectively; $|a_{ik}|^2$ and $|a_{jk}|^2$ are probabilities of the $k$-th
Q-bit in the $i$-th and $j$-th Q-bit individuals, respectively. $D_{qa}$ in (3.24) is the
average value of the Q-bit distance between $n(n-1)$ pairs of Q-bit individuals.
It is verified that $0 \leq D_{qa} \leq 1$. A larger distance between each pair of Q-bit
individuals in a population takes on a larger value of $D_{qa}$.

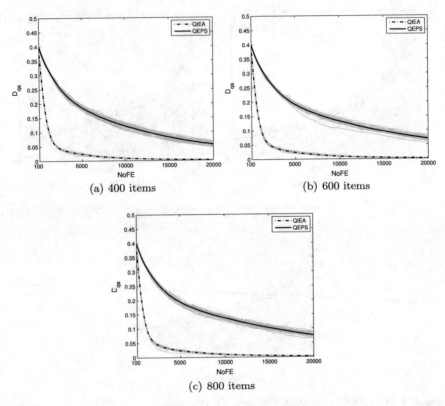

(a) 400 items

(b) 600 items

(c) 800 items

**Fig. 3.27** $D_{qa}$ of QEPS and QIEA with items 400, 600 and 800

(3) $D_{hbw}$ and $D_{hm}$: These are two distances in the binary space. Hamming distance $D_{hbw}$ between the best and worst binary individuals in a population and mean Hamming distance $D_{hm}$ of all binary individuals in a population are defined as

$$D_{hbw} = \sum_{i=1}^{m} (x_{bi} \oplus x_{wi}) \tag{3.25}$$

$$D_{hm} = \frac{2}{n(n-1)} \sum_{i=1}^{n} \sum_{j=i+1}^{n} \left\{ \frac{1}{m} \sum_{k=1}^{m} (x_{ik} \oplus x_{jk}) \right\} \tag{3.26}$$

where $m$ and $n$ are the numbers of bits in a binary solution and individuals in a population; $x_{bi}$ and $x_{wi}$ are the $i$-th bits in the best and worst binary solutions, respectively; symbol $\oplus$ is exclusive OR operator; $x_{ik}$ and $x_{jk}$ are the $k$-th bits in the $i$-th and $j$-th binary solutions, respectively. It is verified that $D_{hbw}$ and $D_{hm}$ are in the range $[0, m]$. More varieties between the best and worst binary

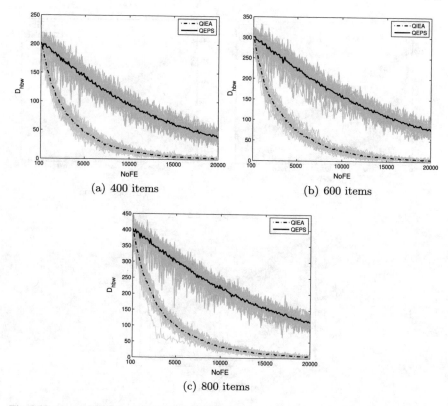

Fig. 3.28  $D_{hbw}$ of QEPS and QIEA with items 400, 600 and 800

individuals, and each pair of binary individuals in a population, respectively, show the larger values of $D_{hbw}$ and $D_{hm}$.

(4) $D_{bc}$ and $D_{ba}$: Two diversity measures based on dispersion statistical measures including the diversity $D_{bc}$ between chromosomes and the diversity $D_{ba}$ between the alleles [3] are defined as

$$D_{bc} = \frac{1}{n-1}\left(\frac{\sum_i S_i^2}{L} - \frac{S^2}{L*n}\right) \qquad (3.27)$$

$$D_{ba} = \frac{1}{L-1}\left(\frac{\sum_j S_j^2}{n} - \frac{S^2}{L*n}\right) \qquad (3.28)$$

where $n$ and $L$ are population size and the length of a chromosome; $S$ is the sum of genes '1'; $S_i$ and $S_j$ are the sum over a row $i$ and the sum over a column $j$, respectively. $D_{bc}$ and $D_{ba}$ have the following properties [25]:

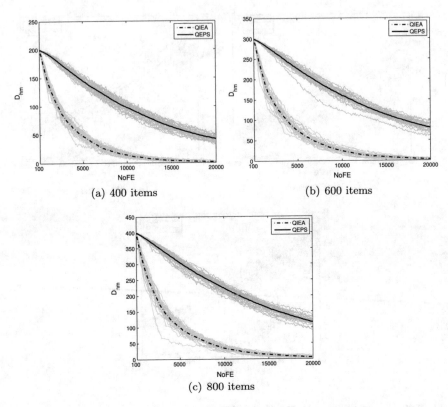

(a) 400 items

(b) 600 items

(c) 800 items

**Fig. 3.29** $D_{hm}$ of QEPS and QIEA with items 400, 600 and 800

(a) $D_{bc}$ and $D_{ba}$ equals zero when the population is homogeneous in either 0 or 1;

(b) If all the chromosomes in the population are identical, $D_{bc}$ is zero and $D_{ba}$ holds a constant value depending on how many genes '1' are in a chromosome.

The two diversity measures, $D_{qbw}$ and $D_{qa}$, in Q-bit space, can be considered as genotypic diversity measures. The rest four diversity measures, $D_{hbw}$, $D_{hm}$, $D_{bc}$ and $D_{ba}$, fall into phenotypic diversity measures.

To intuitively show the changes of the six population diversities in the process of evolution, knapsack problems described in Sect. 3.2.2.2 are used to conduct the experiments. Population size is set to 20. The maximal number 20000 of function evaluations (NoFE) is used as the halting condition. The comparisons of QEPS and QIEA for three instances of the knapsack problem with 400, 600 and 800 items are shown in Figs. 3.26, 3.27, 3.28, 3.29, 3.30 and 3.31, where each subfigure shows the results of 30 independent random runs (green solid lines for QEPS and cyan solid lines for QIEA) and the mean values over 30 runs (black bold solid lines for QEPS and black bold dash-dot lines for QIEA).

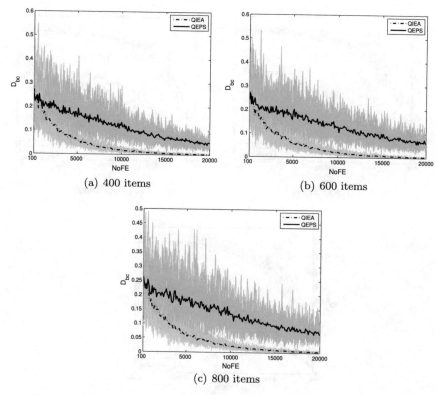

**Fig. 3.30** $D_{bc}$ of QEPS and QIEA with items 400, 600 and 800

The comparisons of population diversity between QEPS and QIEA, which are shown in Figs. 3.26, 3.27, 3.28, 3.29, 3.30 and 3.31, indicate the following observations: a consistent trend for QEPS and QIEA is observed from the three subfigures corresponding to the instances of the knapsack problem with 400, 600 and 800 items, in each of Figs. 3.26, 3.27, 3.28, 3.29, 3.30 and 3.31, respectively. This observation indicates the reasonableness of the six diversity measures to a certain degree. Better population diversity than QIEA in Q-bit space can be obtained by QEPS according to Figs. 3.26 and 3.27. At the beginning of evolution, QEPS and QIEA have almost the same value, of about 0.4, for $D_{qbw}$ and $D_{qa}$. After the evolution starts, the values of $D_{qbw}$ and $D_{qa}$ of Q-bit individuals in QEPS decrease gradually to around 0.05 at the value 20000 of NoFE, whereas the values in QIEA rapidly fall below 0.05 at the value 3000 of NoFE and approximate 0 at the value 20000 of NoFE. $D_{hbw}$ and $D_{hm}$ in Figs. 3.28 and 3.29 indicate that QEPS has a greater potential to preserve the population diversity than QIEA. When NoFE increases to 20000, QIEA loses the population diversity and therefore has not any exploration capability. While QEPS still has one-fourth of initial values of $D_{hbw}$ and $D_{hm}$, which indicates that QEPS still has a certain degree of exploration capability. Figures 3.30 and 3.31 show that

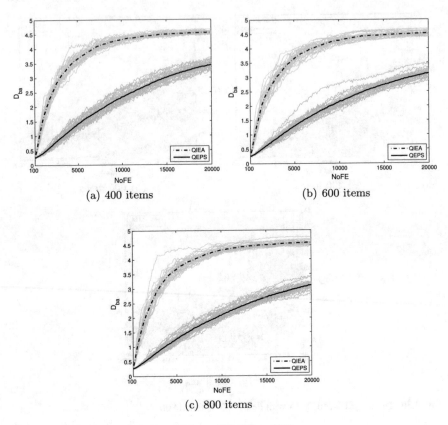

(a) 400 items                                          (b) 600 items

(c) 800 items

**Fig. 3.31** $D_{ba}$ of QEPS and QIEA with items 400, 600 and 800

QEPS is superior to QIEA with respect to $D_{bc}$ and $D_{ba}$. It is noting that the loss of population diversity implies that the algorithm fails to further explore the solution space.

### 3.8.2 Convergence Analysis

Another one of the most factors determining the performance of a population-based search method is convergence at the algorithmic level. Convergence can indicates how fast an algorithm finds a satisfactory solution to an optimization problem. In what follows, four measures: best Q-bit individual convergence ($C_{qb}$), average Q-bit individual convergence ($C_{qa}$), the best fitness convergence ($C_{fb}$) and the average fitness convergence ($C_{fa}$), are used to qualitatively analyze the convergence of QEPS and QIEA. It is worth pointing out that $C_{qb}$ and $C_{qa}$, calculated in the Q-bit space, measure how much Q-bits approach 0 or 1 in the searching process. Thus, $C_{qb}$ and

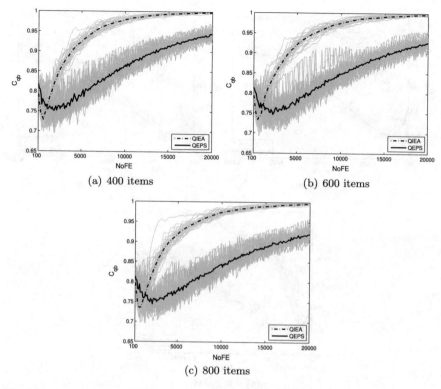

Fig. 3.32 $C_{qb}$ of QEPS and QIEA with items 400, 600 and 800

$C_{qa}$ can be considered as genotypic convergence measures. While $C_{fb}$ and $C_{fa}$, which have a direct relationship to the quality of solutions, can be regarded as phenotypic convergence measures.

The definitions of the four convergence measures are as follows:

(1) $C_{qb}$: This is a convergence measure in the Q-bit space. The best Q-bit individual convergence is defined as

$$C_{qb} = \frac{1}{m} \sum_{j=1}^{m} \max \left\{ |\alpha_{bj}|^2, |\beta_{bj}|^2 \right\} \qquad (3.29)$$

where $m$ is the number of Q-bits in a Q-bit individual; $[\alpha_{bj} \ \beta_{bj}]^T$ is the $j$-th Q-bit in the best Q-bit individual corresponding to the best fitness in a population. It is verified that $0.5 \leq C_{qb} \leq 1$.

(2) $C_{qa}$: This is also a convergence measure in the Q-bit space. The average Q-bit individual convergence is

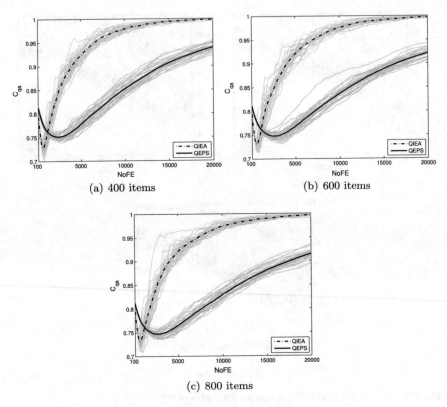

**Fig. 3.33** $C_{qa}$ of QEPS and QIEA with items 400, 600 and 800

$$C_{qa} = \frac{1}{n} \sum_{i=1}^{n} \left\{ \frac{1}{m} \sum_{j=1}^{m} \max \left\{ |\alpha_{ij}|^2, |\beta_{ij}|^2 \right\} \right\} \tag{3.30}$$

where $m$ is the number of Q-bits in a Q-bit individual; $[\alpha_{ij} \ \beta_{ij}]^T$ is the $j$-th Q-bit in the $i$-th Q-bit individual in the population with $n$ individuals. It is verified that $0.5 \leq C_{qa} \leq 1$.

(3) $C_{fb}$ and $C_{fa}$: The two convergence measures $C_{fb}$ and $C_{fa}$ are described as

$$C_{fb} = \max\{f_i(x) \mid 1 \leq i \leq n\} \tag{3.31}$$

$$C_{fa} = \frac{1}{n} \sum_{i=1}^{n} f_i(x) \tag{3.32}$$

where $f_i(x)$ is the fitness of the $i$-th individual. It is noting that the description of $C_{fb}$ in (3.31) is based on a maximization optimization problem.

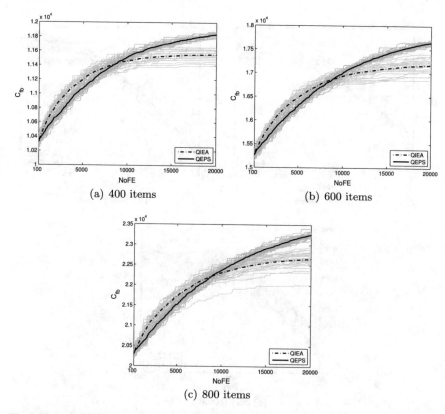

(a) 400 items

(b) 600 items

(c) 800 items

**Fig. 3.34** $C_{fb}$ of QEPS and QIEA with items 400, 600 and 800

Three instances of the knapsack problem with 400, 600 and 800 items are used to compare the convergence performances of QEPS and QIEA. Population size is set to 20. The halting criterion considers the value 20000 of NoFE. The independent runs is 30. Figures 3.32, 3.33, 3.34 and 3.35 show the changes of $C_{qb}$, $C_{qa}$, $C_{fb}$ and $C_{fa}$. Each subfigure shows the results of 30 independently random runs (green solid lines for QEPS and cyan solid lines for QIEA) and the mean values over 30 runs (black bold solid lines for QEPS and black bold dash-dot lines for QIEA).

The experimental results in Figs. 3.32, 3.33, 3.34 and 3.35 indicate that there are similar changes for QEPS and QIEA in the three subfigures corresponding to the instances of the knapsack problem with 400, 600 and 800 items, respectively. To be specific, Figs. 3.32 and 3.33 show that $C_{qb}$ and $C_{qa}$ take on similar tendencies. QEPS converges much slower in Q-bit space than QIEA, which indicates that QEPS has much higher possibility to find better solutions than QIEA. Figures 3.34 and 3.35 show that QEPS has a slower increase of $C_{fb}$ and $C_{fa}$ than QIEA. When the value of NoEF goes up, QIEA stays at a relatively flat level, while QEPS can continue its ascending trend. QEPS is superior to QIEA due to better solutions.

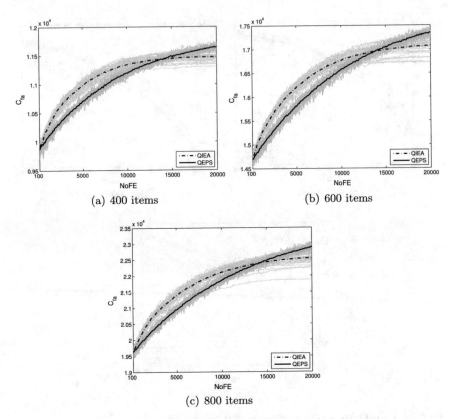

**Fig. 3.35** $C_{fa}$ of QEPS and QIEA with items 400, 600 and 800

Population diversity and convergence are the most important two conflicting factors affecting the algorithm performance. As usual, too high population diversity will result in a slow convergence, vice versa, too low population also result in a bad convergence. So the most important consideration for a population-based search method is how to balance the population diversity and convergence, that is, how to trade off the exploration and exploitation capabilities. Better balance indicates better performance. The analysis of dynamic behavior in this section show the advantage of a MIEA over its counterpart method with respect to the balance capability of exploration and exploitation.

## 3.9   Conclusions

This chapter presented six classes of membrane algorithms from the perspective of membrane structures. We discussed the principles, algorithm steps and experimental results with respect to widely used and well-known benchmark problems. Following the introduction of various membrane algorithms, the roles of P systems in membrane algorithms was also discussed. This chapter highlighted the algorithm principles and their effectiveness verification with benchmark problems, as the fundamentals of the engineering applications of membrane algorithms in the next chapter.

# References

1. Becerra, R.L., and C.A.C. Coello. 2006. Cultured differential evolution for constrained optimization. *Computer Methods in Applied Mechanics and Engineering* 195 (33–36): 4303–4322.
2. Bernardini, F., and M. Gheorghe. 2008. Population P systems. *Journal of Universal Computer Science* 10 (5): 509–539.
3. Burke, E., S. Gustafson, and G. Kendall. 2004. Diversity in genetic programming: an analysis of measures and correlation with fitness. *IEEE Transactions on Evolutionary Computation* 8 (1): 47–62.
4. Chen, H., and J. Lu. 2012. A constrained optimization evolutionary algorithm based on membrane computing. *Journal of Digital Information Management* 10 (2): 121–125.
5. Cheng, J., G. Zhang, and X. Zeng. 2011. A novel membrane algorithm based on differential evolution for numerical optimization. *International Journal of Unconventional Computing* 7 (3): 159–183.
6. Cheng, J., G. Zhang, and T. Wang. 2015. A membrane-inspired evolutionary algorithm based on population P systems and differential evolution for multi-objective optimization. *Journal of Computational and Theoretical Nanoscience* 12 (7): 1150–1160.
7. Coello, C.A.C., and N.C. Cortés. 2004. Hybridizing a genetic algorithm with an artificial immune system for global optimization. *Engineering Optimization* 36 (5): 607–634.
8. Coello, C.A.C., G.B. Lamont, and D.A.V. Veldhuizen. 2007. *Evolutionary algorithms for solving multi-objective problems*, 2nd ed. New York: Springer.
9. Deb, K. 2000. An efficient constraint handling method for genetic algorithm. *Computer Methods in Applied Mechanics and Engineering* 186 (2–4): 311–338.
10. Deb, K., M. Mohan, and S. Mishra. 2005. Evaluating the $\epsilon$-domination based multi-objective evolutionary algorithm for a quick computation of Pareto-optimal solutions. *Evolutionary Computation* 13 (4): 501–525.
11. Deb, K., A. Pratap, S. Agarwal, and T. Meyarivan. 2002. A fast and elitist multiobjective genetic algorithm: NSGA-II. *IEEE Transactions on Evolutionary Computation* 6 (2): 182–197.
12. Elias, S., V. Gokul, K. Krithivasan, M. Gheorghe, and G. Zhang. 2012. A variant of the distributed P system for real time cross layer optimization. *Journal of Universal Computer Science* 18 (13): 1760–1781.
13. Escuela, G., and M.A. Gutiérrez-Naranjo. 2010. An application of genetic algorithms to membrane computing. In *Proceedings of the Eighth Brainstorming Week on Membrane Computing*, 101–108.
14. Folino, G., C. Pizzuti, and G. Spezzano. 2001. Parallel hybrid method for SAT that couples genetic algorithms and local search. *IEEE Transactions on Evolutionary Computation* 5 (4): 323–334.
15. Gao, H., and J. Cao. 2012. Membrane-inspired quantum shuffled frog leaping algorithm for spectrum allocation. *Journal of Systems Engineering and Electronics* 23 (5): 679–688.
16. Gao, H., J. Cao, and Y. Zhao. 2012. Membrane quantum particle swarm optimisation for cognitive radio spectrum allocation. *International Journal of Computer Applications in Technology* 43 (4): 359–365.
17. García, S., D. Molina, M. Lozano, and F. Herrera. 2009. A study on the use of non-parametric tests for analyzing the evolutionary algorithms' behaviour: a case study on the CEC'2005 special session on real parameter optimization. *Journal of Heuristics* 15: 617–644.

18. Garey, M., and D. Johnson. 1979. *Computers and intractability: a guide to the theory of NP-completeness*. New York: W. H. Freeman & Co.
19. Glover, F., E. Taillard, and D. de Werra. 1993. A users guide to tabu search. *Annals of Operations Research* 41 (1): 3–28.
20. Gottlieb, J., E. Marchiori, and C. Rossi. 2001. Evolutionary algorithms for the satisfiability problem. *Evolutionary Computation* 10 (1): 35–50.
21. Hajela, P., and J.S. Yoo. 1999. Immune network modelling in design optimization. In *New Ideas in Optimization*, ed. D. Corne, M. Dorigo, and F. Glover, 167–183. New York: McGraw-Hill.
22. Han, K., and J. Kim. 2000. Genetic quantum algorithm and its application to combinatorial optimization problem. In *Proceedings of IEEE Congress on Evolutionary Computation*, 1354–1360.
23. Han, K., and J. Kim. 2002. Quantum-inspired evolutionary algorithm for a class of combinatorial optimization. *IEEE Transactions on Evolutionary Computation* 6 (6): 580–593.
24. Han, K., and J. Kim. 2004. Quantum-inspired evolutionary algorithms with a new termination criterion, $H_\epsilon$ gate, and two-phase scheme. *IEEE Transactions on Evolutionary Computation* 8 (2): 156–169.
25. Herrera, F., and M. Lozano. 1996. Adaptation of genetic algorithm parameters based on fuzzy logic controllers. In F. Herrera, J.L. Verdegay (eds.), *Genetic Algorithms and Soft Computing*, Physica-Verlag, pages 95–125,
26. Huang, L., and I.H. Suh. 2009. Controller design for a marine diesel engine using membrane computing. *International Journal of Innovative Computing, Information and Control* 5 (4): 899–912.
27. Huang, L., X. He, N. Wang, and Y. Xie. 2007. P systems based multi-objective optimization algorithm. *Progress in Natural Science* 17 (4): 458–465.
28. Huang, L., L. Sun, N. Wang, and X. Jin. 2007. Multiobjective optimization of simulated moving bed by a kind of tissue P system. *Chinese Journal of Chemical Engineering* 15 (5): 683–690.
29. Huang, F., L. Wang, and Q. He. 2007. An effective co-evolutionary differential evolution for constrained optimization. *Applied Mathematics and Computation* 186 (1): 340–356.
30. Huang, L., N. Wang, and J. Zhao. 2008. Multiobjective Optimization for Controller Design. *Acta Automatica Sinica* 34 (4): 472–477.
31. Huang, L., I.H. Suh, and A. Abraham. 2011. Dynamic multi-objective optimization based on membrane computing for control of time-varying unstable plants. *Information Sciences* 181 (11): 2370–2391.
32. Huang, X., G. Zhang, H. Rong, and F. Ipate. 2012. Evolutionary design of a simple membrane system. In *Membrane Computing (CMC 2011)*, ed. M. Gheorghe, G. Păun, G. Rozenberg, A. Salomaa, and S. Verlan, 203–214. Lecture Notes in Computer Science Berlin: Springer.
33. Karaboga, D., and B. Basturk. 2007. Artificial bee colony (ABC) optimization algorithm for solving constrained optimization problems. In *Foundations of Fuzzy Logic and Soft Computing (IFSA 2007)*, ed. P. Melin, O. Castillo, L.T. Aguilar, J. Kacprzyk, and W. Pedrycz, 789–798. Lecture Notes in Computer Science Berlin: Springer.
34. Krasnogor, N., and J. Smith. 2005. A tutorial for competent memetic algorithms: model, taxonomy, and design issues. *IEEE Transactions on Evolutionary Computation* 9 (5): 474–488.
35. Kukkonen, S., and J. Lampinen. 2005. GDE3: the third evolution step of generalized differential evolution. In *Proceedings of IEEE Congress on Evolutionary Computation*, 443–450.
36. Leporati, A., and D. Pagani. 2006. A membrane algorithm for the min storage problem. In *Membrane Computing (WMC 7)*, vol. 4361, ed. H.J. Hoogeboom, G. Păun, G. Rozenberg, and A. Salomaa, 443–462. Lecture Notes in Computer Science Berlin: Springer.
37. Li, H., and Q.F. Zhang. 2009. Multiobjective optimization problems with complicated Pareto sets. *MOEA/D and NSGA-II, IEEE Transactions on Evolutionary Computation* 13 (2): 284–302.
38. Li, B., and Z. Zhuang. 2002. Genetic algorithm based on quantum probability representation. In *Intelligent Data Engineering and Automated Learning (IDEAL 2002)*, vol. 2412, ed. H. Yin, N. Allinson, R. Freeman, J. Keane, and S. Hubbard, 500–505. Lecture Notes in Computer Science Berlin: Springer.

39. Liu, C., G. Zhang, X. Zhang, and H. Liu. 2009. A memetic algorithm based on P systems for IIR digital filter design. In *Proceedings of the Eighth IEEE International Conference on Dependable, Autonomic and Secure Computing*, 330–334.
40. Liu, C., G. Zhang, Y. Zhu, C. Fang, and H. Liu. 2009. A quantum-inspired evolutionary algorithm based on P systems for radar emitter signals. In *Proceedings of the 8th IEEE International Conference on Dependable, Autonomic and Secure Computing*, 24–28.
41. Liu, H., Z. Cai, and Y. Wang. 2010. Hybridizing particle swarm optimization with differential evolution for constrained numerical and engineering optimization. *Applied Soft Computing* 10 (2): 629–640.
42. Liu, C., G. Zhang, H. Liu, M. Gheorghe, and F. Ipate. 2010. An improved membrane algorithm for solving time-frequency atom decomposition. In *Membrane Computing (WMC 2009)*, vol. 5957, ed. M.J. Gh Păun, A. Pérez-Jiménez, G.Rozenberg Riscos-Núñez, and A. Salomaa, 371–384. Lecture Notes in Computer Science Berlin: Springer.
43. Liu, C., M. Han, and X. Wang. 2011. A multi-objective evolutionary algorithm based on membrane systems. In *Proceedings of the 4th International Workshop on Advanced Computational Intelligence*, 103–109.
44. Liu, C., M. Han, and X. Wang. 2012. A novel evolutionary membrane algorithm for global numerical optimization. In *Proceedings of the 3rd International Conference on Intelligent Control and Information Processing*, 727–732.
45. Mezura-Montes, E., and C.A.C. Coello. 2005. A simple multimembered evolution strategy to solve constrained optimization problems. *IEEE Transactions on Evolutionary Computation* 9 (1): 1–17.
46. Nelder, J., and R. Mead. 1965. A simplex method for function minimization. *The Computer Journal* 7 (4): 308–313.
47. Nishida, T. 2004. An application of P systems: a new algorithm for NP-complete optimization problems. In *Proceedings of the 8th World Multi-Conference on Systems, Cybernetics and Informatics*, Vol. 5, 109–112.
48. Nishida, T. 2005. Membrane algorithm: an approximate algorithm for NP-complete optimization problems exploiting P-systems. In *Proceedings of 6th International Workshop on Membrane Computing*, 26–43.
49. Nishida, T. 2006. Membrane algorithms. In *Membrane Computing (WMC 2005)*, vol. 3850, ed. R. Freund, Gh. Păun, G. Rozenberg, and A. Salomaa, 55–66. Lecture Notes in Computer Science Berlin: Springer.
50. Nishida, T. 2006. Membrane algorithms: approximate algorithms for NP-complete optimization problems. In *Applications of Membrane Computing, Chapter 11*, ed. G. Ciobanu, Gh Păun, and M.J. Pérez-Jiménez, 303–314. Natural Computing Series Berlin: Springer.
51. Nishida, T. 2007. Membrane algorithm with brownian subalgorithm and genetic subalgorithm. *International Journal of Foundations of Computer Science* 18 (6): 1353–1360.
52. Nishida, T., T. Shiotani, and Y. Takahashi. 2008. Membrane algorithm solving job-shop scheduling problems. In *Proceedings of the 9th International Workshop on Membrane Computing*, 363–370.
53. Păun, G., G. Rozenberg, and A. Salomaa. 2010. *The Oxford Handbook of Membrane Computing*. New York: Oxford University Press.
54. Peng, H., J. Shao, B. Li, J. Wang, M.J. Pérez-Jiménez, Y. Jiang, and Y. Yang. 2012. Image thresholding with cell-like P systems. In *Proceedings of the Tenth Brainstorming Week on Membrane Computing*, 75–87.
55. Peng, H., J. Wang, M.J. Pérez-Jiménez, and P. Shi. 2013. A novel image thresholding method based on membrane computing and fuzzy entropy. *Journal of Intelligent and Fuzzy Systems* 24 (2): 229–237.
56. Rao, R., V. Savsani, and D. Vakharia. 2011. Teaching-learning-based optimization: A novel method for constrained mechanical design optimization problems. *Computer-Aided Design* 43 (3): 303–315.
57. Robic, T., and B. Filipic. 2005. DEMO: differential evolution for multiobjective optimization. In *Proceedings of 3rd International Conference on Evolutionary Multi-Criterion Optimization*, 520–533.

58. Sun, Y., L. Zhang, and X. Gu. 2010. Membrane computing based particle swarm optimization algorithm and its application. In *Proceedings of the 5th International Conference on Bio-Inspired Computing: Theories and Applications*, 631–636.
59. Traveling salesman problems. http://www.iwr.uni-heidelberg.de/groups/comopt/software/TSPLIB95/tsp/.
60. Vlachogiannis, J., and K. Lee. 2008. Quantum-inspired evolutionary algorithm for real and reactive power dispatch. *IEEE Transactions on Power Systems* 23 (4): 1627–1636.
61. Wang, F., Y. Huang, M. Shi, and S. Wu. 2012. Membrane computing optimization method based on catalytic factor. In *Advances in Brain Inspired Cognitive Systems (BICS 2012)*, vol. 7366, ed. H. Zhang, A. Hussain, D. Liu, and Z. Wang, 129–137. Lecture Notes in Artificial Intelligence Berlin: Springer.
62. Wang, H., H. Peng, J. Shao, and T. Wang. 2012. A thresholding method based on P systems for image segmentation. *ICIC Express Letters* 6 (1): 221–227.
63. Wang, T., J. Wang, H. Peng, and M. Tu. 2012. Optimization of PID controller parameters based on PSOPS algorithm. *ICIC Express Letters* 6 (1): 273–280.
64. Wang, X., G. Zhang, J. Zhao, H. Rong, F. Ipate, and R. Lefticaru. 2015. A modified membrane-inspired algorithm based on particle swarm optimization for mobile robot path planning. *International Journal of Computers, Communications and Control* 10 (5): 732–745.
65. Xiao, J., X. Zhang, and J. Xu. 2012. A membrane evolutionary algorithm for DNA sequence design in DNA computing. *Chinese Science Bulletin* 57 (6): 698–706.
66. Xiao, J., Y. Huang, and Z. Cheng. 2013. A bio-inspired algorithm based on membrane computing for engineering design problem. *International Journal of Computer Science Issues* 10 (1): 580–588.
67. Xiao, J., Y. Huang, Z. Cheng, J. He, and Y. Niu. 2014. A hybrid membrane evolutionary algorithm for solving constrained optimization problems. *Optik* 125 (2): 897–902.
68. Xing, J., and H. Yang. 2012. An optimization algorithm based on evolution rules on cellular system. In *Computational Intelligence and Intelligent Systems (ISICA 2012)*, vol. 316, ed. Z. Li, X. Li, Y. Liu, and Z. Cai, 314–320. Communications in Computer and Information Science Berlin: Springer.
69. Yang, S., and N. Wang. 2012. A novel P systems based optimization algorithm for parameter estimation of proton exchange membrane fuel cell model. *International Journal of Hydrogen Energy* 37 (10): 8465–8476.
70. Yao, X., Y. Liu, and G.M. Lin. 1999. Evolutionary programming made faster. *IEEE Transactions on Evolutionary Computation* 3 (2): 82–101.
71. Yıldız, A. 2009. An effective hybrid immune-hill climbing optimization approach for solving design and manufacturing optimization problems in industry. *Journal of Materials Processing Technology* 209: 2773–2780.
72. Yin, X., L. Qiu, and H. Zhang. 2008. A distributed approach inspired by membrane computing for optimizing bijective S-boxes. In *Proceedings of the 27th Chinese Control Conference*, 60–64.
73. Zaharie, D., and G. Ciobanu. 2006. Distributed evolutionary algorithms inspired by membranes in solving continuous optimization problems. In *Membrane Computing (WMC 7)*, vol. 4361, ed. H.J. Hoogeboom, Gh. Păun, G. Rozenberg, and A. Salomaa, 536–553. Lecture Notes in Computer Science Berlin: Springer.
74. Zavala, A., A. Aguirre, and E. Diharce. 2005. Constrained optimization via evolutionary particle swarm optimization algorithm (PESO). In *Proceedings of the Genetic and Evolutionary Computation Conference*, 209–216.
75. Zhang, R., and H. Gao. 2007. Improved quantum evolutionary algorithm for combinatorial optimization problem. In *International Conference on Machine Learning and Cybernetics*, 3501–3505.
76. Zhang, Y., and L. Huang. 2009. A variant of P systems for optimization. *Neurocomputing* 72 (4–6): 1355–1360.
77. Zhang, J., and A. Sanderson. 2009. JADE: adaptive differential evolution with optional external archive. *IEEE Transactions on Evolutionary Computation* 13 (5): 945–958.

78. Zhang, G., M. Gheorghe, and C. Wu. 2008. A quantum-inspired evolutionary algorithm based on P systems for knapsack problem. *Fundamenta Informaticae* 87 (1): 93–116.
79. Zhang, G., C. Liu, M. Gheorghe, and F. Ipate. 2009. Solving satisfiability problems with membrane algorithm. In *Proceedings of the 4th International Conference on Bio-Inspired Computing: Theories and Applications*, 29–36.
80. Zhang, G., L. Hu, and W. Jin. 2010. Resemblance coefficient and a quantum genetic algorithm for feature selection. In *Discovery Science (DS 2004)*, vol. 3245, ed. E. Suzuki, and S. Arikawa, 155–168. Lecture Notes in Artificial Intelligence Berlin: Springer.
81. Zhang, G., Y. Li, and M. Gheorghe. 2010. A multi-objective membrane algorithm for knapsack problems. In *Proceedings of the 5th International Conference on Bio-Inspired Computing: Theories and Applications*, 604–609.
82. Zhang, G., C. Liu, and H. Rong. 2010. Analyzing radar emitter signals with membrane algorithms. *Mathematical and Computer Modelling* 52 (11–12): 1997–2010.
83. Zhang, G., J. Cheng, and M. Gheorghe. 2011. A membrane-inspired approximate algorithm for traveling salesman problems. *Romanian Journal of Information Science and Technology* 14 (1): 3–19.
84. Zhang, G., M. Gheorghe, and Y. Li. 2012. A membrane algorithm with quantum-inspired subalgorithms and its application to image processing. *Natural Computing* 11 (4): 701–717.
85. Zhang, G., F. Zhou, X. Huang, J. Cheng, M. Gheorghe, F. Ipate, and R. Lefticaru. 2012. A novel membrane algorithm based on particle swarm optimization for solving broadcasting problems. *Chinese Journal of Electronics* 13 (18): 1821–1841.
86. Zhang, G., J. Cheng, M. Gheorghe, and Q. Meng. 2013. A hybrid approach based on differential evolution and tissue membrane systems for solving constrained manufacturing parameter optimization problems. *Applied Soft Computing* 13 (3): 1528–1542.
87. Zhang, G., J. Cheng, and M. Gheorghe. 2014. Dynamic behavior analysis of membrane-inspired evolutionary algorithms. *International Journal of Computers, Communications and Control* 9 (2): 235–250.
88. Zhang, G., M. Gheorghe, L. Pan, and M.J. Pérez-Jiménez. 2014. Evolutionary membrane computing: a comprehensive survey and new results. *Information Sciences* 279: 528–551.
89. Zhang, G., H. Rong, J. Cheng, and Y. Qin. 2014. A population membrane system-inspired evolutionary algorithm for distribution network reconfiguration. *Chinese Journal of Electronics* 23 (3): 437–441.
90. Zhang, G., J. Cheng, M. Gheorghe, F. Ipate, and X. Wang. 2015. QEAM: an approximate algorithm using P systems with active membranes. *International Journal of Computers, Communications and Control* 10 (2): 263–279.
91. Zhao, J., and N. Wang. 2011. Hybrid optimization method based on membrane computing. *Industrial and Engineering Chemistry Research* 50 (3): 1691–1704.
92. Zhao, J., and N. Wang. 2011. A bio-inspired algorithm based on membrane computing and its application to gasoline blending scheduling. *Computers and Chemical Engineering* 35 (2): 272–283.
93. Zhao, J., N. Wang, and P. Zhou. 2012. Multiobjective bio-inspired algorithm based on membrane computing. In *Proceedings of International Conference on Computer Science and Information Processing*, 473–477.
94. Zhou, F., G. Zhang, H. Rong, M. Gheorghe, J. Cheng, F. Ipate, and R. Lefticaru. 2010. A particle swarm optimization based on P systems. In *Proceedings of the 6th International Conference on Natural Computation*, 3003–3007.
95. Zitzler, E., K. Deb, and L. Thiele. 2000. Comparison of multiobjective evolutionary algorithms: empirical results. *Evolutionary Computation* 8 (2): 173–195.

# Chapter 4
# Engineering Optimization with Membrane Algorithms

**Abstract** In this chapter are described engineering applications of the membrane algorithms introduced in Chap. 3. The engineering problems we consider are the following: radar emitter signal analysis, digital image processing, controller design, mobile robot path planning, constrained manufacturing parameter optimization problems, distribution network reconfiguration and electric power system fault diagnosis.

## 4.1 Introduction

Numerous complex engineering problems are intractable and they can be effectively solved by formulating them as optimization problems providing approximate acceptable solutions rather than optimal solutions. As usual, an engineering optimization problem may have continuous, discrete or mixed variables, non-linear constraints and non-linear objective functions with unimodal, multimodal, separable or non-separable characteristics. Thus, engineering optimization problems are typical representatives of difficult or (computationally) hard optimization problems. A heuristic approach is a good choice for this sort of problems.

Of various heuristic methods, membrane algorithms, as introduced in Chap. 3, are alternatives to solve engineering optimization problems. Given the rigor and sound theoretical foundation, the parallel distributed framework and flexible evolution rules of membrane computing models, membrane algorithms can achieve better balance between exploration and exploitation than their counterpart heuristic approaches [47, 49].

The general steps for the use of membrane algorithms to solve an engineering problem include the formulation of an appropriate evaluation function, i.e., a minimization or maximization problem, the encoding of the parameters and the design of a membrane algorithm.

This chapter describes the engineering applications of membrane algorithms with cell-like, tissue-like and neural-like P systems. We will omit the details on the membrane algorithms used in this chapter, due to the description of various membrane algorithms in Chap. 3. The engineering problems we consider in this chapter are the following: radar emitter signal analysis, digital image processing, controller design,

© Springer International Publishing AG 2017

G. Zhang et al., *Real-life Applications with Membrane Computing*,
Emergence, Complexity and Computation 25, DOI 10.1007/978-3-319-55989-6_4

mobile robot path planning, constrained manufacturing parameter optimization problems, distribution network reconfiguration and electric power system fault diagnosis. We also briefly review the applications of membrane algorithms in other engineering fields.

## 4.2   Engineering Optimizations with Cell-Like P Systems

In this section, we will describe how to apply membrane algorithms with cell-like membrane structures to analyze radar signals and digital images, and to optimize the controller design and mobile robot paths.

### 4.2.1   Signal Analysis

The signal analysis with cell-like P systems in this subsection refers to the use of the membrane algorithm based on quantum-inspired evolutionary algorithms (QIEA), introduced in Sect. 3.3.2, to analyze radar emitter signals. To be specific, the modified QIEA based on P systems (MQEPS) in [53] is applied to optimize the time-frequency atom decomposition process of radar emitter signals. MQEPS is a modified variant of QIEA based on P systems (QEPS) by inserting a local search, the tubu search. The related work was introduced in several papers [7, 25, 27, 53].

The time-frequency atom decomposition (TFAD) of radar emitter signals is the process by which a radar emitter signal is decomposed into elementary components with good time and frequency resolution and hence the radar emitter signal can be represented by a linear superposition of a series of waveforms adapting to the local structures of the signal. The waveforms and their collection are called time-frequency atoms and time-frequency atom dictionary, respectively. As usual, the dictionary is redundant or over-complete. The problem of selecting a series of atoms from a redundant time-frequency atom dictionary to optimally approximate a signal has been proved to be an NP-hard problem [8]. The classic method for implementing TFAD of a signal is the greedy algorithm [32]. But a relatively large time-frequency dictionary will result in the impossibility to decompose a signal in a reasonable finite time.

By using TFAD, a radar emitter signal $S(t)$ can be expanded into a linear sum of time-frequency atoms $g_\gamma(t)$, which are elements of an appropriate countable subset of $g_\gamma(t)$ with $\gamma \in \Gamma = \mathcal{R}^+ \times \mathcal{R}^\in$ selecting from a redundant time-frequency atom dictionary $\mathcal{D} = (g_\gamma(t))_{\gamma \in \Gamma}$. Thus, $S(t)$ can be described as

$$S(t) = \sum_{h=1}^{+\infty} a_h g_{\gamma_h}(t), \tag{4.1}$$

where $a_h$ is the expansion coefficient of atom $g_{\gamma_h}(t)$ providing explicit information on certain types or properties of $S(t)$ [32]. $\Gamma = \mathcal{R}^+ \times \mathcal{R}^\in$ is a set of indices for the elements of the dictionary. TFAD is implemented by an iterative procedure starting by projecting $S(t)$ on an atom $g_{\gamma_0} \in \mathcal{D}$ and computing the residual $RS$:

$$S = \langle S, g_{\gamma_0} \rangle g_{\gamma_0} + RS, \tag{4.2}$$

where $RS$ is the residual vector after approximating $S$ in the direction of $g_{\gamma_0}$. Then, the iterative algorithm subdecomposes the residual $RS$ in a sequential manner by projecting it on an atom of $\mathcal{D}$. To maximize the module $|\langle R^h S, g_{\gamma_h} \rangle|$, a nearly best time-frequency atom $g_{\gamma_h}$ is searched at the $h$th iteration from a time-frequency atom dictionary $\mathcal{D}$, where $R^0 S = S$ and $R^h S$ is the $h$th order residual of the radar emitter signal $S(t)$, for $h \geq 1$. Finally, the signal $S$ can be denoted as

$$S = \sum_{h=0}^{H-1} \langle R^h S, g_{\gamma_h} \rangle g_{\gamma_h} + R^H S, \tag{4.3}$$

where $H$ is the maximal number of iterations. As $R^H S$ is orthogonal to $g_{\gamma_h}$, the module of $S$ is

$$\|S\|^2 = \sum_{h=0}^{H-1} |\langle R^h S, g_{\gamma_h} \rangle|^2 + \|R^H S\|^2, \tag{4.4}$$

where $\|R^H S\|$ converges exponentially to 0 when $H$ tends to infinity.

The iterative procedure described above is a local and time-consuming technique. The problem that the best atoms and their corresponding best expansion coefficients in (4.1) are found in a redundant time-frequency dictionary to optimally approximate a radar emitter signal, is an NP-hard problem [8]. In this subsection, a membrane algorithm, MQEPS in [53], is used to search the suboptimal time-frequency atoms from redundant time-frequency atom dictionaries to reduce the computational load of TFAD. The pseudocode algorithm for MQEPS-based TFAD is shown in Fig. 4.1.

In what follows, MQEPS is used to analyze 16 radar emitter signals. Their labels and lengths are listed in Table 4.1. Figure 4.4a shows a signal with the length of 1024 points. To draw a comparison, we consider four algorithms: greedy algorithm (GrA), QIEA, QEPS and MQEPS in the experiments. We evaluate the performances of the four algorithms by five criteria: a correlation rate $C_r$, elapsed time, a decaying rate $D_r$, a final value $D_{rf}$ of the decaying rate $D_r$ and a time-frequency energy distribution (TFED). The correlation rate $C_r = \langle S, S_r \rangle / (\|S\| . \|S_r\|)$ is used to measure the resemblance between the original radar emitter signal $S$ and the restored radar emitter signal $S_r$. Signal $S_r$ is obtained by using TFAD from the corresponding signal $S$. The decaying rate $D_r = \log_{10}(\|R^h S\| / \|S\|)$ is used to measure the decaying speed of signal energy in the process of decomposing a radar emitter signal into a certain number of time-frequency atoms, where $h$ is the number of iterations and $R^h S$ is the $h$th order residual of the original signal $S$ [32]. The final value $D_{rf}$ of

```
Begin
    R⁰S=S ;   h=0 ;
    While (not termination condition) do
        Set parameters of time-frequency atom;
(•)     Search the suboptimal time-frequency atom gγₕ in 𝒟 using MQEPS;

        Compute |⟨RʰS, gγₕ⟩gγₕ| ;

        RʰS ←(RʰS−⟨RʰS, gγₕ⟩gγₕ);

        h←h+1 ;
    End
End
```

**Fig. 4.1** Pseudocode algorithm for MQEPS-based TFAD

**Table 4.1** Comparisons of GrA, QIEA, QEPS and MQEPS in terms of $C_r$ and $D_{rf}$. NoP represents the number of points. Symbol/means that it cannot be obtained within a tolerable computing time

| No. | NoP | GrA | | QIEA | | QEPS | | MQEPS | |
|-----|-----|-----|-----|-----|-----|-----|-----|-----|-----|
| | | $C_r$ | $D_{rf}$ | $C_r$ | $D_{rf}$ | $C_r$ | $D_{rf}$ | $C_r$ | $D_{rf}$ |
| 1 | 512 | 0.9991 | −1.3691 | 0.9965 | −1.0753 | 0.9984 | −1.2511 | 0.9993 | −1.4297 |
| 2 | 1024 | 0.9992 | −1.3944 | 0.9971 | −1.1161 | 0.9986 | −1.2752 | 0.9991 | −1.3677 |
| 3 | 1087 | 0.9991 | −1.3638 | 0.9968 | −1.0988 | 0.9986 | −1.2827 | 0.9989 | −1.3236 |
| 4 | 1631 | 0.9983 | −1.2301 | 0.9950 | −1.0025 | 0.9973 | −1.1367 | 0.9977 | −1.1709 |
| 5 | 2767 | 0.9956 | −1.0272 | 0.9883 | −0.8164 | 0.9929 | −0.9243 | 0.9940 | −0.9598 |
| 6 | 3212 | / | / | 0.9506 | −0.5080 | 0.9678 | −0.5988 | 0.9753 | −0.6559 |
| 7 | 3529 | / | / | 0.9611 | −0.5589 | 0.9740 | −0.6445 | 0.9780 | −0.6811 |
| 8 | 4007 | / | / | 0.8450 | −0.2718 | 0.8745 | −0.3143 | 0.8933 | −0.3472 |
| 9 | 4550 | / | / | 0.9541 | −0.5234 | 0.9717 | −0.6264 | 0.9800 | −0.7013 |
| 10 | 5434 | / | / | 0.9641 | −0.5754 | 0.9803 | −0.7041 | 0.9821 | −0.7247 |
| 11 | 6114 | / | / | 0.9236 | −0.4162 | 0.9566 | −0.5354 | 0.9653 | −0.5829 |
| 12 | 7082 | / | / | 0.8398 | −0.2652 | 0.8759 | −0.3159 | 0.8919 | −0.3446 |
| 13 | 8906 | / | / | 0.8930 | −0.3462 | 0.9509 | −0.5090 | 0.9618 | −0.5625 |
| 14 | 9506 | / | / | 0.9786 | −0.6808 | 0.9952 | −1.0074 | 0.9971 | −1.1181 |
| 15 | 13354 | / | / | 0.9128 | −0.3887 | 0.9531 | −0.5190 | 0.9607 | −0.5563 |
| 16 | 15642 | / | / | 0.9041 | −0.3685 | 0.9794 | −0.6943 | 0.9879 | −0.8093 |

decaying rates is obtained when the number $h$ of iterations reaches its prescribed maximum. The TFED is used to intuitively show the time-frequency characteristics and time-frequency concentration of the time-frequency atoms.

The parameter setting for the experiments is as follows: the maximal number $H$ of iterations for TFAD, the population size $n$ for MQEPS and the number of Q-bits for each parameter are set to 200, 20, and 10, respectively; $I_{max} = 9$ and the number

**Fig. 4.2**  Elapsed time of
QIEA, QEPS and MQEPS

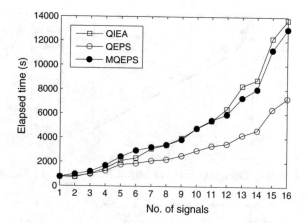

$m = 15$ of elementary membranes in QEPS and MQEPS suggested in [25, 50]; the
maximal number of iterations and tabu length for tabu search are set to 50 and 5,
respectively. We consider the following Gabor time-frequency atom

$$g_\gamma(t) = \frac{1}{\sqrt{s}} g \left( \frac{t - u}{s} \right) \cos{(vt + w)}. \tag{4.5}$$

The index $\gamma = (s, u, v, w)$ is a set of parameters representing the scale, translation,
frequency and phase, respectively, and $g(.)$ is a Gauss-modulated window function,
$g(t) = e^{-\pi t^2}$.

Table 4.1 lists the correlation rates $C_r$ and the final values $D_{rf}$ of the decaying rates
of GrA, QIEA, QEPS and MQEPS. Figure 4.2 shows the elapsed time of MQEPS and
QIEA and QEPS, excluding GrA, for decomposing the 16 signals. Figure 4.3 shows
the comparisons of the progresses of the decaying rates of the four algorithms. Since
GrA is a very time-consuming approach, Table 4.1 and Fig. 4.3 show only the first
five results. GrA consumes 14543, 38874, 61305, 125797 and 262953 (equivalent
to about 73 h) seconds for the signals with 512, 1024, 1087, 1631 and 2767 points,
respectively, which is about 19, 39, 50, 73 and 108 times as much as that in MQEPS,
respectively. Even the elapsed time (14543 s) of GrA for the signal with 512 points is
much more than that (12969 s) of MQEPS for the signal with 15642 points. To further
demonstrate the performance differences of the four algorithms, Fig. 4.4 shows the
TFEDs of the original signal with a length of 1024 points and the signals reconstructed
using time-frequency atoms.

Figure 4.2 shows that MQEPS consumes nearly equivalent computing time to
QIEA and QEPS requires smaller elapsed time for 16 signals than MQEPS and QIEA.
MQEPS has advantages over QIEA for processing long signals. According to the $C_r$
and $D_{rf}$ results in Table 4.1, MQEPS is superior to QEPS and QIEA. The progresses
of the decaying rates of the 16 signals in Fig. 4.3 indicate the MQEPS advantage over
the other two algorithms. The superiority of MQEPS can be also observed in Fig. 4.4,

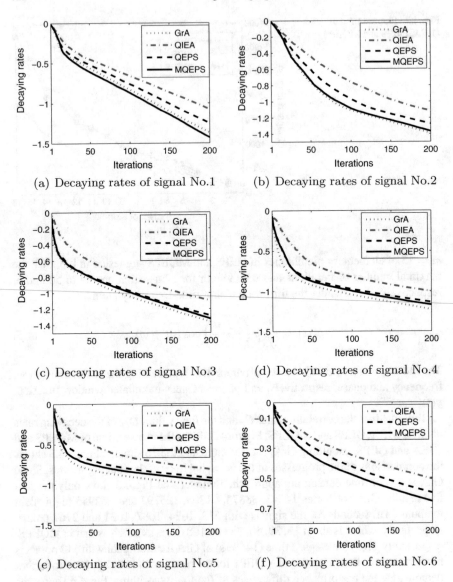

**Fig. 4.3** Decaying rates obtained by GrA, QIEA, QEPS and MQEPS

where TFED obtained by MQEPS can effectively reduce the effect of interference terms in the signal with two linear frequency-modulated components and has good time-frequency concentration.

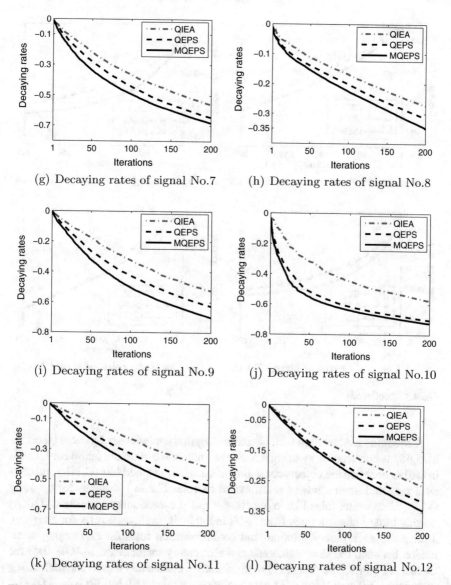

(g) Decaying rates of signal No.7       (h) Decaying rates of signal No.8

(i) Decaying rates of signal No.9       (j) Decaying rates of signal No.10

(k) Decaying rates of signal No.11      (l) Decaying rates of signal No.12

**Fig. 4.3** (continued)

### 4.2.2  *Image Processing*

Image processing is very useful and necessary in various engineering applications
such as robotics, communication systems, biomedical systems and remote sensing.
The image processing discussed here will focus on the use of the membrane algorithm, MAQIS [48], to solve the image sparse decomposition problem.

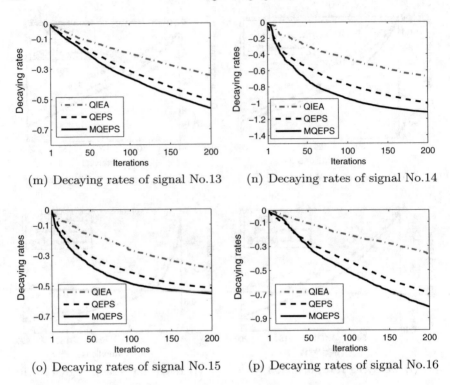

(m) Decaying rates of signal No.13          (n) Decaying rates of signal No.14

(o) Decaying rates of signal No.15          (p) Decaying rates of signal No.16

**Fig. 4.3**  (continued)

A Membrane Algorithm with Quantum-Inspired evolutionary Systems (short for MAQIS) is constructed by using a one-level membrane structure (short for OLMS) in [50] to organize the objects consisting of quantum-inspired bits (Q-bits) and classical bits, and three kinds of rules: evolution rules like in P systems for updating Q-bits, observation rules like in quantum-inspired evolutionary algorithms (QIEA) making binary bits solutions from Q-bit individuals, evaluation rules for assigning a fitness to each binary solution, and communication rules for exchanging information between the skin membrane and elementary membranes. MAQIS uses the OLMS in Fig. 4.5. The skin membrane contains $m$ elementary membranes delimiting $m$ regions. Different types of Q-gate evolutionary rules, $QRG_1$, $QRG_2$, ..., $QRG_m$, are put into different regions. A Q-bit individual consisting of a certain number of Q-bits and a binary string are dealt with as multisets of objects. The transformation of a Q-bit individual into a binary string is implemented by using a probabilistic observation process. In MAQIS, a binary string corresponds to a candidate solution of a problem. The set of rules are responsible for making the system evolve; the initialization, observation and evaluation are executed in the skin membrane; Q-gate update processes for generating offspring are performed in elementary membranes. More details about MAQIS can be referred to [48, 50].

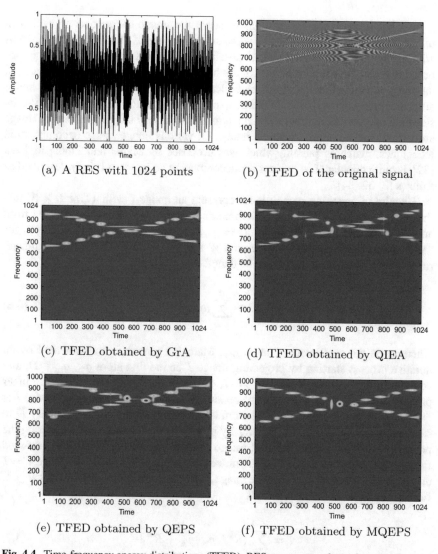

(a) A RES with 1024 points

(b) TFED of the original signal

(c) TFED obtained by GrA

(d) TFED obtained by QIEA

(e) TFED obtained by QEPS

(f) TFED obtained by MQEPS

**Fig. 4.4** Time-frequency energy distributions (TFED). RES represents radar emitter signal

**Fig. 4.5** The OLMS used in MAQIS

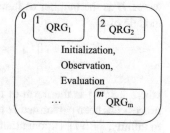

In what follows, we use MAQIS to solve the image sparse decomposition in image processing. The aim is to find as small number of elementary components with the key features of an image as possible to represent or reconstruct the image or to obtain a sparse representation of the image [3]. The elementary components of an image are called image atoms. A collection of image atoms is called an image atom dictionary. The problem of image sparse decomposition can be described as the selection of a certain number of image atoms from a very large or redundant image atom dictionary to approximate an image. Of various image sparse decomposition techniques, matching pursuit, which was proposed by Mallat and Zhang in 1993 [32], is quite popular. The process for decomposing an image is briefly described as follows [3, 32, 34].

The aim of matching pursuit is to represent an image $I$ with a size $L \times W$ in a linear superposition by using a certain number $H$ of image atoms $g_\gamma$ selected from an over-complete atom dictionary $D = \{g_\gamma\}_{\gamma \in \Gamma}$, where $\Gamma$ is a set of indices for the elements of the dictionary $D$, $L$ and $W$ are the length and the width of image, respectively. Thus, an image $I$ can be denoted as

$$I = \sum_{h=1}^{H} a_h g_{\gamma_h} \tag{4.6}$$

where $a_h$ is the coefficient of atom $g_{\gamma_h}$. Matching pursuit is implemented by an iterative process starting by projecting image $I$ on the first atom $g_{\gamma_0}$, $g_{\gamma_0} \in D$, and computing the residual $RI$, i.e., $I = \langle I, g_{\gamma_0} \rangle g_{\gamma_0} + RI$, where $\langle I, g_{\gamma_0} \rangle$ is the inner product of image $I$ and $g_{\gamma_0}$; $RI$ is the residual vector after approximating image $I$ at the direction of $g_{\gamma_0}$. Then, the next atom is selected from the atom dictionary $D$ to subdecompose the residual vector $RI$ by projecting it on an atom of $D$. During the process of decomposition, the best image atom $g_{\gamma_h}$ at the $h$th iteration is searched from image atom dictionary $D$ to maximize the module $|\langle R^h I, g_{\gamma_h} \rangle|$, where $R^0 I = I$ and $R^h I$ is the residual vector of the image $I$ after $h$ atoms found matches, for $h \geq 1$. Finally, we represent the image $I$ as

$$I = \sum_{h=0}^{H-1} \langle R^h I, g_{\gamma_h} \rangle g_{\gamma_h} + R^H I \tag{4.7}$$

where $H$ is the number of iterations. $R^H I$ is orthogonal to $g_{\gamma_h}$, which leads to the module of $I$

$$||I||^2 = \sum_{h=0}^{H-1} |\langle R^h I, g_{\gamma_h} \rangle|^2 + ||R^H I||^2 \tag{4.8}$$

where $\|R^H I\|$ is the norm of the residual vector of the image $I$ after the iterative process has been performed $H$ times, i.e., there are $H$ atoms found. As $H$ approaches to infinity, $\|R^H I\|$ exponentially converges to 0.

**Fig. 4.6** Pseudocode algorithm for image sparse decomposition

The optimal approximation of an image by using a sequence of best atoms and their corresponding best coefficients in (4.6), which are obtained from a redundant or over-complete atom dictionary by applying matching pursuit, is an NP-hard problem [8]. As usual, the iterative process for matching pursuit is very time-consuming [58], so MAQIS is used to search the suboptimal atoms from an over-complete dictionary in the following description. The pseudocode algorithm is shown in Fig. 4.6.

The over-complete image atom dictionary is constructed by using a non-symmetric atom [34] with the following basic form without normalization processing

$$g(x, y) = (4x^2 - 2)e^{-(x^2+y^2)} \qquad (4.9)$$

where $x$ and $y$ are the coordinates in a plane representing the length and width of an image, respectively. The updated version $g_\gamma$, which is used in the experiments, of the basic non-symmetric atom by rotating, translating and scaling is

$$g_\gamma = g\left(\frac{x-u}{s_x}, \frac{y-v}{s_y}\right) \qquad (4.10)$$

where $\theta, u, v, s_x, s_y$ are the rotation factors, translation factors at the direction of $x$ and $y$, scale factors at the direction of $x$ and $y$, respectively, and they form a vector $\gamma = (\theta, u, v, s_x, s_y)$.

In the experiments, four benchmark images widely used in the community of image processing are considered. The four images are Lena$_1$ with $128 \times 128$ pixels, Lena$_2$ with $256 \times 256$ pixels, Cameraman$_1$ with $128 \times 128$ pixels and Cameraman$_2$ with $256 \times 256$ pixels, respectively. We consider two versions of MAQIS, MAQIS$_1$ and MAQIS$_2$ [48, 58], and QEPS and five types of QIEAs constructed by using five types of Q-gate evolutionary approaches, QRG$_1$ [13], QRG$_2$ [14], QRG$_3$ [52], QRG$_4$ [54], QRG$_5$ [11], respectively. According to the description on matching pursuit, We use the evaluation function

$$f(\gamma) = |\langle R^h I, g_{\gamma_h}\rangle| \qquad (4.11)$$

where $\gamma = (\theta, u, v, s_x, s_y)$; $R^h I$ is the residual vector of image $I$ after $h$ atoms are found; $g_{\gamma_h}$ is the atom found at iteration $h$. The population size is set to 20. The number 1000 of atoms and the maximal number 100 of evolutionary generations are used as the stopping condition. The algorithm performance is evaluated by three criteria: the elapsed time, the reconstructed images and the quantitative measure for the quality of reconstructed images, which uses the peak signal-to-noise ratio (PSNR)

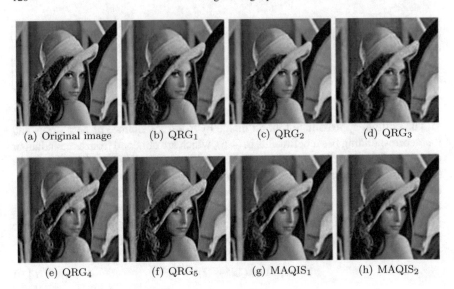

(a) Original image     (b) QRG$_1$     (c) QRG$_2$     (d) QRG$_3$

(e) QRG$_4$     (f) QRG$_5$     (g) MAQIS$_1$     (h) MAQIS$_2$

**Fig. 4.7** Reconstructed Lena$_1$ images

$$PSNR = 10 \log \left( \frac{255^2}{LW \sum\limits_{j=0}^{L-1} \sum\limits_{i=0}^{W-1} (x_{ij} - \widehat{x}_{ij})} \right) \qquad (4.12)$$

where $L$ and $W$ are the length and width of images, respectively; $x_{ij}$ and $\widehat{x}_{ij}$ are the pixels of the original image and the reconstructed image, respectively.

The original four images, Lena$_1$, Lena$_2$, Cameraman$_1$ and Cameraman$_2$, are shown in Figs. 4.7a, 4.8a, 4.9a and 4.10a, respectively. The reconstructed images obtained by using QRG$_1$, QRG$_2$, QRG$_3$, QRG$_4$, QRG$_5$, MAQIS$_1$ and MAQIS$_2$, are shown in Figs. 4.7b–h, 4.8b–h, 4.9b–h and 4.10b–h, respectively. The PSNR of the reconstructed images and mean of elapsed time (MET) per iteration are listed in Table 4.2.

The images reconstructed by using 1000 image atoms of seven algorithms, which are shown in Figs. 4.7, 4.8, 4.9 and 4.10, have a good visual quality close to the original images. Table 4.2 shows different PSNR values of the reconstructed images from the seven algorithms at the cost of nearly equal elapsed time in the process of decomposing each of the four images with different sizes. An order of seven algorithms with respect to PSNR values from the worst to the best, i.e., QRG$_3$, QRG$_4$, QRG$_1$, QRG$_2$, MAQIS$_1$, QRG$_5$ and MAQIS$_2$, as indicated in Table 4.2.

**Fig. 4.8** Reconstructed
Lena$_2$ images

(a) Original image            (b) QRG$_1$

(c) QRG$_2$                   (d) QRG$_3$

(e) QRG$_4$                   (f) QRG$_5$

(g) MAQIS$_1$                 (h) MAQIS$_2$

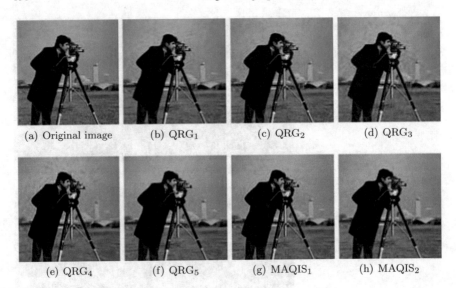

(a) Original image      (b) QRG$_1$      (c) QRG$_2$      (d) QRG$_3$

(e) QRG$_4$      (f) QRG$_5$      (g) MAQIS$_1$      (h) MAQIS$_2$

**Fig. 4.9** Reconstructed Cameraman$_1$ images

### 4.2.3 Controller Design

A controller is used to monitor and alter the operating conditions of a control system, which may be a social system, an ecology system, an industrial process, an aircraft system, or an anti-lock brake system, and so on. Controller design is not only the central work of control theory, but also an important task in various disciplines. This subsection discusses the use of the membrane algorithm, particle swarm optimization based on P systems (PSOPS) [57, 59], to optimize the design of a proportional-integral-derivative (PID) controller [41].

By combining P systems and particle swarm optimization, PSOPS uses the representation of individuals, evolutionary rules of particle swarm optimization, one-level membrane structure (OLMS) [50] and transformation or communication-like rules in P systems to design its algorithm. To be specific, PSOPS was designed by using particle swarm optimization algorithms as algorithm-in-membrane (AIM) organized in OLMS. The details on PSOPS can be referred to [57, 59].

The block diagram of a PID control system is shown in Fig. 4.11, where $r(t)$, $y(t)$, $e(t)$ and $u(t)$ represent the input, output, control error and control signal [23]. Their relationship can be depicted by the following formula

$$u(t) = K_p \left[ e(t) + \frac{1}{T_i} \int_0^t e(t)dt + T_d \frac{de(t)}{dt} \right] \qquad (4.13)$$

**Fig. 4.10** Reconstructed
Cameraman$_2$ images

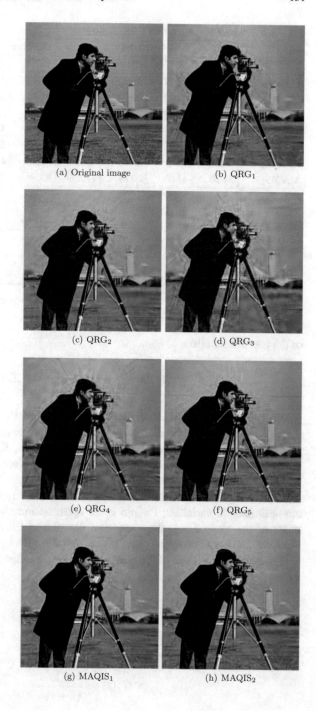

(a) Original image                  (b) QRG$_1$

(c) QRG$_2$                         (d) QRG$_3$

(e) QRG$_4$                         (f) QRG$_5$

(g) MAQIS$_1$                       (h) MAQIS$_2$

**Table 4.2** Experimental results for seven algorithms

| Algorithms | Lena$_1$ | | Lena$_2$ | | Cameraman$_1$ | | Cameraman$_2$ | |
|---|---|---|---|---|---|---|---|---|
| | PSNR | MET (s) | PSNR | MET (s) | PSNR | MET (s) | PSNR | MET (s) |
| QRG$_1$ | 33.81 | 38.27 | 29.17 | 106.18 | 32.72 | 37.54 | 28.25 | 105.83 |
| QRG$_2$ | 34.18 | 43.79 | 29.50 | 112.78 | 33.16 | 39.71 | 28.57 | 111.80 |
| QRG$_3$ | 30.78 | 39.42 | 27.00 | 99.94 | 29.62 | 39.05 | 25.87 | 101.41 |
| QRG$_4$ | 31.64 | 42.49 | 27.71 | 113.77 | 30.35 | 42.17 | 26.63 | 113.71 |
| QRG$_5$ | 34.55 | 40.01 | 29.81 | 106.97 | 33.55 | 38.86 | 29.02 | 106.92 |
| MAQIS$_1$ | 34.53 | 41.47 | 29.65 | 110.30 | 33.33 | 41.14 | 28.75 | 108.72 |
| MAQIS$_2$ | 34.90 | 43.00 | 29.89 | 110.39 | 33.68 | 41.80 | 29.18 | 110.01 |

**Fig. 4.11** Block diagram of a PID control system

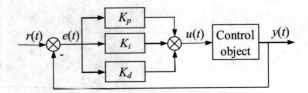

or the transfer function

$$C(s) = \frac{U(s)}{E(s)} = K_p + \frac{K_i}{s} + K_d s \qquad (4.14)$$

where $T_i$ and $T_d$ are integral and derivative time, respectively; $K_p$, $K_i = K_p/T_i$ and $K_d = K_p T_d$ are proportional, integral and derivative gains, respectively. The design of a PID controller is to tune the three parameters, $K_p$, $K_i$ and $K_d$, to obtain good control performance, such as quick-response, small overshoot, short settling time and high stability precision.

In the design of a PID controller with PSOPS, the three parameters, $K_p$, $K_i$ and $K_d$, are optimization variables; PSOPS is the optimization algorithm; the performance index of the controller is the evaluation function of PSOPS. In what follows, we illustrate the design procedure and experimental results by using four examples.

The transfer functions of four single-input single-output control objects [6] are as follows:

$$G_1(s) = \frac{1}{s+1}e^{-0.2s} \qquad (4.15)$$

$$G_2(s) = \frac{0.33}{134s^2 + 18.5s + 1}e^{-18.25s} \qquad (4.16)$$

$$G_3(s) = \frac{1 - 1.4s}{(1+s)^3} \qquad (4.17)$$

$$G_4(s) = \frac{1}{(s+1)(s/6+1)^3} \tag{4.18}$$

In the experiments, the results obtained by the three methods, Z-N, SGA and dsDNA-MC [6], are used as benchmarks. The integral of absolute error of control performance index (IAE) is considered as the fitness function. The parameters of PSOPS are set as follows: population size is set to 30; the number of elementary membranes is set to 16; the maximal generations both in the skin membrane and in each elementary membrane are set to 10; the inertia weight is set to 0.6; the two acceleration factors are set to 2 and the dimension is set to 3. The values of PID controller parameters, $K_p$, $K_i$ and $K_d$, and the control performance indexes, adjustment time $t_s$, the overshoot $\sigma$ and the IAE value, are shown in Table 4.3. The step responses of the closed loop systems for the four control objects are shown in Fig. 4.12.

The results in Table 4.3 and the curves in Fig. 4.12 show that PSOPS achieves better results than the other three methods in [6]. For the objects $G_1$, $G_2$ and $G_4$, the PID control systems tuned by PSOPS have shorter settling time, smaller overshoot and performance index than the other three methods. For object $G_3$, the result in Fig. 4.12c show that the unit step response curves of the former three methods are slightly better than PSOPS, but the performance index of the fourth method is far better than others with faster response speed.

**Table 4.3** Comparisons of four methods with respect to PID controllers

| Objects | Methods | $K_p$ | $K_i$ | $K_d$ | $t_s$ (s) | $\sigma$ (%) | IAE |
|---------|---------|-------|-------|-------|-----------|--------------|-----|
| $G_1$ | Z-N | 5.1070 | 13.7190 | 0.4750 | 1.18 | 57.60 | 0.5400 |
| | SGA | 4.8911 | 3.1814 | 0.5513 | 1.03 | 6.31 | 0.3550 |
| | dsDNA-MC | 4.3118 | 3.2737 | 0.3218 | 0.62 | 8.09 | 0.3330 |
| | PSOPS | 3.8884 | 3.1098 | 0.2675 | 0.48 | 2.39 | 0.3312 |
| $G_2$ | Z-N | 2.9010 | 0.0800 | 26.4300 | 103.00 | 11.30 | 41.9400 |
| | SGA | 2.4446 | 0.1010 | 30.8800 | 122.50 | 13.00 | 39.7500 |
| | dsDNA-MC | 2.3757 | 0.0882 | 26.8500 | 66.10 | 7.00 | 38.3100 |
| | PSOPS | 2.3226 | 0.0863 | 25.6560 | 65.53 | 5.99 | 38.2715 |
| $G_3$ | Z-N | 0.9230 | 0.2700 | 0.7880 | 8.50 | 2.75 | 3.7450 |
| | SGA | 0.8714 | 0.2718 | 0.7990 | 4.55 | 6.10 | 3.7130 |
| | dsDNA-MC | 0.8890 | 0.2895 | 0.8295 | 4.31 | 2.13 | 3.6580 |
| | PSOPS | 0.9527 | 0.3019 | 0.8946 | 5.54 | 7.71 | 3.4787 |
| $G_4$ | Z-N | 4.6050 | 5.9040 | 0.8470 | 2.35 | 52.10 | 0.9830 |
| | SGA | 8.3690 | 2.9330 | 2.5870 | 2.74 | 10.60 | 0.5850 |
| | dsDNA-MC | 6.6190 | 2.0280 | 1.9980 | 1.54 | 4.10 | 0.5430 |
| | PSOPS | 4.9104 | 1.8422 | 1.3680 | 1.59 | 3.08 | 0.5706 |

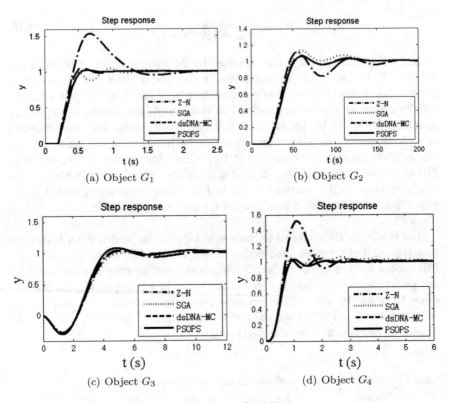

**Fig. 4.12** Step responses of closed loop systems for four objects

### 4.2.4  Mobile Robot Path Planning

This subsection uses a modified membrane-inspired algorithm based on particle swarm optimization (mMPSO) to solve a multi-objective mobile robot path planning problem (MR3P) in a challenging environment with various static and dynamic obstacles. We will first describe MR3P and then present the procedure and results. The mMPSO is briefly discussed. More details can be referred to [43].

MR3P is proven to be an NP-complete problem [31]. Its aim is to use an optimization technique to find a short, smooth and safe path for a mobile robot from a starting point to the ending point in an environment with static or dynamic obstacles. The requirements for the path are characterized by: the shortest distance, smoothness and safety. So the MR3P objective function $f$ can be defined as

$$f = K_d \cdot Dis + K_f \cdot S + K_s \cdot SD \qquad (4.19)$$

where $Dis$, $S$ and $SD$ are path length, smoothness and safety degree, respectively, and their weighing factors are $K_d$, $K_s$ and $K_v$, respectively.

Path length $Dis$ is defined as the sum of distances between $n$ nodes from the starting point to the ending point, i.e.,

$$Dis = \sum_{i=0}^{n-1} L(i, i+1) \tag{4.20}$$

where $L(i, i+1) = \sqrt{(x_{i+1} - x_i)^2 + (y_{i+1} - y_i)^2}$ is the distance between nodes $i = (x_i, y_i)$ and $i+1 = (x_{i+1}, y_{i+1})$.

Smoothness is defined as the sum of the reflection angles formed by any arbitrary three neighboring nodes of a path. Usually, the direct calculation of the smoothness is a time-consuming process, thus, we use an indirect approach, i.e., it uses the ratio $S_c$ of the number of deflection angles which are less than the given expected value over the total number of deflection angles and the ratio $S_p$ of the number of path segments which are greater than the number of the segments in the path with the smallest number of path segments in a group over the total number of path segments to evaluate the smoothness of a path. Smoothness is computed by the following formula:

$$S = \alpha \cdot S_c + \beta \cdot S_p \tag{4.21}$$

where

$$S_c = 1 - \frac{DA_l}{N_f - 1} \tag{4.22}$$

$$S_p = 1 - \frac{S_{min}}{N_f} \tag{4.23}$$

where $\alpha$ and $\beta$ are two weighing coefficients; $DA_l$ is the number of deflection angles greater than the expected value; $N_f$ the total number of path segments; $S_{min}$ is the number of the segments in the path with the smallest number of path segments in a group.

Safety degree (SD) refers to the sum of deviation degrees $C_i$ $(i = 1, 2, \ldots, N)$ between any segment in a path and its nearest obstacle, i.e.,

$$SD = \sum_{i=1}^{n-1} C_i = \begin{cases} 0, & d_i \geq \lambda \\ \sum_{i=1}^{n-1} e^{\lambda - d_i}, & d_i < \lambda \end{cases} \tag{4.24}$$

where $d_i$ is the minimal distance between the $i$th segment and its nearest obstacle, and $\lambda$ is the threshold of safety degree.

A grid (occupancy cell) environment is used and can be represented by two ways: a $X$–$Y$ coordinates plane [39] and an orderly numbered grid. We use the latter technique. A square environment is evenly divided into a certain number of squares, i.e., the $x$-axis and $y$-axis are divided equally into $m$ segments. Thus, we obtain $m \times m$ grids, where several grids are used to represent the obstacles. The grid-based

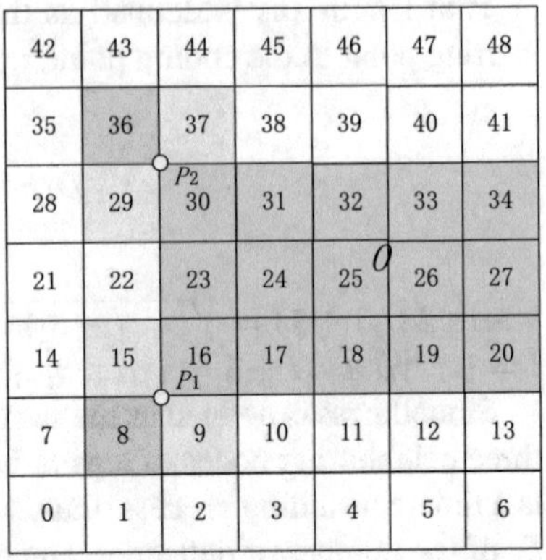

**Fig. 4.13** A 7 × 7 grid
environment

| 42 | 43 | 44 | 45 | 46 | 47 | 48 |
|----|----|----|----|----|----|----|
| 35 | 36 | 37 | 38 | 39 | 40 | 41 |
| 28 | 29 | $P_2$ 30 | 31 | 32 | 33 | 34 |
| 21 | 22 | 23 | 24 | 25 $O$ | 26 | 27 |
| 14 | 15 | 16 $P_1$ | 17 | 18 | 19 | 20 |
| 7 | 8 | 9 | 10 | 11 | 12 | 13 |
| 0 | 1 | 2 | 3 | 4 | 5 | 6 |

environment considered is dynamic and therefore moving obstacles or dangerous
sources may appear or disappear. Figure 4.13 shows an example of 7 × 7 grids. In
this figure, the grid map is encoded by using Matlab and the grey shadow grids denote
obstacles. The mapping relations between coordinates $(x, y)$ and the serial number
$p$ starting from one is described as

$$p = fix(y/SoG) \cdot NoC + fix(x/SoG) \tag{4.25}$$

where $SoG$ is the grid size; $NoC$ is the number of columns; the function $fix(t)$
rounds $t$ to its nearest integer towards zero.

Thus, MR3P can be described as a minimization problem with the objective
function (4.19) in an orderly numbered grid environment with moving obstacles or
dangerous sources. In what follows, we use mMPSO to solve MR3P.

The modified membrane-inspired algorithm based on particle swarm optimization (mMPSO) is designed by using a variable dimension PSO and a dynamic membrane structure with membrane division and dissolution. In mMPSO, a point repair
algorithm, a smoothness approach and a moving direction adjustment technique are
considered. Their details and mMPSO can be referred to [43].

In what follows, several experiments are used to show the application. We consider
three grid models, 16 × 16, 32 × 32 and 64 × 64, and five environments: 16 × 16
with 9 static obstacles ($O_s = 9$), 16 × 16 with $O_s = 9$ and one dynamic obstacle
($O_d = 1$), 32 × 32 with $O_s = 20$, 32 × 32 with $O_s = 20$ and $O_d = 1$, 64 × 64 with
$O_s = 20$. The 16 × 16 grid model environment with 9 static obstacles is first applied
to compare mMPSO with its counterpart particle swarm optimization with variable
dimensions (vPSO) (when $m = 1$, mMPSO becomes vPSO) and genetic algorithm
(GA) [39]. Complex environments, 32 × 32 and 64 × 64 grid model environments
with 20 static obstacles, are applied to further show the application. One dynamic
obstacle representing a moving obstacle or a dangerous source occurring suddenly
is considered. The position for the possible occurrence of the dynamic obstacle is
close to the near center because it may block the feasible paths. Parameter setting is
as follows: the population size is set to 100; the proportion coefficients $\delta_1 = 0.65$,

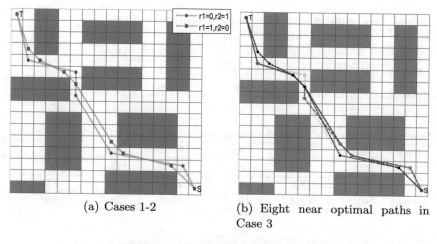

(a) Cases 1-2

(b) Eight near optimal paths in Case 3

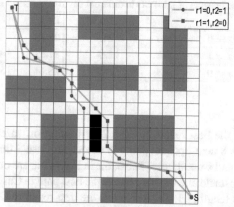

(c) Cases 1-2 with one dynamic obstacle

**Fig. 4.14** Experimental results of mMPSO in the environments $16 \times 16$ grids, $O_s = 9$ and $O_d = 1$

$\delta_2 = 0.35$; $\rho$ is defined as a variable, which varies from 0.246 to 0.157 along the logarithm function $\log_{10}(y)$; the number $m$ of elementary membranes inside the skin membrane is set to 13; in (4.21), $\alpha = 0.6$, $\beta = 0.4$; in (4.24), $\lambda$ is set to the robot radius. The termination condition is designated as the maximal number 2000 of iterations.

In the experiments on the model with $16 \times 16$ grids, we consider three cases for $K_d, K_s, K_f$ as follows:

(1) Case 1: $K_d = 1$, $K_s = K_f = 0$, $\gamma_1 = 1$, $\gamma_2 = 0$;
(2) Case 2: $K_d = 0.6$, $K_s = K_f = 0.2$, $\gamma_1 = 0$, $\gamma_2 = 1$;
(3) Case 3: $K_d = 0.8$, $K_s + K_f = 0.2$, $\gamma_1 + \gamma_2 = 1$.

Figure 4.14 shows the experimental results of mMPSO. Figure 4.14a indicates that the blue line is the best path in Case 1 considering only one objective, path length,

**Table 4.4**  Comparisons of three methods in the environment (Fig. 4.14a)

| Method | NoO | NoNO | NoI | Fv | Gn | St |
|--------|-----|------|-----|-------|-----|------|
| GA [36] | 9 | 78 | 13 | 24.68 | 16 | 1.68 |
| vPSO | 83 | 108 | 0 | 24.95 | 65 | 2.97 |
| mMPSO | 94 | 239 | 0 | 24.26 | 27 | 0.84 |

**Table 4.5**  Comparisons of three methods in the environment (Fig. 4.14c)

| Method | NoO | NoNO | NoI | Fv | Gn | St |
|--------|-----|------|-----|-------|-----|------|
| GA [36] | 32 | 68 | 0 | 24.71 | 12 | 0.69 |
| vPSO | 81 | 103 | 0 | 28.56 | 73 | 3.12 |
| mMPSO | 92 | 235 | 0 | 27.43 | 34 | 0.97 |

**Table 4.6**  Experimental results of mMPSO in different environments in Fig. 4.15

| Environment | NoO | NoNo | Fv | Gn | St |
|-------------|-----|------|-------|-----|------|
| $32 \times 32$, $O_s = 20$, $O_d = 0$ | 86 | 242 | 28.79 | 36 | 1.72 |
| $32 \times 32$, $O_s = 20$, $O_d = 1$ | 82 | 225 | 31.53 | 45 | 1.93 |
| $64 \times 64$, $O_s = 20$, $O_d = 0$ | 83 | 247 | 28.14 | 59 | 2.68 |

and the red line is the best result in Case 2 by trading-off safety and smoothness. Figure 4.14b shows 8 near optimal paths (8 colors) in Case 3 through balancing the path length, safety and smoothness. The paths in Fig. 4.14c are obtained by considering one dynamic obstacle and the blue line is the best path in Case 1 considering only one objective, path length, and the red line is the best result in Case 2 by trading-off safety and smoothness.

Let $K_d = 1$. We perform the experiment for 100 independent runs to draw a comparison with GA in [39] and vPSO. The experimental results of GA, vPSO and mMPSO in the environments with static obstacles and the environments with static and dynamic obstacles are shown in Tables 4.4 and 4.5. NoO, NoNO, NoI, Fv, Gn and St in Tables 4.4, 4.5 and 4.6 are the number of optimal solutions, the number of near optimal solutions, the number of infeasible solutions and the fitness value in 100 trials, the average generations for finding the optimal solution and the mean of the elapsed time (s) in each trial, respectively.

Table 4.4 and Fig. 4.14 shows that mMPSO takes smaller elapsed time to find much more optimal paths and near optimal paths than GA. The solutions found by vPSO and mMPSO are feasible, while GA finds some infeasible solutions. mMPSO is better than vPSO and GA with respect to optimal and near optimal solutions and the elapsed time. Similar conclusions are indicated in Tables 4.5 and 4.4.

In the following experiments, complex environments with $32 \times 32$ and $64 \times 64$ grids containing 20 or 21 obstacles are considered and shown in Fig. 4.15a–d. Figure 4.15a shows the environment with $32 \times 32$ grids and 20 static obstacles.

**Fig. 4.15** Experimental
results of mMPSO in the
environments $32 \times 32$,
$64 \times 64$ grids, $O_s = 20$ and
$O_d = 1$

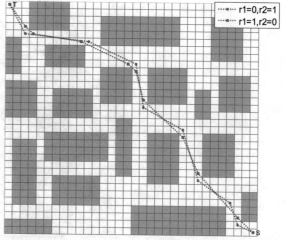

(a) Cases 1-2 in the environment $32 \times 32$

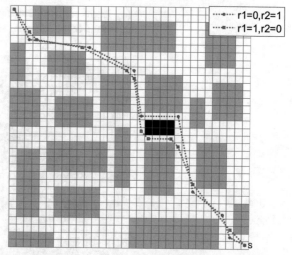

(b) Cases 1-2 in the environment $32 \times 32$

The environment with 20 static obstacles and one dynamic obstacle is shown in
Fig. 4.15b–c. The three objectives, path length, smoothness and safety, are considered in Fig. 4.15c. We use the same parameters as the previous experiment except
for the population size 150 and $m = 15$. 100 independent runs are executed for all
the tests. Experimental results are shown in Table 4.6. Tables 4.4, 4.5 and 4.6 show
that the optimal solutions obtained by mMPSO decrease from 94 to 83, the elapsed
time goes up from 0.84 to 2.68 and the average generations vary from 27 to 59 as

**Fig. 4.15** (continued)

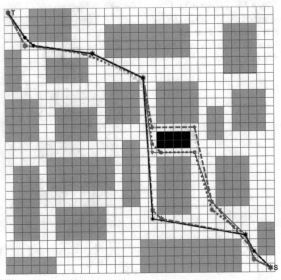

(c) Eight near optimal paths in Case 3 in the environment $32 \times 32$ with one dynamic obstacle

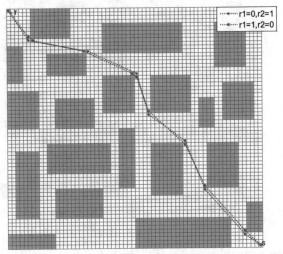

(d) Cases 1-2 in the environment with $64 \times 64$

the number of grids increases from $16 \times 16$ to $64 \times 64$ and the static obstacles go up from 9 to 20. The dynamic obstacle slightly increases the elapsed time and average generations.

### 4.2.5  Other Applications

In addition to the engineering applications discussed above, membrane algorithms were also applied to solve a wider range of engineering problems. For instance, in [56], gasoline blending scheduling problems were discussed. In [25], infinite impulse response digital filters were designed by using membrane algorithms. In [45], the parameters of proton exchange membrane fuel cell models were estimated. The cognitive radio spectrum allocation problem was solved in [9, 10].

## 4.3  Engineering Optimization with Tissue-Like P Systems

### 4.3.1  Manufacturing Parameter Optimization Problems

This subsection uses the membrane algorithm designed with a tissue P system and differential evolution (DETPS) [46], which is discussed in Sect. 3.6, to solve manufacturing parameter optimization problems.

Manufacturing is a complex process that transforms raw materials, components or parts into finished products so as to satisfy expectations or constraints imposed by customers' specifications. To design and produce a suitable final product, lots of conditions or factors or parameters are considered in the formulation of the manufacturing process. A set of parameters corresponding to a set of manufacturing conditions is a solution of the problem. To obtain an efficient manufacturing process to produce a high-quality product with low cost, a manufacturing parameter optimization problem is usually solved by using an optimization technique to find the optimal condition. Considering various parameters (or decision variables) and various constraints, a manufacturing parameter optimization problem can be formulated as a mixed discrete-continuous constrained minimization problem:

$$\min f(x), \quad x = (x_1, x_2, \ldots, x_n) \tag{4.26}$$

subject to:

$$g_i(x) \leq 0, \quad i = 1, 2, \ldots, p \tag{4.27}$$
$$h_j(x) = 0, \quad j = 1, 2, \ldots, q \tag{4.28}$$

where $x = (x_1, \ldots, x_n)$ represents a solution to the manufacturing parameter optimization problem; $x_i \in R, (i = 1, 2, \ldots, n)$ is a parameter or a variable to be optimized; $n$ is the number of parameters; $f(x)$ is the objective function; $p$ and $q$ are the number of inequality constraints and equality constraints, respectively. The two kinds of constraints, $g_i(x) \leq 0$ and $h_j(x) = 0$, could be linear or nonlinear. Parameter $x_i(i = 1, 2, \ldots, n)$ is one of the four types of variables: type I, type II, type III and type IV [21]. Type I refers to a binary variable, 0 or 1. Type II is an integer

**Fig. 4.16** The welded beam
design problem [2, 46]

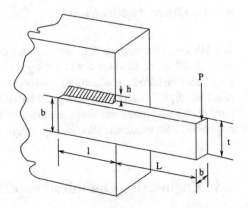

variable confined within the uniformly spaced values in a bounded range. Type III
is a real-valued parameter consisting of uniformly spaced values within a bounded
range. Type IV could be a discrete or continuous variable that varies within a bounded
range.

In what follows, we use DETPS to solve four manufacturing parameter optimization problems [2, 15, 17, 26, 29, 35, 36]: welded beam design problem, pressure
vessel design problem, tension/compression string design problem and speed reducer
design problem.

A welded beam design problem is designed for the minimum cost subject to
constraints on shear stress ($\tau$), bending stress in the beam ($\theta$), buckling load on the
bar ($P_c$), end deflection of the beam ($\delta$), and side constraints. There are four design
variables as shown in Fig. 4.16 [2, 46], i.e. $h(x_1), l(x_2), t(x_3)$ and $b(x_4)$. This problem
can be mathematically formulated as follows:

$$\min f(x) = 1.10471x_1^2 x_2 + 0.04811 x_3 x_4 (14.0 + x_2)$$

subject to

(1) $g_1(x) = \tau(x) - \tau_{\max} \leq 0,$
(2) $g_2(x) = \sigma(x) - \sigma_{\max} \leq 0,$
(3) $g_3(x) = x_1 - x_4 \leq 0,$
(4) $g_4(x) = 0.10471x_1^2 + 0.04811 x_3 x_4 (14.0 + x_2) - 5.0 \leq 0,$
(5) $g_5(x) = 0.125 - x_1 \leq 0,$
(6) $g_6(x) = \delta(x) - \delta_{\max} \leq 0,$
(7) $g_7(x) = P - P_c(x) \leq 0.$

where

$$\tau(x) = \sqrt{(\tau')^2 + 2\tau'\tau'' \frac{x_2}{2R} + (\tau'')^2},$$
$$\tau' = \frac{P}{\sqrt{2}x_1 x_2},$$
$$\tau'' = \frac{MR}{J},$$

$$M = P\left(L + \frac{x_2}{2}\right),$$

$$R = \sqrt{\frac{x_2^2}{4} + \left(\frac{x_1 + x_3}{2}\right)^2},$$

$$J = 2\sqrt{2}x_1 x_2 \left[\frac{x_2^2}{12} + \left(\frac{x_1 + x_3}{2}\right)^2\right],$$

$$\sigma(x) = \frac{6PL}{x_4 x_3^2},$$

$$\delta(x) = \frac{4PL^3}{Ex_3^3 x_4},$$

$$P_c(x) = \frac{4.013E\sqrt{\frac{x_3^2 x_4^6}{36}}}{L^2}\left(1 - \frac{x_3}{2L}\sqrt{\frac{E}{4G}}\right).$$

where $P = 6000$ lb, $L = 14$ in, $E = 30 \times 10^6$ psi, $G = 12 \times 10^6$ psi, $\tau_{max} = 13{,}600$ psi, $\sigma_{max} = 30{,}000$ psi, $\delta_{max} = 0.25$ in, $0.1 \leq x_1 \leq 2, 0.1 \leq x_2 \leq 10, 0.1 \leq x_3 \leq 10, 0.1 \leq x_4 \leq 2$.

A pressure vessel design problem is for minimizing the total cost $f(x)$ of a pressure vessel considering the cost of material, forming and welding. A cylindrical vessel is capped at both ends by hemispherical heads as shown in Fig. 4.17 [2, 46]. There are four design variables: $T_s$ ($x_1$, thickness of the shell), $T_h$ ($x_2$, thickness of the head), $R$ ($x_3$, inner radius) and $L$ ($x_4$, length of the cylindrical section of the vessel, not including the head). Among the four variables, $T_s$ and $T_h$, which are integer multiples of 0.0625, are the available thicknesses of rolled steel plates, and $R$ and $L$ are continuous variables. This problem can be formulated as follows:

$$\min f(x) = 0.6224x_1 x_3 x_4 + 1.7781x_2 x_3^2 + 3.1661x_1^2 x_4 + 19.84x_1^2 x_3$$

subject to

(1) $g_1(x) = -x_1 + 0.0193x_3 \leq 0,$
(2) $g_2(x) = -x_2 + 0.00954x_3 \leq 0,$
(3) $g_3(x) = -\pi x_3^2 x_4 - \frac{4}{3}\pi x_3^3 + 1296000 \leq 0,$
(4) $g_4(x) = x_4 - 240 \leq 0,$

where $1 \leq x_1 \leq 99, 1 \leq x_2 \leq 99, 10 \leq x_3 \leq 200, 10 \leq x_4 \leq 200$.

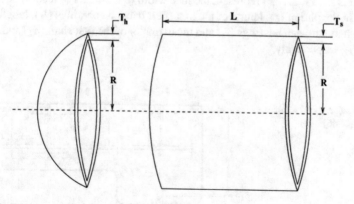

**Fig. 4.17** Center and end section of pressure vessel design problem [2, 46]

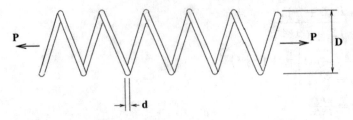

**Fig. 4.18** Tension/compression string design problem [2, 46]

A tension/compression string design problem is to minimize the weight ($f(x)$) of a tension/compression spring (as shown in Fig. 4.18 [2, 46]) subject to constraints on minimum deflection, shear stress, surge frequency, limits on outside diameter and on design variables. The design variables are the mean coil diameter $D(x_2)$, the wire diameter $d(x_1)$ and the number of active coils $P(x_3)$. The mathematical formulation of this problem can be described as follows:

$$\min f(x) = (x_3 + 2) x_2 x_1^2$$

subject to

(1) $g_1(x) = 1 - \frac{x_2^3 x_3}{71785 x_1^4} \leq 0,$

(2) $g_2(x) = \frac{4x_2^2 - x_1 x_2}{12566(x_2 x_1^3 - x_1^4)} + \frac{1}{5108 x_1^2} - 1 \leq 0,$

(3) $g_3(x) = 1 - \frac{140.45 x_1}{x_2^2 x_3} \leq 0,$

(4) $g_4(x) = \frac{x_1 + x_2}{1.5} - 1 \leq 0,$

where $0.05 \leq x_1 \leq 2, 0.25 \leq x_2 \leq 1.3, 2 \leq x_3 \leq 15$.

A speed reducer design problem is shown in Fig. 4.19 [2, 46] for minimizing the weights of the speed reducer subject to constraints on bending stress of the gear teeth, surface stress, transverse deflections of the shafts and stresses in the shafts. The parameters $x_1, x_2, \ldots, x_7$ represent the face width ($b$), module of teeth ($m$), number of teeth in the pinion ($z$), length of the first shaft between bearings ($l_1$), length of the second shaft between bearings ($l_2$) and the diameter of the first shaft ($d_1$) and second shaft ($d_2$), respectively.

**Fig. 4.19** The speed reducer problem [2, 46]

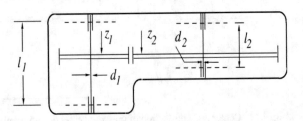

$$\min f(x) = 0.7854x_1x_2^2 \left(3.3333x_3^2 + 14.9334x_3 - 43.0934\right)$$
$$- 1.508x_1\left(x_6^2 + x_7^2\right) + 7.4777\left(x_6^3 + x_7^3\right)$$

subject to

(1)  $g_1(x) = \frac{27}{x_1x_2^2x_3} - 1 \le 0,$

(2)  $g_2(x) = \frac{397.5}{x_1x_2^2x_3^2} - 1 \le 0,$

(3)  $g_3(x) = \frac{1.93x_4^3}{x_2x_3x_6^4} - 1 \le 0,$

(4)  $g_4(x) = \frac{1.93x_5^3}{x_2x_3x_7^4} - 1 \le 0,$

(5)  $g_5(x) = \frac{\left(\left(\frac{745x_4}{x_2x_3}\right)^2 + 16.9 \times 10^6\right)^{1/2}}{110.0x_6^3} - 1 \le 0,$

(6)  $g_6(x) = \frac{\left(\left(\frac{745x_4}{x_2x_3}\right)^2 + 157.5 \times 10^6\right)^{1/2}}{85.0x_7^3} - 1 \le 0,$

(7)  $g_7(x) = \frac{x_2x_3}{40} - 1 \le 0,$

(8)  $g_8(x) = \frac{5x_2}{x_1} - 1 \le 0,$

(9)  $g_9(x) = \frac{x_1}{12x_2} - 1 \le 0,$

(10)  $g_{10}(x) = \frac{1.5x_6 + 1.9}{x_4} - 1 \le 0,$

(11)  $g_{11}(x) = \frac{1.1x_7 + 1.9}{x_5} - 1 \le 0,$

where $2.6 \le x_1 \le 3.6, 0.7 \le x_2 \le 0.8, 17 \le x_3 \le 28, 7.3 \le x_4 \le 8.3, 7.8 \le x_5 \le 8.3, 2.9 \le x_6 \le 3.9, 5.0 \le x_7 \le 5.5$.

DETPS is designed by using a network membrane structure, evolution and communication rules like in a tissue P system and five widely used differential evolution variants put inside five cells of the tissue P system. In the experiments, a population with 20 individuals that are equally divided into 5 groups, i.e., $NP = 20$, $NP_1 = NP_2 = NP_3 = NP_4 = NP_5 = 4$. A prescribed number of 10,000 function evaluations is regarded as the stoping criterion. Table 4.7 lists the statistical results over 30 independent runs. We consider seven optimization algorithms as benchmarks. They are TLBO [36], $(\mu + \lambda)$-evolutionary strategy $((\mu + \lambda)$-ES) [29], unified particle swarm optimization (UPSO) [35], co-evolutionary particle swarm optimization (CPSO) [15], CoDE [17], hybridizing particle swarm optimization with differential evolution (PSO-DE) [26] and artificial bee colony algorithm (ABCA) [2], respectively. In Table 4.7, the statistical results of the seven optimization algorithms are also shown. Table 4.7 indicates that DETPS is better than the other seven algorithms for the welded beam and pressure vessel problems in terms of the best, mean, worst solutions, standard deviations and the number of function evaluations. DETPS is competitive to TLBO with respect to the best and mean and the number of function evaluations on the tension/compression spring problem; DETPS uses a much smaller number of function evaluations to obtain competitive results compared with PSO-DE; the rest of five optimization algorithms are inferior to DETPS in terms of the

**Table 4.7**  Statistical results of seven algorithms for test problems 6–9. The results of $(\mu + \lambda)$-ES, UPSO, CPSO, CoDE, PSO-DE, ABCA and TLBO are referred from [2, 15, 17, 26, 29, 35, 36], respectively. BT, MN, WT and FE represent best solution, mean best solution, worst solution and the mean number of function evaluations over independent 30 runs, respectively. Symbol '-' means that no result can be referred. Pm, WB, PV, TS and SR represent problems, welded beam, pressure vessel, tension/compression string and speed reducer design problems, respectively

| Pm | | DETPS | $(\mu+\lambda)$ $-ES$ | UPSO | CPSO | CoDE | PSO-DE | ABCA | TLBO |
|---|---|---|---|---|---|---|---|---|---|
| WB | BT | **1.724852** | **1.724852** | 1.92199 | 1.728024 | 1.733462 | 1.724853 | **1.724852** | **1.724852** |
| | MN | **1.724852** | 1.777692 | 2.83721 | 1.748831 | 1.768158 | 1.724858 | 1.741913 | 1.728447 |
| | WT | **1.724853** | 2.074562 | 4.88360 | 1.782143 | 1.824105 | 1.724881 | - | - |
| | SD | **2.1e−7** | 8.8e−2 | 6.8e−1 | 1.3e−2 | 2.2e−2 | 4.1e−6 | 3.1e−2 | - |
| | FE | **10,000** | 30,000 | 100,000 | 200,000 | 240,000 | 33,000 | 30,000 | **10,000** |
| PV | BT | **5885.3336** | 6059.7016 | 6544.27 | 6061.0777 | 6059.7340 | 6059.7143 | 6059.7147 | 6059.7143 |
| | MN | **5887.3161** | 6379.9380 | 9032.55 | 6147.1332 | 6085.2303 | 6059.7143 | 6245.3081 | 6059.7143 |
| | WT | **5942.3234** | 6820.3975 | 11638.20 | 6363.8041 | 6371.0455 | 6059.7143 | - | - |
| | SD | **1.0e+1** | 2.1e+2 | 9.9e+2 | 8.6e+1 | 4.3e+1 | 1.0e−10 | 2.1e+2 | - |
| | FE | **10,000** | 30,000 | 100,000 | 200,000 | 240,000 | 42,100 | 30,000 | **10,000** |
| TS | BT | **0.012665** | 0.012689 | 0.013120 | 0.012675 | 0.012670 | **0.012665** | **0.012665** | **0.012665** |
| | MN | 0.012680 | 0.013165 | 0.022948 | 0.012730 | 0.012703 | **0.012665** | 0.012709 | 0.012666 |
| | WT | 0.012769 | 0.014078 | 0.050365 | 0.012924 | 0.012790 | **0.012665** | - | - |
| | SD | 2.7e−5 | 3.9e−4 | 7.2e−3 | 5.2e−5 | 2.7e−5 | **1.2e−8** | 1.3e−2 | - |
| | FE | **10,000** | 30,000 | 100,000 | 200,000 | 240,000 | 24,950 | 30,000 | **10,000** |
| SR | BT | **2996.348** | 2996.348 | - | - | - | **2996.348** | 2997.058 | **2996.348** |
| | MN | **2996.348** | 2996.348 | - | - | - | **2996.348** | 2997.058 | **2996.348** |
| | WT | **2996.348** | 2996.348 | - | - | - | **2996.348** | - | - |
| | SD | 5.2e−5 | **0.0** | - | - | - | 6.4e−6 | **0.0** | - |
| | FE | **10,000** | 30,000 | - | - | - | 54,350 | 30,000 | **10,000** |

quality of solutions and the number of function evaluations. DETPS is competitive to $(\mu + \lambda)$-ES, PSO-DE, ABCA and TLBO in the results of the speed reducer problem.

## 4.3.2  Distribution Network Reconfiguration

This subsection discusses the application of a population-membrane-system-inspired evolutionary algorithm (PMSIEA, for short) described in Sect. 3.7.2 to solve the distribution network reconfiguration problem in power systems.

The reconfiguration of distribution network in a power system is an important process that uses remote-controlled switches to improve operating conditions and planning studies, service restoration and distribution automation of a power system [1, 30]. There exist lots of candidate-switching combinations in a distribution system, so the distribution system reconfiguration is a complex combinatorial problem with a large number of integer and continuous variables and various constraints such as power flow equations, upper and lower bounds of nodal voltages, upper and lower bounds of line currents, feasible conditions in terms of network topology. Thus, how to obtain the optimal distribution system reconfiguration is an ongoing issue in a power system. The problem can be formulated as a minimization cost function $f$ to minimize the power loss of the system by changing the topology of distribution systems through altering the open/closed status of sectionalizing switches [1, 37], i.e.,

$$\min f = \sum_{i=1}^{L} r_i \frac{P_i^2 + Q_i^2}{V_i^2} \tag{4.29}$$

subject to

(1)  $g(x) = 0$,
(2)  $V_{min} < V_n < V_{max}$,
(3)  $I_i^{min} < I_i < I_i^{max}$,
(4)  $\det(A) = 1$ or $-1$ (for radial systems),
(5)  $\det(A) = 0$ (for not radial systems),

where

- $f$ is the objective function (kW);
- $L$ is the number of branches;
- $P_i$ is the active power at the sending end of branch $i$;
- $Q_i$ is the reactive power at sending end of branch $i$;
- $V_n$ is the voltage at node $n$;
- $I_i$ is the line current at branch $i$;
- $g(x)$ is the power flow equations;
- $V_{min}$ and $V_{max}$ are the lower and upper voltage limits, respectively;
- $I_i^{min}$ and $I_i^{max}$ are the lower and upper current limits, respectively;
- $A$ is the bus incidence matrix;
- $r_i$ is the resistance of branch $i$.

In what follows, the distribution system reconfiguration problems will be solved by using PMSIEA described in Sect. 3.7.2 of Chap. 3. We use IEEE 33-bus and PG&E 69-bus distribution systems as examples to show the application. Figures 4.20 and 4.21 show the IEEE 33-bus and PG&E 69-bus systems. In Fig. 4.20, the IEEE 33-bus system has 33 buses, 37 branches and 5 tie-lines. Figure 4.21 shows that the PG&E 69-bus system consists of 69 buses, 68 sectionalizing switches and 5 tie switches. In the IEEE 33-bus system, the normally open switches are 33, 34, 35, 36 and 37, and the initial real power losses (before reconfiguration) are 202.68 kW. The normally open switches are 69, 70, 71, 72 and 73, and the initial real power losses (before reconfiguration) are 226.4419 kW in the PG&E 69-bus system. The objective function or fitness function is shown in (4.29). The population size and the value of $p_c$ are set to 10 and 0.9, respectively. The maximal number 100 of evolutionary generations is considered as the determination criterion, which is the

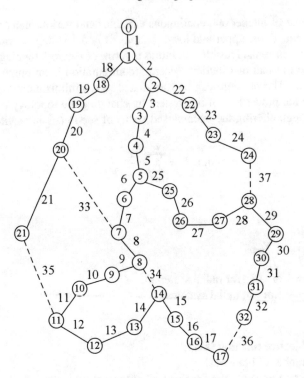

**Fig. 4.20**   IEEE 33-bus system

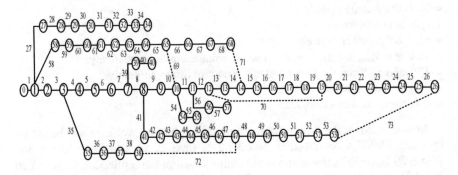

**Fig. 4.21**   PG&E 69-bus system

same as reported in the literature. Table 4.8 shows the experimental results of the IEEE 33-bus system. The results obtained by five optimization approaches reported in the recent literature, a heuristic approach (HeAp) [30], SA+TS [18], MTS [28], PSO [1] and ACO [37], are also listed in Table 4.8 to make a comparison. Table 4.9 lists the experimental results of the PG&E 69-bus system. To bring into a comparison, the results obtained by ACS [12], HPSO [20], VSHDE [33] and ACO [37] are also

**Table 4.8** Results of IEEE 33-bus test system

| Methods | OpCo | RPL (kW) | MNV (pu) |
|---|---|---|---|
| BeRe | 33, 34, 35, 36, 37 | 202.68 | 0.9378 |
| HeAp [30] | 7, 9, 14, 32, 37 | 139.55 | 0.9378 |
| SA + TS [18] | 7, 9, 14, 32, 37 | 139.55 | 0.9378 |
| MTS [28] | 7, 9, 14, 32, 37 | 139.55 | 0.9378 |
| PSO [1] | 7, 9, 14, 32, 37 | 139.55 | 0.9378 |
| ACO [37] | 7, 9, 14, 32, 37 | 139.55 | 0.9378 |
| PMSIEA | 7, 9, 14, 32, 37 | 139.55 | 0.9378 |

**Table 4.9** Results of the PG&E 69-bus test system

| Methods | OpCo | RPL (kW) | MNV (pu) |
|---|---|---|---|
| BeRe | 69, 70, 71, 72, 73 | 226.4419 | 0.9089 |
| ACS [12] | 61, 69, 14, 70, 55 | 99.519 | 0.943 |
| HPSO [20] | 69, 12, 14, 47, 50 | 99.6704 | 0.9428 |
| VSHDE [33] | 11, 24, 28, 43, 56 | 99.6252 | 0.9427 |
| IIGA [40] | 69, 14, 70, 47, 50 | 99.618 | 0.9427 |
| PMSIEA | 47, 12, 50, 14, 69 | 99.4944 | 0.9441 |

shown in Table 4.9. BeRe, OpCo, RPL and MNV represent before reconfiguration, optimal configuration, real power loss and minimum node voltage, respectively.

It can be seen from Table 4.8 that PMSIEA is competitive to the nine optimization approaches, a heuristic approach, SA+TS, MTS, PSO and ACO, in terms of the optimal solutions. Table 4.9 shows that PMSIEA achieves lower real power losses and higher minimum node voltage than ACS, HPSO, VSHDE and IIGA.

## 4.4 Engineering Optimization with Neural-Like P Systems

This section discusses the use of an optimization spiking neural P system (OSNPS) to solve the power system fault diagnosis problem, specifically fault section estimation problem.

The power system fault diagnosis consists of five processes: fault detection, fault section estimation, fault type identification, failure isolation and recovery [42, 44]. As a very important process in the fault diagnosis of a power system [38, 42], fault section estimation uses the status information of protective relays and circuit breakers (CBs) obtained from supervisor control and data acquisition (SCADA) systems to identify faulty sections [16]. Until now, some methods have been reported in the literature to diagnose faults of a power system. These methods include fuzzy logic (FL) [5], fuzzy Petri nets (FPN) [38], expert systems (ES) [19], multi agent systems

(MAS) [51], artificial neural networks (ANN) [4], and optimization methods (OM) [16, 22–44]. Each method has its own pros and cons [42]. Therefore, the continuous efforts for better solving FSE is a hot topic in the research field of electrical power systems.

The problem of fault section estimation in a power system can be formulated into a 0–1 integer programming problem that can be effectively solved by an optimization method such as OSNPS [55]. This section discusses how to use OSNPS to solve the power system fault section estimation problem. When we input the status information of protective relays and circuit breakers read from a supervisory control and data acquisition system, the results on fault sections will be output by OSNPS. We will use different cases with single fault, multiple faults or multiple faults with incomplete and uncertain information to show the application of OSNPS for fault sections estimation of power systems.

The problem description of the power system fault section estimation is as follows. Fault section estimation in a power system is to derive a fault hypothesis explaining warning signals (status information) in the maximum degree. To be specific, fault section estimation can be described as a 0–1 programming problem, where an error function obtained according to the causality between a fault and the statuses of protection devices (protective relays and CBs), is regarded as an objective function, as shown in (4.30) [22]. Thus, an optimization approach can be used to find the fault hypothesis, i.e. the minimal value of a status function $E(S)$ in (4.30) of all the sections in a power system.

$$E(S) = \sum_{j=1}^{n_c} \left| c_j - c_j^*(S, R) \right| + \sum_{k=1}^{n_r} \left| r_k - r_k^*(S) \right|, \tag{4.30}$$

where $n_c$ is the number of circuit breakers (CBs) in a protection system; $n_r$ is the number of protective relays; $S = (S_1, \ldots, S_n)$ is an $n$ vector representing the status of sections in a power system and $n$ represents the number of sections: if a section is faulty, then $S_i = 1$, otherwise, $S_i = 0$ $(1 \leq i \leq n)$; $c = (c_1, \ldots, c_{n_c})$ is an $n_c$-vector representing the real status of CBs in a protection system. If a CB trips, then $c_j = 1$, otherwise, $c_j = 0$ $(1 \leq j \leq n_c)$; $c^*(S, R) = (c_1^*, \ldots, c_{n_c}^*)$ is an $n_c$-vector representing the expected status of CBs in a protection system. If a CB should trip, then $c_j^* = 1$, otherwise, $c_j^* = 0$ $(1 \leq j \leq n_c)$; $r = (r_1, \ldots, r_{n_r})$ is an $n_r$-vector representing the real status of protective relays in a protection system and $n_r$ represents the number of protective relays: if a protective relays operates, then $r_k = 1$, otherwise, $r_k = 0$ $(1 \leq k \leq n_r)$; $r^*(S) = (r_1^*, \ldots, r_{n_r}^*)$ is an $n_r$-vector representing the expected status of protective relays in a protection system and $n_r$ represents the number of protective relays: if a protective relay should operate, then $r_k^* = 1$, otherwise, $r_k^* = 0$ $(1 \leq k \leq n_r)$.

When we use OSNPS to minimize $E(S)$ in (4.30) to solve the fault section estimation problem in a power system, the real status of protective relays and CBs are normally read from a power SCADA system and the expected status of protective relays and CBs can be obtained according to their operation principles and the

**Fig. 4.22** The ESNPS
structure

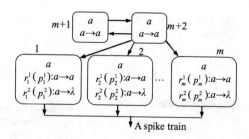

A spike train

protection structure of a power system. After all the expected and real statuses of
protection are obtained, OSNPS is used to find the minimal value of $E(S)$ in (4.30).
The aim of fault section estimation is to obtain vector elements of $S$ corresponding
to the minimum value of (4.30), where $S_i = 1$ is the faulty section (diagnosis target),
$i = 1, \ldots, n$.

An extended spiking neural P system (ESNPS) of degree $m \geq 1$, as shown in
Fig. 4.22, is described in [55] as the following tuple

$$\Pi = (O, \sigma_1, \ldots, \sigma_{m+2}, syn, I_0), \tag{4.31}$$

where:

- $O = \{a\}$ is the singleton alphabet ($a$ is called *spike*);
- $\sigma_1, \ldots, \sigma_m$ are neurons of the form $\sigma_i = (1, R_i, P_i)$, $1 \leq i \leq m$, such that $R_i = \{r_i^1, r_i^2\}$ is a set of rules with $r_i^1 = \{a \to a\}$ and $r_i^2 = \{a \to \lambda\}$, $P_i = \{p_i^1, p_i^2\}$ is a finite set of probabilities, where $p_i^1$ and $p_i^2$ are the selection probabilities of rules $r_i^1$ and $r_i^2$, respectively, satisfying $p_i^1 + p_i^2 = 1$; and $\sigma_{m+1} = \sigma_{m+2} = (1, \{a \to a\})$.
- $syn = \{(i, j) \mid (i = m + 2 \land 1 \leq j \leq m + 1) \lor (i = m + 1 \land j = m + 2)\}$
- $I_0 = \{1, 2, \ldots, m\}$ is a finite set of *output* neurons, i.e., the output is a spike train formed by concatenating the outputs of $\sigma_1, \sigma_2, \ldots, \sigma_m$.

This system contains the subsystem consisting of two identical neurons: $\sigma_{m+1}$
and $\sigma_{m+2}$, each of which fires at each moment of time and sends a spike to each of
neurons $\sigma_1, \ldots, \sigma_m$, and reloads each other continuously. At each time unit, each of
neurons $\sigma_1, \ldots, \sigma_m$ performs the firing rule $r_i^1$ by probability $p_i^1$ and the forgetting
rule $r_i^2$ by probability $p_i^2$, $i = 1, 2, \ldots, m$. If the $i$th neuron spikes, its output is 1,
i.e., we obtain 1 by probability $p_i^1$, otherwise, we obtain its output 0, i.e., we obtain 0
by probability $p_i^2$, $i = 1, 2, \ldots, m$. Thus, this system outputs a spike train consisting
of 0 and 1 at each moment of time. If the probabilities $p_1^1, \ldots, p_m^1$ can be adjusted,
the outputted spike train can be controlled. In what follows, a method for adjusting
the probabilities $p_i^1, \ldots, p_m^1$ is described by introducing a family of ESNPS.

A certain number of ESNPS can be arranged into a family of ESNPS (called
OSNPS) by introducing a guider to adjust the selection probabilities of rules inside
each neuron of each ESNPS. Figure 4.23 shows the OSNPS structure. In this figure,
OSNPS is composed of $H$ ESNPS: $ESNPS_1, ESNPS_2, \ldots, ESNPS_H$. Each ESNPS
is identical with the one in Fig. 4.22 and the pseudocode algorithm of the guider

**Fig. 4.23** The OSNPS
structure

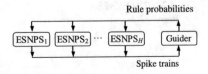

**Fig. 4.24** Guider algorithm

**Input:** Spike train $T_s$, $p_j^a$, $\Delta$, $H$ and $m$
1: Rearrange $T_s$ as matrix $P_R$
2:   $i = 1$
3: **while** $(i \leq H)$ **do**
4:     $j=1$
5:     **while** $(j \leq m)$ **do**
6:       **if** $(rand < p_j^a)$ **then**
7:           $k_1, k_2 = ceil(rand * H), k_1 \neq k_2 \neq i$
8:           **if** $(f(C_{k_1}) > f(C_{k_2}))$ **then**
9:               $b_j = b_{k_1}$
10:          **else**
11:              $b_j = b_{k_2}$
12:          **end if**
13:          **if** $(b_j > 0.5)$ **then**
14:              $p_{ij}^1 = p_{ij}^1 + \Delta$
15:          **else**
16:              $p_{ij}^1 = p_{ij}^1 - \Delta$
17:          **end if**
18:      **else**
19:          **if** $(b_j^{max} > 0.5)$ **then**
20:              $p_{ij}^1 = p_{ij}^1 + \Delta$
21:          **else**
22:              $p_{ij}^1 = p_{ij}^1 - \Delta$
23:          **end if**
24:      **end if**
25:      **if** $(p_{ij}^1 > 1)$ **then**
26:          $p_{ij}^1 = p_{ij}^1 - \Delta$
27:      **else**
28:          **if** $(p_{ij}^1 < 0)$ **then**
29:              $p_{ij}^1 = p_{ij}^1 + \Delta$
30:          **end if**
31:      **end if**
32:      $j = j + 1$
33:    **end while**
34:    $i = i + 1$
35: **end while**
**Output:**  Rule probability matrix $P_R$

algorithm is described in Fig. 4.24. For details about the guider and more information about ESNPS and OSNPS, please see [55].

The sketch map of the use of OSNPS to solve the fault section estimation problem is shown in Fig. 4.25. Details on the steps are described as follows:

**Fig. 4.25**  The sketch map of fault section estimation using OSNPS

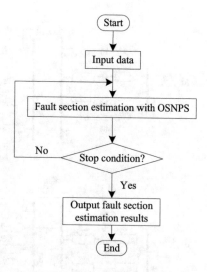

(1)  Step 1. *Input data*
SCADA data, parameter setting of OSNPS and the initial value of the fitness function need to be prepared. Thus, the input data consist of three parts:

- Read SCADA data including the status information including the status of protective relays and CBs, the topological connection of a given power system and its protection system structure information;
- Set OSNPS parameters: the number of ESNPS ($H$), the dimension of each ESNPS ($m$), the learning probabilities, the learning rate, the rule probability matrix and the maximal number of iterations;
- Initialize the fitness function of the fault section estimation problem according to (4.30) using the above data.

(2)  Step 2. *Fault section estimation with OSNPS*
OSNPS is used to find the minimum value of (4.30). Each ESNPS produces a spike train for storing the binary results. $H$ ESNPS are arranged into an OSNPS by a guider to adjust the selection probabilities of rules inside each neuron of each ESNPS. The guider algorithm, as shown in Fig. 4.24 and described in [55] in detail, is applied to help OSNPS obtaining the minimum value of (4.30).

(3)  Step 3. *Termination*
The algorithm stops when the maximal number of iterations is obtained.

(4)  Step 4. *Output fault section estimation results*
The spike train corresponding to the minimum value of (4.30) is output in an $n$ vector $S = (S_1, \ldots, S_n)$ and $S_i = 1$ is the $i$th faulty section, $i = 1, \ldots, n$, where $n$ represents the number of sections in the given power system.

A typical 4-substation system is shown in Fig. 4.26, where there are 28 system sections, 40 CBs and 84 protective relays [42, 44]. The protective relays are composed of main protective relays (MPRs), first backup protective relays (FBPRs) and second

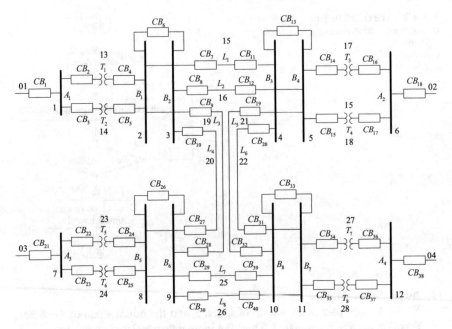

**Fig. 4.26** A local sketch map of the protection system of an EPS

**Table 4.10** Status information about protective relays and CBs

| Cases | Status information | |
|---|---|---|
| | Operated relays | Tripped CBs |
| 1 | $B_{1m}, L_{2Rs}, L_{4Rs}$ | $CB_4, CB_5, CB_7$ |
| | | $CB_9, CB_{12}, CB_{27}$ |
| 2 | $B_{1m}, L_{1Sm}, L_{1Rp}$ | $CB_4, CB_5, CB_6$ |
| | $B_{2m}, L_{2Sp}, L_{2Rm}$ | $CB_7, CB_8, CB_9$ |
| | | $CB_{10}, CB_{11}, CB_{12}$ |
| 3 | $T_{7m}, T_{8P}, B_{7m}$ | $CB_{19}, CB_{20}, CB_{29}, CB_{30}$ |
| | $B_{8m}, L_{5Sm}, L_{5Rp}$ | $CB_{32}, CB_{33}, CB_{34}, CB_{35}$ |
| | $L_{6Ss}, L_{7Sp}, L_{7Rm}, L_{8Ss}$ | $CB_{36}, CB_{37}, CB_{39}$ |

backup protective relays (SBPRs) in a power system. The detailed operational rules of protective relays for main sections in a power system can be seen in [42, 44].

Three cases of the local power system in Fig. 4.26 are used to show the fault section estimation application of OSNPS. Table 4.10 lists the status information about protective relays and CBs of these cases. *Case 1* has a single fault. *Case 2* has multiple faults. *Case 3* has multiple faults with incompleteness and uncertainty. The fault section estimation results of OSNPS for the three cases are given in Table 4.11, where the results of other three methods are also listed. The symbol "−" means that this case was not considered in the corresponding literature.

**Table 4.11** Comparisons between OSNPS and three fault diagnosis methods

| Cases | Diagnosis results | | | | |
|---|---|---|---|---|---|
| | OSNPS | FL [5] | GA [44] | FDSNP [42] | GATS [22] |
| 1 | $B_1$ | $B_1$ | $B_1$ | $B_1$ | - |
| 2 | $B_1, B_2$ | $B_1, B_2$ | $B_1, B_2$ | $B_1, B_2$ | - |
| | $L_1, L_2$ | $L_1, L_2$ | $L_1, L_2$ | $L_1, L_2$ | |
| 3 | $L_5, L_7$ | $L_5, L_7$ | (1) $L_5, L_7, B_7, B_8$ | $L_5, L_7$ | $L_5, L_7$ |
| | $B_7, B_8$ | $B_8, T_7$ | $T_7, T_8$ | $B_7, B_8$ | $B_7, B_8$ |
| | $T_7, T_8$ | $T_8$ | (2) $L_5, L_7, T_7, B_8$ | $T_7, T_8$ | $T_7, T_8$ |

Table 4.11 shows that OSNPS obtains the same results in *Cases 1* and *2* as those by fuzzy logic [FL], genetic algorithm (GA) and FDSNP in [5, 42, 44], respectively. This indicates the success of OSNPS for solving the fault section estimation problem of a power system with a single or multiple faults. The results of OSNPS in *Case 3* are different from those in [5, 44]. Referring to [22, 42], we confirm that OSNPS obtains the correct results. Thus, OSNPS is able to correctly estimate the fault sections for a power system with a single fault, multiple faults or multiple faults with incomplete and uncertain alarm information.

## 4.5 Conclusions

This chapter discussed the engineering applications of several membrane algorithms designed with cell-like, tissue-like and neural-like P systems. These engineering applications consist of radar emitter signal decomposition with MQEPS, image processing with MAQIS, PID controller design with PSOPS, mobile robot path planning with mMPSO, manufacturing parameter optimization problems with DETPS, distribution network reconfiguration with PMSIEA and power system fault diagnosis problem with OSNPS.

The engineering optimization of membrane algorithms discussed in this chapter show the good potential of the approach. More and wider practical application problems should be investigated in order to prove the impact of these membrane algorithms.

## References

1. Abdelaziz, A., F. Mohammed, S. Mekhamer, and M. Badr. 2009. Distribution systems reconfiguration using a modified particle swarm optimization algorithm. *Electric Power Systems Research* 79 (11): 1521–1530.
2. Akay, B., and D. Karaboga. 2012. Artificial bee colony algorithm for large-scale problems and engineering design optimization. *Journal of Intelligent Manufacturing* 23 (4): 1001–1014.

3. Bergeau, F., and S. Mallat. 1994. Matching pursuit of Images. In *Proceedings of IEEE International Conference on Signal Processing*, 330–333.

4. Cardoso, G., J.G. Rolim, and H.H. Zurn. 2008. Identifying the primary fault section after contingencies in bulk power systems. *IEEE Transactions on Power Delivery* 23 (3): 1335–1342.

5. Chang, C.S., J.M. Chen, D. Srinivasan, F.S. Wen, and A.C. Liew. 1997. Fuzzy logic approach in power system fault section identification. *IEEE Proceedings–Part C, Generation, Transmission and Distribution* 144 (5): 406–414.

6. Chen, J.W. 2008. *Optimal design of control system based on membrane computing optimization method*, Master dissertation, Zhejiang University, Hangzhou.

7. Cheng, J., G. Zhang, and X. Zeng. 2011. A novel membrane algorithm based on differential evolution for numerical optimization. *International Journal of Unconventional Computing* 7 (3): 159–183.

8. Davis, G., S. Mallat, and M. Avellaneda. 1997. Adaptive greedy approximation. *Journal of Constructive Approximation* 13 (1): 57–98.

9. Gao, H., and J. Cao. 2012. Membrane-inspired quantum shuffled frog leaping algorithm for spectrum allocation. *Journal of Systems Engineering and Electronics* 23 (5): 679–688.

10. Gao, H., J. Cao, and Y. Zhao. 2012. Membrane quantum particle swarm optimisation for cognitive radio spectrum allocation. *International Journal of Computer Applications in Technology* 43 (4): 359–365.

11. Gao, H., G.H. Xu, and Z.R. Wang. 2006. A novel quantum evolutionary algorithm and its application. In *Proceedings of the Sixth World Congress on Intelligent Control and Automation*, 3638–3642.

12. Ghorbani, M.A., S.H. Hosseinian, and B. Vahidi. 2008. Application of ant colony system algorithm to distribution networks reconfiguration for loss reduction. In *Proceedings of International Conference on Optimization of Electrical and Electronic Equipment*, 269–273.

13. Han, K.H., and J.H. Kim. 2000. Genetic quantum algorithm and its application to combinatorial optimization problem. In *Proceedings of IEEE Congress on Evolutionary Computation*, 1354–1360.

14. Han, K.H., and J.H. Kim. 2002. Quantum-inspired evolutionary algorithm for a class of combinatorial optimization. *IEEE Transactions on Evolutionary Computation* 6 (6): 580–593.

15. He, Q., and L. Wang. 2007. An effective co-evolutionary particle swarm optimization for constrained engineering design problems. *Engineering Applications of Artificial Intelligence* 20 (1): 89–99.

16. Huang, S.J., and X.Z. Liu. 2013. Application of artificial bee colony-based optimization for fault section estimation in power systems. *International Journal of Electrical Power & Energy Systems* 44 (1): 210–218.

17. Huang, F.Z., L. Wang, and Q. He. 2007. An effective co-evolutionary differential evolution for constrained optimization. *Applied Mathematics and Computation* 186 (1): 340–356.

18. Jeon, Y.J., and J.C. Kim. 2004. Application of simulated annealing and tabu search for loss minimization in distribution systems. *International Journal of Electrical Power & Energy Systems* 26 (1): 9–18.

19. Lee, H.J., B.S. Ahn, and Y.M. Park. 2000. A fault diagnosis expert system for distribution substations. *IEEE Transactions on Power Delivery* 15 (1): 92–97.

20. Li, Z.K., X.Y. Chen, K. Yu, Y. Sun, and H.M. Liu. 2008. A hybrid particle swarm optimization approach for distribution network reconfiguration problem. In *Proceedings of Power and Energy Society General Meeting*, 1–7

21. Liao, T.W. 2010. Two hybrid differential evolution algorithms for engineering design optimization. *Applied Soft Computing* 10 (4): 1188–1199.

22. Lin, X.N., S.H. Ke, Z.T. Li, H.L. Weng, and X.H. Han. 2010. A fault diagnosis method of power systems based on improved objective function and genetic algorithm-tabu search. *IEEE Transactions on Power Delivery* 25 (3): 1268–1274.

23. Liu, J.K. 2004. *Advanced PID control and Matlab simulation*, 2nd ed. Beijing: PHEI Press.

24. Liu, C., G. Zhang, X. Zhang, and H. Liu. 2009. A memetic algorithm based on P systems for IIR digital filter design. In *Proceedings of the Eighth IEEE International Conference on Dependable, Autonomic and Secure Computing*, 330–334.
25. Liu, C., G. Zhang, Y. Zhu, C. Fang, and H. Liu. 2009. A quantum-inspired evolutionary algorithm based on P systems for radar emitter signals. In *Proceedings of the 8th IEEE International Conference on Dependable, Autonomic and Secure Computing*, 24–28.
26. Liu, H., Z. Cai, and Y. Wang. 2010. Hybridizing particle swarm optimization with differential evolution for constrained numerical and engineering optimization. *Applied Soft Computing* 10 (2): 629–640.
27. Liu, C., G. Zhang, H. Liu, M. Gheorghe, and F. Ipate. 2010. An improved membrane algorithm for solving time-frequency atom decomposition. In *Membrane Computing. WMC 2009*, vol. 5957, ed. G. Păun, M.J. Pérez-Jiménez, A. Riscos-Núñez, G. Rozenberg, and A. Salomaa, 371–384. Lecture Notes in Computer Science. Berlin: Springer.
28. Mekhamer, S., A. Abdelaziz, F. Mohammed, and M. Badr. 2008. A new intelligent optimization technique for distribution systems reconfiguration. In *Proceedings of International Middle-East Power System Conference*, 397–401.
29. Mezura-Montes, E., and C.A.C. Coello. 2005. Useful infeasible solutions in engineering optimization with evolutionary algorithms. In *MICAI 2005: Advances in Artificial Intelligence*, vol. 3789, ed. A. Gelbukh, A. de Albornoz, and H. Terashima-Marín, 652–662. Lecture Notes in Artificial Intelligence. Berlin: Springer.
30. Martín, J.A., and A.J. Gil. 2008. A new heuristic approach for distribution systems loss reduction. *Electric Power Systems Research* 78 (11): 1953–1958.
31. Masehian, E., and D. Sedighizadeh. 2007. Classic and heuristic approaches in robot motion planning-a chronological review. *International Journal of Mechanical, Aerospace, Industrial, Mechatronic and Manufacturing Engineering* 1 (5): 228–233.
32. Mallat, S.G., and Z.F. Zhang. 1993. Matching pursuits with time-frequency dictionaries. *IEEE Transactions on Signal Processing* 41 (12): 3397–3415.
33. Nournejad, F., R. Kazemzade, and A.S. Yazdankhah. 2011. A multiobjective evolutionary algorithm for distribution system reconfiguration. In *Proceedings of the 16th Conference on Electrical Power Distribution Networks*, 1–7.
34. Pierre, V., and F. Pascal. 2001. Efficient image representation by anisotropic refinement in matching pursuit. In *Proceedings of IEEE International Conference on Acoustics, Speech, and Signal Processing*, 1757–1760.
35. Parsopoulos, K.E., and M.N. Vrahatis. 2005. Unified particle swarm optimization for solving constrained engineering optimization problems. In *Advances in Natural Computation (ICNC 2005)*, vol. 3612, ed. L. Wang, K. Chen, and Y.S. Ong, 582–591. Lecture Notes in Computer Science. Berlin: Springer.
36. Rao, R.V., V.J. Savsani, and D.P. Vakharia. 2011. Teaching-learning-based optimization: A novel method for constrained mechanical design optimization problems. *Computer-Aided Design* 43 (3): 303–315.
37. Swarnkar, A., N. Gupta, and K. Niazi. 2011. Efficient reconfiguration of distribution systems using ant colony optimization adapted by graph theory. In *Proceedings of Power and Energy Society General Meeting*, 1–8.
38. Sun, J., S.Y. Qin, and Y.H. Song. 2004. Fault diagnosis of electric power systems based on fuzzy Petri nets. *IEEE Transactions on Power Systems* 19 (4): 2053–2059.
39. Tuncer, A., and M. Yildirim. 2012. Dynamic path planning of mobile robots with improved genetic algorithm. *Computers & Electrical Engineering* 38 (6): 1564–1572.
40. Wang, C.X., A.J. Zhao, H. Dong, and Z.J. Li. 2009. An improved immune genetic algorithm for distribution network reconfiguration. In *Proceedings of International Conference on Information Management, Innovation Management and Industrial Engineering*, 218–223.
41. Wang, T., J. Wang, H. Peng, and M. Tu. 2012. Optimization of PID controller parameters based on PSOPS algorithm. *ICIC Express Letters* 6 (1): 273–280.
42. Wang, T., G.X. Zhang, J.B. Zhao, Z.Y. He, J. Wang, and M.J. Pérez-Jiménez. 2015. Fault diagnosis of electric power systems based on fuzzy reasoning spiking neural P systems. *IEEE Transactions on Power Systems* 30 (3): 1182–1194.

43. Wang, X., G. Zhang, J. Zhao, H. Rong, F. Ipate, and R. Lefticaru. 2015. A modified membrane-inspired algorithm based on particle swarm optimization for mobile robot path planning. *International Journal of Computers, Communications and Control* 10 (5): 732–745.

44. Wen, F.S., and Z.X. Han. 1995. Fault section estimation in power systems using a genetic algorithm. *Electric Power Systems Research* 34 (3): 165–172.

45. Yang, S., and N. Wang. 2012. A novel P systems based optimization algorithm for parameter estimation of proton exchange membrane fuel cell model. *International Journal of Hydrogen Energy* 37 (10): 8465–8476.

46. Zhang, G., J. Cheng, M. Gheorghe, and Q. Meng. 2013. A hybrid approach based on differential evolution and tissue membrane systems for solving constrained manufacturing parameter optimization problems. *Applied Soft Computing* 13 (3): 1528–1542.

47. Zhang, G., J. Cheng, and M. Gheorghe. 2014. Dynamic behavior analysis of membrane-inspired evolutionary algorithms. *International Journal of Computers, Communications and Control* 9 (2): 235–250.

48. Zhang, G., M. Gheorghe, and Y. Li. 2012. A membrane algorithm with quantum-inspired subalgorithms and its application to image processing. *Natural Computing* 11 (4): 701–717.

49. Zhang, G., M. Gheorghe, L. Pan, and M.J. Pérez-Jiménez. 2014. Evolutionary membrane computing: a comprehensive survey and new results. *Information Sciences* 279: 528–551.

50. Zhang, G., M. Gheorghe, and C. Wu. 2008. A quantum-inspired evolutionary algorithm based on P systems for knapsack problem. *Fundamenta Informaticae* 87 (1): 93–116.

51. Zhu, Y.L., L.M. Huo, and J.L. Liu. 2006. Bayesian networks based approach for power systems fault diagnosis. *IEEE Transactions on Power Delivery* 21 (2): 634–639.

52. Zhang, G., N. Li, W. Jin, and L. Hu. 2006. Novel quantum genetic algorithm and its application. *Frontiers of Electrical and Electronic Engineering in China* 1 (1): 31–36.

53. Zhang, G., C. Liu, and H. Rong. 2010. Analyzing radar emitter signals with membrane algorithms. *Mathematical and Computer Modelling* 52 (11–12): 1997–2010.

54. Zhang, G., and H. Rong. 2007. Real-observation quantum-inspired evolutionary algorithm for a class of numerical optimization problems. In *Computational Science-ICCS 2007*, vol. 4490, ed. Y. Shi, G.D. van Albada, J. Dongarra, and P.M.A. Sloot, 989–996. Lecture Notes in Computer Science. Berlin: Springer.

55. Zhang, G.X., H.N. Rong, F. Neri, and M.J. Pérez-Jiménez. 2014. An optimization spiking neural P system for approximately solving combinatorial optimization problems. *International Journal Neural Systems*, 24 (5), Article no. 1440006, 16 p.

56. Zhao, J., and N. Wang. 2011. A bio-inspired algorithm based on membrane computing and its application to gasoline blending scheduling. *Computers and Chemical Engineering* 35 (2): 272–283.

57. Zhang, G., F. Zhou, X. Huang, J. Cheng, M. Gheorghe, F. Ipate, and R. Lefticaru. 2012. A novel membrane algorithm based on particle swarm optimization for solving broadcasting problems. *Journal of Universal Computer Science* 13 (18): 1821–1841.

58. Zhang, H., G. Zhang, H. Rong, and J. Cheng. 2010. Comparisons of quantum rotation gates in quantum-inspired evolutionary algorithms. In *Proceedings of the 6th International Conference on Natural Computation*, 2306–2310.

59. Zhou, F., G. Zhang, H. Rong, M. Gheorghe, J. Cheng, F. Ipate, and R. Lefticaru. 2010. A particle swarm optimization based on P systems. In *Proceedings of the 6th International Conference on Natural Computation*, 3003–3007.

# Chapter 5
# Electric Power System Fault Diagnosis with Membrane Systems

**Abstract** Spiking Neural P systems (SN P systems, for short) are used in electric power systems fault diagnostics, by expanding their modeling capabilities with fuzzy theory concepts. The following variants of SN P systems are introduced and investigated: fuzzy reasoning spiking neural P systems with real numbers, weighted fuzzy reasoning spiking neural P systems and fuzzy reasoning spiking neural P systems with trapezoidal fuzzy numbers.

## 5.1 Introduction

Currently, there are three basic types of P systems: cell-like P systems, tissue-like P systems and neural-like P systems. In recent years, the research on neural-like P systems mainly focused on spiking neural P systems (SN P systems, for short), which was introduced in [17]. This computational paradigm is inspired by the neurophysiological behavior of neurons sending electrical impulses (spikes) along axons from presynaptic neurons to postsynaptic neurons in a distributed and parallel manner. An SN P system can be considered as a set of nodes representing neurons in a directed graph whose arcs express synaptic connections among neurons. The contents of each neuron are composed of several copies of a single object type. Likewise, each neuron has a finite set of firing (spiking) and forgetting rules. Firing rules send information between neurons in the form of spikes and forgetting rules remove spikes from neurons. The rules associated with each neuron are used in a sequential manner, but neurons communicate with each other in parallel. Recently, SN P systems have become a hot topic in membrane computing [2, 7, 12, 26–28, 30–42, 47]. The features of SN P systems, such as inherent parallelism, understandability, dynamics, synchronization/asynchronization, non-linearity and non-determinism, are suitable for solving various engineering problems [27, 35].

Since a power system usually is composed of a large number of parts such as generators, transmission lines, bus bars and transformers, fault diagnosis is a complicated process because these parts are protected by a defence system consisting of protective relays, circuit breakers (CBs) and communication equipments [22]. Fault diagnosis is always an important and attractive research topic in power systems. Various methods

© Springer International Publishing AG 2017
G. Zhang et al., *Real-life Applications with Membrane Computing*,
Emergence, Complexity and Computation 25, DOI 10.1007/978-3-319-55989-6_5

have been reported in literature, such as artificial neural networks (ANNs) [1, 33], multi-agent systems (MAS) [11, 16], fuzzy logic (FL) [3–6], expert systems (ES) [18, 23], Bayesian networks (BNs) [10, 48], Petri nets (PNs) [22, 28, 46], information theory (IT) [20, 32], cause-effect networks (CE-Nets) [5, 6, 9] and optimization methods (OM) [19, 43]. Each method has its own pros and cons. ANNs have good tolerance and powerful learning ability, but their network structures and parameters are usually designed in an empirical manner [9]. A large number of samples are required to train ANNs and they suffer from premature convergence. MAS combine effectively several agents representing different methods to cooperatively diagnose faults. However, there are difficulties in defining the way these agents cooperate [16, 20]. FL can process imprecision and uncertainty and is often combined with other methods in studying fault diagnosis [9, 20]. As the earliest artificial intelligence method for power system fault diagnosis, ES can fully use the knowledge of experts, but it has a slow inference engine, due to its sequential search approach, and the difficulty of designing and maintaining a rule-based knowledge system [9, 23]. BNs-based fault diagnosis methods can find relationships of causality between data, while it is difficult to gain accurate prior probabilities and model a complex power grid [20, 48]. PN-based methods have graphical knowledge representation and parallel information processing, but they cannot avoid bad tolerance and combinatorial explosion [5]. IT-based fault diagnosis techniques emerging with the informatization of power systems can deal with the uncertainty in failure processes with fast diagnosis speed to a certain extent, while there are difficulties in the dynamic description of fault information needed [20]. As a graphical tool for knowledge representation, CE-Nets have the features of easy algebraic reasoning and parallel information processing. However, CE-Nets are required to improve their fault tolerance and are not capable to visually describe all possible combinations of main, first and second protections [9, 45], due to their forward reasoning strategy. OM formulates a complex fault diagnosis problem as a minimization optimization problem, which is conveniently solved by applying various optimization techniques, but the difficult issues are the construction of an objective function reflecting the discrepancy between the expected and actual states of protective devices and the parameter adjustment of optimization models [9, 15]. Hence, the improvement of these methods and the exploration of new ones to solve fault diagnosis problems in a power system is of great interest and very necessary.

As a class of distributed and parallel computing models, SN P systems have good understandability and dynamics [27, 35, 39]. In an electric power system, the fault occurrence is a discrete and dynamical process [3–6]. Thus, SN P systems can be used for diagnosing faults in power systems. This chapter focuses on the application of several variants of SN P systems to fault diagnosis of electric power systems, called fuzzy reasoning spiking neural P systems (FRSN P systems, for short). These variants are fuzzy reasoning spiking neural P systems with real numbers (rFRSN P systems) [27], weighted fuzzy reasoning spiking neural P systems (WFRSN P systems) [40] and fuzzy reasoning spiking neural P systems with trapezoidal fuzzy numbers (tFRSN P systems) [42], respectively.

This chapter is structured as follows. Section 5.2 introduces some concepts and essentials that will be used throughout this chapter. The definitions of FRSN P systems for fault diagnosis, including models and reasoning algorithms, are presented in Sect. 5.3. In Sect. 5.4, case studies are used to show how to use three kinds of FRSN P systems to solve fault diagnosis problems. Finally, a comparison between FRSN P systems and several fault diagnosis approaches, and future work are given in Sect. 5.5.

## 5.2 Preliminaries

In this section, we introduce some concepts and essential notations which will be used throughout this chapter. Fuzzy set theory allows us to effectively model and transform imprecise information. Bearing in mind that numbers are the predominant carriers of information, fuzzy numbers play a significant role among all fuzzy entities [21]. First, trapezoidal fuzzy numbers and fuzzy production rules for fuzzy knowledge representation and reasoning are described. Then, essential concepts of electric power systems fault diagnosis and principles of model-based fault diagnosis methods are presented to help readers understand the operating principle of protection devices and the framework of fault diagnosis in power systems using reasoning model-based methods.

### 5.2.1 Fuzzy Knowledge Representation and Reasoning

In order to properly represent real world knowledge, trapezoidal fuzzy numbers and fuzzy production rules have been used for knowledge representation and reasoning to process uncertain, incomplete, imprecise and ambiguous knowledge [21]. A fault diagnosis process consisting of fault information acquisition, processing and representation often contains plenty of incompleteness, imprecision or uncertainty. Thus, trapezoidal fuzzy numbers and fuzzy production rules are used in fault diagnosis.

#### 5.2.1.1 Trapezoidal Fuzzy Numbers

**Definition 1** ([13]) A fuzzy set $A$ of the real line $\mathcal{R}$ with membership function $\mu_A$ is a mapping from $\mathcal{R}$ to $[0, 1]$, associating with each real number $x$ its grade of membership $\mu_A(x)$.

**Definition 2** A fuzzy number is a fuzzy set $A$ of the real line $\mathcal{R}$ with membership function $\mu_A$ such that:

(1) $A$ is *normal*, that is, there exists an element $x_0$ such that $\mu_A(x_0) = 1$;
(2) $A$ is *fuzzy convex*, that is, $\mu_A(\lambda x_1 + (1 - \lambda)x_2) \geq \mu_A(x_1) \wedge \mu_A(x_2)$, for all $x_1, x_2 \in \mathcal{R}$ and $\lambda \in [0, 1]$;
(3) $\mu_A$ is *upper semicontinuous*, that is, for all $x \in \mathcal{R}$ and $\alpha \in [0, 1]$ such that $\mu_A(x) < \alpha$, there exists a $\delta > 0$ such that $x' \in \mathcal{R} \wedge |x' - x| \geq \delta \implies \mu_A(x') < \alpha$;
(4) The set *suppA* is *bounded*, where $suppA = cl(\{x \in \mathcal{R} : \mu_A(x) > 0)$, and $cl$ is the closure operator.

Roughly speaking, a fuzzy number is a quantity whose value is imprecise, rather than exact as is the case with ordinary (single-valued) numbers. There are different kinds of fuzzy numbers, such as rectangular fuzzy numbers, trapezoidal fuzzy numbers, triangular fuzzy numbers, Cauchy fuzzy numbers, bell-shaped fuzzy numbers and so on, among which trapezoids (triangular fuzzy numbers are specific forms of trapezoids) are a good class of functions for representing imprecision and uncertainty, and for modeling granular information by fuzzy sets.

As shown in Fig. 5.1, we use a 4-tuple $\tilde{A} = (a_1, a_2, a_3, a_4)$ to parameterize a *trapezoidal fuzzy number*. In Fig. 5.1, $a_1, a_2, a_3$ and $a_4$ are real numbers representing four horizontal axis values of the trapezoid and satisfy $a_1 < a_2 < a_3 < a_4$. We define the membership function $\mu_{\tilde{A}}(x)$ of the trapezoidal fuzzy number $\tilde{A}$ as follows.

$$\mu_{\tilde{A}}(x) = \begin{cases} 0, & x \leq a_1 \\ \frac{x-a_1}{a_2-a_1}, & a_1 < x \leq a_2 \\ 1, & a_2 < x \leq a_3 \\ \frac{a_4-x}{a_4-a_3}, & a_3 < x \leq a_4 \\ 0, & x > a_4 \end{cases} \tag{5.1}$$

In what follows, we describe arithmetic operations of two trapezoidal fuzzy numbers $\tilde{A}$ and $\tilde{B}$, $\tilde{A} = (a_1, a_2, a_3, a_4)$ and $\tilde{B} = (b_1, b_2, b_3, b_4)$. More operations can be referred in [4, 5].

(1) Addition $\oplus$:
$\tilde{A} \oplus \tilde{B} = (a_1, a_2, a_3, a_4) \oplus (b_1, b_2, b_3, b_4) = (a_1 + b_1, a_2 + b_2, a_3 + b_3, a_4 + b_4)$;

**Fig. 5.1** A trapezoidal fuzzy number ($a_1, a_2, a_3$ and $a_4$ are real numbers)

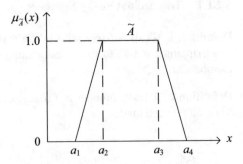

(2) Subtraction $\ominus$:

$\tilde{A} \ominus \tilde{B} = (a_1, a_2, a_3, a_4) \ominus (b_1, b_2, b_3, b_4) = (a_1 - b_1, a_2 - b_2, a_3 - b_3, a_4 - b_4)$;

(3) Multiplication $\otimes$:

$\tilde{A} \otimes \tilde{B} = (a_1, a_2, a_3, a_4) \otimes (b_1, b_2, b_3, b_4) = (a_1 \times b_1, a_2 \times b_2, a_3 \times b_3, a_4 \times b_4)$;

(4) Division of trapezoidal fuzzy numbers $\oslash$:

$\tilde{A} \oslash \tilde{B} = (a_1, a_2, a_3, a_4) \oslash (b_1, b_2, b_3, b_4) = (a_1/b_1, a_2/b_2, a_3/b_3, a_4/b_4)$.

Four logic operations are listed as follows, where $\tilde{A}$ and $\tilde{B}$ are trapezoidal fuzzy numbers, and $a, b$ are real numbers [42]:

(1) *Minimum operator* $\wedge$: $a \wedge b = min\{a, b\}$;
(2) *Maximum operator* $\vee$: $a \vee b = max\{a, b\}$;
(3) *and* $\bigotimes$: $\tilde{A} \bigotimes \tilde{B} =$
    $(a_1, a_2, a_3, a_4) \bigotimes (b_1, b_2, b_3, b_4) = ((a_1 \wedge b_1), (a_2 \wedge b_2), (a_3 \wedge b_3), (a_4 \wedge b_4))$;
(4) *or* $\bigvee$: $\tilde{A} \bigvee \tilde{B} =$
    $(a_1, a_2, a_3, a_4) \bigvee (b_1, b_2, b_3, b_4) = ((a_1 \vee b_1), (a_2 \vee b_2), (a_3 \vee b_3), (a_4 \vee b_4))$.

In addition, a scalar multiplication operation of a real number $b$ by a trapezoidal number $\tilde{A}$ is defined as follows [42]:

$$b \cdot \tilde{A} = b \cdot (a_1, a_2, a_3, a_4) = (b \cdot a_1, b \cdot a_2, b \cdot a_3, b \cdot a_4).$$

The usual order on real numbers can be extended to trapezoidal numbers in a natural way, as follows: $\tilde{A} < \tilde{B}$ if and only if $a_1 < b_1, a_2 < b_2, a_3 < b_3, a_4 < b_4$, being $\tilde{A} = (a_1, a_2, a_3, a_4)$ and $\tilde{B} = (b_1, b_2, b_3, b_4)$. Similarly, $\tilde{A} \leq \tilde{B}, \tilde{A} > \tilde{B}$ and $\tilde{A} \geq \tilde{B}$ can be defined.

### 5.2.1.2 Fuzzy Production Rules

Fuzzy production rules are usually presented in the form of a fuzzy IF-THEN rule in which both the rule antecedent and the rule consequent have fuzzy concepts denoted by fuzzy sets. If the rule antecedent part or rule consequent part of a fuzzy production rule contains AND or OR logic connectors, then this rule is called a composite fuzzy production rule [8, 21]. Fuzzy production rules are widely used in fuzzy knowledge representation [4, 8, 21, 24]. Fuzzy production rules consist of five types: simple fuzzy production rules (*type 1*), composite fuzzy conjunctive rules in the antecedent (*type 2*), composite fuzzy conjunctive rules in the consequent (*type 3*), composite fuzzy disjunctive rules in the antecedent (*type 4*) and composite fuzzy disjunctive rules in the consequent (*type 5*).

A *simple fuzzy production rule (type 1)* is of the form

$$R_i: \text{IF } p_j \text{ THEN } p_k \ (CF = \tau_i) \tag{5.2}$$

where $R_i$ indicates the $i$th fuzzy production rule; $\tau_i \in [0, 1]$ represents its confidence factor; $p_j$ and $p_k$ represents two propositions each of which has a fuzzy truth value in $[0, 1]$. If fuzzy truth values of propositions $p_j$ and $p_k$ are $\alpha_j$ and $\alpha_k$, respectively, then $\alpha_k = \alpha_j * \tau_i$.

A *composite fuzzy conjunctive rule in the antecedent* (*type 2*) is of the form

$$R_i: \text{IF } p_1 \text{ AND } \cdots \text{ AND } p_{k-1} \text{ THEN } p_k \text{ (CF} = \tau_i) \tag{5.3}$$

where $R_i$ indicates the $i$th fuzzy production rule; $\tau_i \in [0, 1]$ represents its confidence factor; $p_1, \ldots, p_{k-1}$, with $k \geq 3$, are propositions in the antecedent part of the rule. If fuzzy truth values of propositions $p_1, \ldots, p_k$ are $\alpha_1, \ldots, \alpha_k$, respectively, then $\alpha_k = \min\{\alpha_1, \ldots, \alpha_{k-1}\} * \tau_i$.

A *composite fuzzy conjunctive rule in the consequent* (*type 3*) is of the form

$$R_i: \text{IF } p_1 \text{ THEN } p_2 \text{ AND } \cdots \text{ AND } p_k \text{  (CF} = \tau_i) \tag{5.4}$$

where $R_i$ indicates the $i$th fuzzy production rule; $\tau_i \in [0, 1]$ represents its confidence factor; $p_1$ is a proposition in the antecedent part of the rule; $p_2, \ldots, p_k$, with $k \geq 3$, are propositions in the consequent part of the rule. If fuzzy truth values of propositions $p_1, \ldots, p_k$ are $\alpha_1, \ldots, \alpha_k$, respectively, then $\alpha_2 = \cdots = \alpha_k = \alpha_1 * \tau_i$,

A *composite fuzzy disjunctive rule in the antecedent* (*type 4*) is of the form

$$R_i: \text{IF } p_1 \text{ OR } \cdots \text{ OR } p_{k-1} \text{ THEN } p_k \text{ (CF} = \tau_i) \tag{5.5}$$

where $R_i$ indicates the $i$th fuzzy production rule; $\tau_i \in [0, 1]$ represents its confidence factor; $p_1, \ldots, p_{k-1}$, with $k \geq 3$, are propositions in the antecedent part of the rule. If the fuzzy truth values of propositions $p_1, \ldots, p_k$ are $\alpha_1, \ldots, \alpha_k$, respectively, then $\alpha_k = \max\{\alpha_1, \ldots, \alpha_{k-1}\} * \tau_i$.

A *composite fuzzy disjunctive rule in the consequent* (*type 5*) is of the form

$$R_i: \text{IF } p_1 \text{ THEN } p_2 \text{ OR } \cdots \text{ OR } p_k \text{ (CF} = \tau_i) \tag{5.6}$$

where $R_i$ indicates the $i$th fuzzy production rule; $\tau_i \in [0, 1]$ represents its confidence factor; $p_1$ is a proposition in the antecedent part of the rule; $p_2, \ldots, p_k$, with $k \geq 3$, are propositions in the consequent part of the rule. This type of rules is unsuitable for knowledge representation due to the fact that it does not make any specific implication [4]. Thus, this type of rules is not discussed here and will not be considered in fault diagnosis in the following sections.

### 5.2.2   Essentials of Electric Power System Fault Diagnosis

In a strict sense, fault diagnosis consists of fault detection, fault section identification, fault type estimation, failure isolation and recovery [43], of which fault section

**Fig. 5.2** A schematic
illustration of a power
system with sections and
protective relays considered
in this chapter

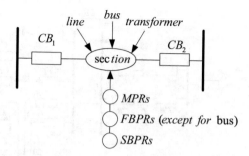

identification is a very important process [22, 28]. When faults occur in a power system, protective relays detect the faults and trip their corresponding circuit breakers (CBs) to isolate faulty sections from the operation of this power system and guarantee the other parts can operate normally. The aim of fault diagnosis in this chapter is to identify the faulty sections by using status information of protective relays and CBs which are read from supervisor control and data acquisition (SCADA) systems.

The protective system of an electric power system is very important in fault diagnosis, as well as the protection configuration of this protective system. Thus, protection devices (protective relays and CBs) and their operating principles are described in detail in this subsection.

The protective relays are composed of main protective relays (MPRs), first backup protective relays (FBPRs) and second backup protective relays (SBPRs). There are no FBPR within buses. A schematic illustration of a power system with sections and protective relays is shown in Fig. 5.2, which is considered in this chapter.

We choose a local sketch map of the protection system of an EPS from [28, 43], which is shown in Fig. 5.3, to demonstrate the operational rules of different types of protections. The local system is composed of 28 system sections, 40 CBs and 84 protective relays.

For the convenience of description, we define the notations as follows. $A$, $B$, $T$ and $L$ represent a single bus, double bus, transformer and line, respectively. The sending and receiving ends of line $L$ are described by $S$ and $R$, respectively. $m$, $p$ and $s$ are used to denote the main protection, the first backup protection and the second backup protection, respectively. The 28 sections ($S_1 \sim S_{28}$) are described by $A_1, \ldots, A_4, T_1, \ldots, T_8, B_1, \ldots, B_8, L_1, \ldots, L_8$. The 40 CBs ($C_1 \sim C_{40}$) are represented by $CB_1, CB_2, \ldots, CB_{40}$. The 84 protective relays consist of 36 main ones, $r_1 \sim r_{36}$ denoted by $A_{1m}, \ldots, A_{4m}, T_{1m}, \ldots, T_{8m}, B_{1m}, \ldots, B_{8m}, L_{1Sm}, \ldots, L_{8Sm}, L_{1Rm}, \ldots, L_{8Rm}$, and 48 backup ones, $r_{37} \sim r_{84}$ described by $T_{1p}, \ldots, T_{8p}, T_{1s}, \ldots, T_{8s}, L_{1Sp}, \ldots, L_{8Sp}, L_{1Rp}, \ldots, L_{8Rp}, L_{1Ss}, \ldots, L_{8Ss}, L_{1Rs}, \ldots, L_{8Rs}$.

We describe the operational rules of the protective relays for the three kinds of sections, transmission lines, buses and transformers in [43], as follows:

(1) *Protective relays of transmission lines*
Protective relays of transmission lines are of two types: sending end protective relays and receiving end protective relays. Both ends of a line have their own main, first and

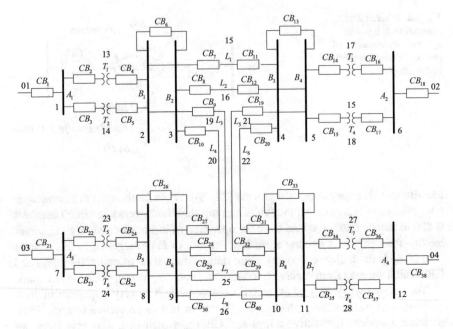

**Fig. 5.3** A local sketch map of the protection system of an EPS

second protections. When the main protective relays of a line operate, CBs connected to the line are tripped. For example, if line $L_7$ fails, MPRs $L_{7Sm}$ and $L_{7Rm}$ are operated to trip $CB_{29}$ and $CB_{39}$, respectively. Likewise, when the main protections of a line fail to operate, the first backup protective relays operate to trip CBs connected to the line. For example, if line $L_7$ fails and main protection relay (MPR) $L_{7Sm}$ fails to operate, first backup protective relay (FBPR) $L_{7Sp}$ operates to trip $CB_{29}$. If line $L_7$ fails and MPR $L_{7Rm}$ fails to operate, FBPR $L_{7Rp}$ operates to trip $CB_{39}$. When the adjacent regions of a line fail and their protections fail to operate, the second backup protections operate to protect the line. For example, if section $B_8$ fails and $CB_{39}$ fails to trip off, second backup protective relay (SBPR) $L_{7Ss}$ operates to trip $CB_{29}$. If section $B_5$ fails and $CB_{29}$ fails to trip off, SBPR $L_{7Rs}$ operates to trip $CB_{39}$.

*(2) Protective relays of buses*
When the main protective relays of a bus operate, all CBs directly connected to the bus will be tripped. For example, if bus $A_1$ fails, MPR $A_{1m}$ operates to trip $CB_1$, $CB_2$ and $CB_3$. Similarly, if bus $B_8$ fails, MPR $B_{8m}$ operates to trip $CB_{32}$, $CB_{33}$ and $CB_{39}$.

*(3) Protective relays of transformers*
When the main protective relays of a transformer operate, all CBs connected to the transformer are tripped. For example, if transformer $T_3$ fails, MPR $T_{3m}$ operates to trip $CB_{14}$ and $CB_{16}$. Likewise, when the main protections of a transformer fail to operate, the first backup protective relays operate to trip CBs connected to the transformer. For example, if transformer $T_3$ fails and MPR $T_{3m}$ fails to operate, FBPR

$T_{3p}$ operates to trip $CB_{14}$ and $CB_{16}$. When the adjacent regions of the transformer fail and their protections fail to operate, the second backup protections operate to protect the transformer. For example, if bus $A_2$ fails and $CB_{16}$ fails to trip off, SBPR $T_{3s}$ operates to trip $CB_{16}$ to protect $T_3$.

### 5.2.3  Principles of Model-Based Fault Diagnosis Methods

Fault diagnosis of power systems based on SN P systems belongs to model-based fault diagnosis methods. So this subsection describes some basic ideas of model-based fault diagnosis methods. The framework of fault diagnosis in power systems using reasoning model-based method is shown as in Fig. 5.4 [14, 45].

There are three important parts in this framework: real-time data, static data and a flowchart of fault sections identification.

(1) *Real-time data*. The real-time data, protective relay operation information and circuit breaker tripping information, are used to estimate the outage areas to obtain candidate faulty sections using a network topology analysis method, so as to reduce the subsequent computational burden [14].

(2) *Static data*. The static data, network topology and protection configuration of a power system, are used to build a fault diagnosis model for each candidate

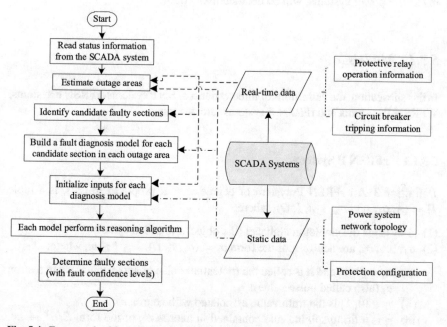

**Fig. 5.4** Framework of fault diagnosis in power systems using reasoning model-based methods

section in each outage area. The inputs of each diagnosis model are initialized by both real-time data and static data.

(3) *Flowchart of fault sections identification*. Each diagnosis model performs a reasoning algorithm to obtain fault confidence levels of the candidate faulty sections to determine faulty sections. The diagnosis results include the faulty sections and their fault confidence levels.

## 5.3   Spiking Neural P Systems for Fault Diagnosis

Up to now, three variants of SN P systems have been proposed for solving fault diagnosis problems [39], which are fuzzy reasoning spiking neural P systems with real numbers (rFRSN P systems) [27], weighted fuzzy reasoning spiking neural P systems (WFRSN P systems) [40] and fuzzy reasoning spiking neural P systems with trapezoidal fuzzy numbers (tFRSN P systems) [42]. This section first describes the definitions of rFRSN P systems, WFRSN P systems and tFRSN P systems, respectively. Then, their reasoning algorithms, called fuzzy reasoning algorithm (FRA), weighted matrix-based reasoning algorithm (WMBRA) and matrix-based fuzzy reasoning algorithm (MBFRA), respectively, are presented. In this section, we focus on the description of SN P systems for fault diagnosis and their reasoning algorithms. How to use the SN P systems and their reasoning algorithms to diagnose the faults in electric power systems, will be detailed in Sect. 5.4.

### 5.3.1   Models

In this subsection, the definitions of three variants of SN P systems (rFRSN P systems, WFRSN P systems and tFRSN P systems) are described.

#### 5.3.1.1   rFRSN P Systems

**Definition 3** An rFRSN P system of degree $m \geq 1$, proposed in [27], is a tuple $\Pi = (A, \sigma_1, \ldots, \sigma_m, syn, I, O)$, where:

(1) $A = \{a\}$ is the singleton alphabet (the object $a$ is called spike);
(2) $\sigma_1, \ldots, \sigma_m$ are neurons, of the form $\sigma_i = (\alpha_i, \tau_i, r_i)$, $1 \leq i \leq m$, where:

  (i) $\alpha_i \in [0, 1]$ and it is called the (potential) value of spike contained in neuron $\sigma_i$ (also called pulse value);
  (ii) $\tau_i \in [0, 1]$ is the truth value associated with neuron $\sigma_i$;
  (iii) $r_i$ is a firing/spiking rule contained in neuron $\sigma_i$, of the form $E/a^\alpha \rightarrow a^\beta$, where $\alpha, \beta \in [0, 1]$;

(3) $syn \subseteq \{1, \ldots, m\} \times \{1, \ldots, m\}$ with $i \neq j$ for all $(i, j) \in syn$, $1 \leq i, j \leq m$ (*synapses* between neurons);

(4) $I, O \subseteq \{1, \ldots, m\}$ indicate the input neuron set and the output neuron set of $\Pi$, respectively.

In what follows, we describe how the rFRSN P systems are extended from SN P systems as follows:

(1) The content of a neuron is denoted by a real number in $[0, 1]$ instead of the number of spikes in SN P systems. From the perspective of biological neuron, this can be interpreted as the (potential) value of spike. If $\alpha_i > 0$ then the neuron $\sigma_i$ contains a spike with (potential) value $\alpha_i$; otherwise, it contains no spike.

(2) Each neuron $\sigma_i$ ($i \in \{1, \ldots, m\}$) of an rFRSN P system has associated either a fuzzy proposition or a fuzzy production rule, and $\tau_i \in [0, 1]$ is used to express the fuzzy truth value of a fuzzy proposition or the confidence factor (CF) of a fuzzy production rule.

(3) Each neuron contains only one spiking (firing) rule, of the form $E/a^\alpha \rightarrow a^\beta$, where $E = a^n$ is the firing condition and $n \geq 1$ is the number of input synapses from other neurons to this neuron. The firing condition $E = a^n$ means that if the neuron receives $n$ spikes, the spiking rule can be applied; otherwise the rule cannot be enabled until $n$ spikes are received. When the number of spikes received by a neuron is less than $n$, the value of the spikes received will be updated according to logical AND or OR operations.

(4) The firing mechanism of neurons in rFRSN P systems is described in detail as follows. For the neuron $\sigma_i$, if its firing rule $E/a^\alpha \rightarrow a^\beta$ can be applied, the neuron fires. This indicates that its pulse value $\alpha > 0$ is consumed (removed) and it produces a spike with value $\beta$. Once a spike with value $\beta$ is excited from neuron $\sigma_i$, all neurons $\sigma_j$ with $(i, j) \in syn$ immediately receive the spike.

(5) Three types of neurons are defined in different ways to handle both $\alpha$ and $\beta$, and $\beta$ relates with both $\alpha$ and $\tau_i$ (see *Definitions 4–6*).

(6) Time delay is ignored in an rFRSN P system, thus all neurons are always open.

In summary, an rFRSN P system has three kinds of neurons: proposition neurons, AND-type rule neurons and OR-type rule neurons. They are defined as follows.

**Definition 4** The *proposition neurons*, shown in Fig. 5.5, are a class of neurons associated with propositions in a fuzzy knowledge base [27].

A *proposition neuron* can be denoted by $\sigma = (\alpha, \tau, r)$, where $\alpha$, $\tau$ and $r$ are its pulse value, the truth value of the proposition associated with it and its spiking rule of the form $E/a^\alpha \rightarrow a^\alpha$, respectively.

If a proposition neuron $\sigma$ is an input neuron in $\Pi$, then $\alpha = \tau$; otherwise, $\alpha$ equals logical "OR" operation of all pulse values received from other neurons. After the neuron updates its content, the truth value of the corresponding proposition will equal its pulse value, i.e., $\tau = \alpha$. When the neuron fires and applies its firing rule, it will generate a spike with pulse value $\alpha$.

**Fig. 5.5** **a** A proposition neuron and **b** its simplified form

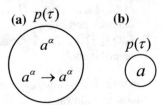

**Fig. 5.6** **a** An AND-type rule neuron and **b** its simplified form

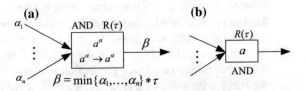

**Fig. 5.7** **a** A OR-type rule neuron and **b** its simplified form

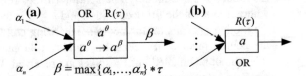

**Definition 5** The *AND-type rule neurons*, shown in Fig. 5.6, are a class of neurons associated with fuzzy production rules with AND-type antecedent part, where confidence factor (CF) of each rule is denoted by $\tau$ [27].

An *AND-type rule neuron* is represented by the symbol AND. If an AND-type rule neuron receives $n$ pulse values from other neurons, $\alpha_1, \ldots, \alpha_n$, it uses logic operator AND to combine all its inputs, i.e., $\alpha = \min\{\alpha_1, \ldots, \alpha_n\}$, where $\alpha_i \in [0, 1]$, $1 \le i \le n$. If its firing condition $E$ is satisfied, it fires and applies its spiking rule $E/a^\alpha \to a^\beta$ to generate a spike with value $\beta = \alpha * \tau$, thus $\beta = \min\{\alpha_1, \ldots, \alpha_n\} * \tau$; otherwise, it only updates its content by using the pulse values of the received spikes.

**Definition 6** The *OR-type rule neurons*, shown in Fig. 5.7, are a class of neurons associated with fuzzy production rules with OR-type antecedent part, where confidence factor (CF) of each rule is denoted by $\tau$ [27].

An *OR-type rule neuron* is denoted by the symbol OR. If an OR-type rule neuron receives $n$ pulse values from other neurons, $\alpha_1, \ldots, \alpha_n$, it uses the logic operator OR to combine all its inputs, i.e., $\alpha = \max\{\alpha_1, \ldots, \alpha_n\}$, where $\alpha_i \in [0, 1]$, $1 \le i \le n$. Similarly, when $E$ is satisfied, it fires and applies its spiking rule $E/a^\alpha \to a^\beta$ to generate a spike with value $\beta = \alpha * \tau$, thus $\beta = \max\{\alpha_1, \ldots, \alpha_n\} * \tau$; otherwise, it only updates its content by using the pulse values of the received spikes.

### 5.3.1.2 WFRSN P Systems

**Definition 7** A WFRSN P system of degree $m \geq 1$, proposed in [40], is a tuple $\Pi = (O, \sigma_1, \ldots, \sigma_m, syn, in, out)$, where:

(1) $O = \{a\}$ is a singleton alphabet ($a$ is called spike);
(2) $\sigma_1, \ldots, \sigma_m$ are neurons, of the form $\sigma_i = (\theta_i, c_i, \overrightarrow{\omega_i}, \lambda_i, r_i)$, $1 \leq i \leq m$, where:

  (i) $\theta_i$ is a real number in $[0, 1]$ representing the potential value of spikes (i.e. value of electrical impulses) contained in neuron $\sigma_i$;
  (ii) $c_i$ is a real number in $[0, 1]$ representing the truth value associated with neuron $\sigma_i$;
  (iii) $\overrightarrow{\omega_i} = (\omega_{i1}, \ldots, \omega_{iN_i})$ is a real number vector in $(0, 1]$ representing the output weight vector of neuron $\sigma_i$, where $\omega_{ij}$ $(1 \leq j \leq N_i)$ represents the weight on the $j$th output arc (synapse) of neuron $\sigma_i$ and $N_i$ is a natural number representing the number of synapses starting from neuron $\sigma_i$;
  (iv) $\lambda_i$ is a real number in $[0, 1)$ representing the firing threshold of neuron $\sigma_i$;
  (v) $r_i$ represents a firing (spiking) rule contained in neuron $\sigma_i$ with the form $E/a^\theta \rightarrow a^\beta$, where $\theta$ and $\beta$ are real numbers in $[0, 1]$, $E = a^n$ is the firing condition. The firing condition means that if and only if neuron $\sigma_i$ receives at least $n$ spikes and the potential value of spikes is with $\theta \geq \lambda_i$, then the firing rule contained in the neuron can be applied, otherwise, the firing rule cannot be applied;

(3) $syn \subseteq \{1, \ldots, m\} \times \{1, \ldots, m\}$ with $i \neq j$ for all $(i, j) \in syn$, $1 \leq i, j \leq m$; that is, $syn$ provides a (weighted) directed graph whose set of nodes is $\{1, \ldots, m\}$;
(4) $in, out \subseteq \{1, \ldots, m\}$ indicate the input neuron set and the output neuron set of $\Pi$, respectively.

In what follows, we describe how a WFRSN P system is extended from a SN P system as follows:

(1) We extend the definition of neurons. WFRSN P systems consist of two kinds of neurons, i.e., proposition neurons and rule neurons, where rule neurons contain three subcategories: *general, and* and *or*.
(2) The pulse value $\theta_i$ contained in each neuron $\sigma_i$ is a real number in $[0, 1]$, which represents the potential value of spikes contained in this neuron instead of the number of spikes in SN P systems.
(3) Each neuron is associated with either a proposition or a production rule, and $c_i \in [0, 1]$ denotes the truth value of this proposition or the confidence factor (CF) of this production rule.
(4) Each weighted directed synapse has an output weight. That is, each synapse in $syn \subseteq \{1, \ldots, m\} \times \{1, \ldots, m\}$ has a weight. The output weights of neurons represent the importance degree of their values in contributing to the computing results in output neurons.
(5) Each neuron contains only one firing (spiking) rule of the form $E/a^\theta \rightarrow a^\beta$. When the firing condition of one neuron is satisfied, the firing rule is applied,

which indicates that the potential value $\theta$ is consumed and then this neuron generates a new spike with potential value of $\beta$. These different types of neurons handle the potential values $\theta$ and $\beta$ in different ways (see Definitions 8–11). If the firing condition of one neuron is not satisfied, the potential value of the spikes received by this neuron is updated by logical operators *and* or *or*.

(6) Time delay is ignored in WFRSN P systems, so all neurons are always open.

The definitions of different types of neurons in WFRSN P systems are described as follows.

**Definition 8** A *proposition neuron* is associated with a proposition in a fuzzy production rule. Such a neuron is represented by a circle and symbol $P$, as shown in Fig. 5.8 [40].

If a *proposition neuron* is an input neuron of a WFRSN P system $\Pi$, its potential value $\theta$ is received from the environment; otherwise, $\theta$ equals the result of logical operation *or* on all weighted potential values received from its presynaptic rule neurons, i.e., $\theta = max\{\theta_1 * \omega_1, \ldots, \theta_k * \omega_k\}$. The firing rule of a *proposition neuron* is of the form $E/a^\theta \rightarrow a^\theta$, that is, the parameter $\beta$ of the firing rule contained in such a neuron is identical to $\theta$. When the firing condition $E$ of a *proposition neuron* is satisfied, the potential value $\theta$ of spikes contained in this neuron is consumed and then a new spike with potential value $\theta$ is generated and emitted.

**Definition 9** A *general rule neuron* is associated with a fuzzy production rule which has only one proposition in the antecedent part of the rule. Such a neuron is represented by a rectangle and symbol $R_{(c, general)}$, as shown in Fig. 5.9 [40], where $c$ is the confidence factor of this rule.

A *general rule neuron* has only one presynaptic proposition neuron and one or more postsynaptic proposition neurons. If a *general rule neuron* $\sigma$ receives a spike from its presynaptic proposition neuron and its firing condition is satisfied, the neuron

**Fig. 5.8** A *proposition neuron* (**a**) and its simplified form (**b**)

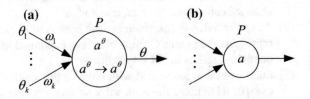

**Fig. 5.9** A *general rule neuron* (**a**) and its simplified form (**b**)

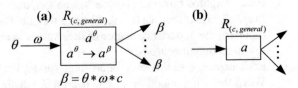

**Fig. 5.10** An *and rule neuron* (**a**) and its simplified form (**b**)

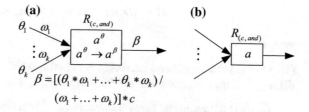

**Fig. 5.11** An *or rule neuron* (**a**) and its simplified form (**b**)

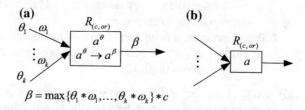

fires and produces a new spike with the potential value $\beta = \theta * \omega * c$, where $\omega$ is the weight of the output arc to $\sigma$.

**Definition 10** An *and rule neuron* is associated with a fuzzy production rule which has more than one propositions with an *and* relationship in the antecedent part of the rule. Such a neuron is denoted by a rectangle and symbol $R_{(c, and)}$, as shown in Fig. 5.10 [40], where $c$ is the confidence factor of this rule.

An *and rule neuron* has more than one presynaptic proposition neuron and only one postsynaptic proposition neuron. If an *and rule neuron* $\sigma$ receives $k$ spikes from its $k$ presynaptic proposition neurons and its firing condition is satisfied, the neuron fires and produces a new spike with the potential value $\beta = [(\theta_1 * \omega_1 + \cdots + \theta_k * \omega_k)/(\omega_1 + \cdots + \omega_k)] * c$, where $\omega$ is the weight of the output arc to $\sigma$

**Definition 11** An *or rule neuron* is associated with a fuzzy production rule with more than one propositions having an *or* relationship in the antecedent part of the rule. Such a neuron is represented by a rectangle and symbol $R_{(c, or)}$, as shown in Fig. 5.11 [40], where $c$ is the confidence factor of this rule.

An *or rule neuron* has more than one presynaptic proposition neuron and only one postsynaptic proposition neuron. If an *or rule neuron* $\sigma$ receives $k$ spikes from its $k$ presynaptic proposition neurons and its firing condition is satisfied, the neuron fires and produces a new spike with the potential value $\beta = max\{\theta_1 * \omega_1, \ldots, \theta_k * \omega_k\} * c$, where $\omega$ is the weight of the output arc to $\sigma$.

### 5.3.1.3 tFRSN P Systems

**Definition 12** A tFRSN P system with trapezoidal fuzzy numbers (with degree $m \geq 1$) is a tuple [42]

$$\Pi = (O, \sigma_1, \ldots, \sigma_m, syn, in, out)$$

where:

(1)  $O = \{a\}$ is a singleton alphabet ($a$ is called spike);
(2)  $\sigma_1, \ldots, \sigma_m$ are neurons of the form $\sigma_i = (\theta_i, c_i, r_i)$, $1 \le i \le m$, where:

  (i)  $\theta_i$ is a trapezoidal fuzzy number in $[0, 1]$ representing the potential value of spikes (i.e., the value of electrical impulses) contained in neuron $\sigma_i$;
  (ii)  $c_i$ is a trapezoidal fuzzy number in $[0, 1]$ representing the fuzzy truth value corresponding to neuron $\sigma_i$;
  (iii)  $r_i$ represents a firing (spiking) rule associated with neuron $\sigma_i$ of the form $E/a^\theta \to a^\beta$, where $E$ is a regular expression, and $\theta$ and $\beta$ are trapezoidal fuzzy numbers in $[0, 1]$.

(3)  $syn \subseteq \{1, \ldots, m\} \times \{1, \ldots, m\}$ with $i \ne j$ for all $(i, j) \in syn$, $1 \le i, j \le m$, is a directed graph of synapses between the linked neurons;
(4)  $in, out \subseteq \{1, 2, \ldots, m\}$ are the input neuron set and the output neuron set of $\Pi$, respectively.

The pulse value contained in each neuron in a tFRSN P system is not the number of spikes denoted by a real number, but a trapezoidal fuzzy number in $[0, 1]$. This can be regarded as the potential value of spikes contained in neuron $\sigma_i$. The introduction of trapezoidal fuzzy numbers are for processing various uncertainties coming from professional knowledge acquisition due to experts' subjectivity, or linguistic terms with a certain degree of uncertainty for expressing human knowledge in the real world such as knowledge of fault diagnosis process, or the knowledge of practical applications with a certain degree of uncertainty. For example, we often use fuzzy concepts (*absolutely-false, very-low, low, medium-low, medium, medium-high, high, very-high, absolutely-high*) to describe a degree of uncertainty. The operation process of protective devices in fault diagnosis usually includes uncertainly protective messages such as maloperation and misinformation.

If the pulse value $\theta_i > (0, 0, 0, 0)$, neuron $\sigma_i$ contains a spike with value $\theta_i$, otherwise, the neuron contains no spike. The meaning of the firing condition $E = a^n$ is that the spiking rule associated with neuron $\sigma_i$ can be applied at an instant $t$ if and only if the number of spikes that neuron $\sigma_i$ receives at that moment equals $n$, otherwise, the firing rule cannot be applied. If the number of spikes that neuron $\sigma_i$ receives is less than $n$, neuron $\sigma_i$ performs the operation $\wedge$ or $\vee$ on the potential values carried by these spikes to update its pulse value.

The neurons in a tFRSN P system are divided into two categories: *proposition neurons* and *rule neurons*. Each neuron $\sigma_i$ corresponds to either a proposition or a fuzzy production rule, which will be described later in this section. Thus, the trapezoidal fuzzy number $c_i$ can be understood as either the fuzzy truth value of a proposition or the confidence factor of a fuzzy production rule.

The definitions of different types of neurons in tFRSN P systems are described as follows.

**Fig. 5.12** A proposition
neuron (**a**) and its simplified
form (**b**)

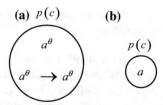

**Definition 13** A *proposition neuron*, as shown in Fig. 5.12 [42], corresponds to a proposition in the fuzzy production rules. Such a neuron is represented by the symbol $p$.

The fuzzy truth value of a proposition neuron equals the fuzzy truth value of its proposition. If such a proposition neuron receives one spike, i.e., $n = 1$, it will fire and emit a spike. The parameter $\beta$ of the firing rule contained in such a proposition neuron is identical to $\theta$. If a proposition neuron is an input, its pulse value $\theta$ equals the fuzzy truth value $c$ of this neuron. Otherwise, if there is only one presynaptic rule neuron, $\theta$ equals the pulse value transmitted from this neuron. In any other case, $\theta$ equals the result of the operation $\otimes$ on all pulse values received from its presynaptic rule neurons.

The three types of *rule neurons*, *general*, *and* and *or*, are represented by the three symbols $\ominus$, $\otimes$ and $\oslash$, respectively. A rule neuron is represented by $R$. If the number of spikes a rule neuron receives equals the number of its presynapses, it will fire and emit a spike. In what follows, each type of rule neurons will be defined.

**Definition 14** A *general rule neuron* $\ominus$, as shown in Fig. 5.13i [42], corresponds to a fuzzy production rule with only one proposition in the antecedent part of the

**Fig. 5.13** Rule neurons.
**i** A *general* rule neuron (*a*)
and its simplified form (*b*);
**ii** An *and* rule neuron (*a*) and
its simplified form (*b*); **iii** An
*or* rule neuron (*a*) and its
simplified form (*b*)

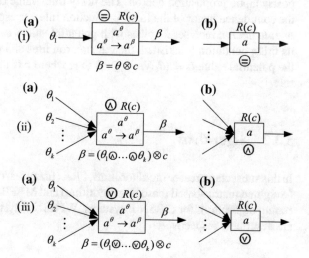

rule. The consequent part of the fuzzy production rule may contain one or more propositions.

A *general rule neuron* has only one presynaptic proposition neuron and one or more postsynaptic proposition neurons. The fuzzy truth value of a *general rule neuron* equals the confidence factor of the fuzzy production rule corresponding to its neuron. If a *general rule neuron* receives a spike with potential value $\theta$ and its firing condition is satisfied, the neuron fires and produces a new spike with potential value $\beta = \theta \otimes c$, where $c$ is the confidence factor of this rule.

**Definition 15** An *and rule neuron* $\bigwedge$, as shown in Fig. 5.13ii [42], corresponds to the fuzzy production rule having more than one proposition with an *and* relationship in the antecedent part of the rule. The consequent part of the fuzzy production rule contains only one proposition.

An *and rule neuron* has more than one presynaptic proposition neuron and only one postsynaptic proposition neuron. The fuzzy truth value of an *and rule neuron* equals the confidence factor of the fuzzy production rule corresponding to its neuron. If an *and rule neuron* receives $k$ spikes with potential values $\theta_1, \ldots, \theta_k$, respectively, and its firing condition is satisfied, then the neuron fires and produces a new spike with the potential value $\beta = (\theta_1 \bigwedge \cdots \bigwedge \theta_k) \otimes c$, where $c$ is the confidence factor of this rule.

**Definition 16** An *or rule neuron* $\bigvee$, as shown in Fig. 5.13iii [42], corresponds to the fuzzy production rule which has more than one proposition with an *or* relationship in the antecedent part of the rule. The consequent part of the fuzzy production rule contains only one proposition.

An *or rule neuron* has more than one presynaptic proposition neuron and only one postsynaptic proposition neuron. The fuzzy truth value of an *or rule neuron* equals the confidence factor of the fuzzy production rule corresponding to its neuron. If an *or rule neuron* receives $k$ spikes with potential values $\theta_1, \ldots, \theta_k$, respectively and its firing condition is satisfied, then the neuron fires and produces a new spike with the potential value $\beta = (\theta_1 \bigvee \cdots \bigvee \theta_k) \otimes c$, where $c$ is the confidence factor of this rule.

## 5.3.2 Algorithms

In this subsection, reasoning algorithms, FRA (fuzzy reasoning algorithm), WMBRA (weighted matrix-based reasoning algorithm) and MBFRA (matrix-based fuzzy reasoning algorithm), for rFRSN P systems, WFRSN P systems and tFRSN P systems are described, respectively.

### 5.3.2.1  FRA

A fuzzy reasoning algorithm (FRA) based on rFRSN P systems was proposed in [27]. The aim of FRA is to use known fuzzy propositions (input neurons) to reason out the fuzzy truth values of unknown fuzzy propositions (proposition neurons associated with output neurons). Assume that all fuzzy production rules in a fuzzy diagnosis knowledge base have been modeled by an rFRSN P system model $\Pi$. The model $\Pi$ consists of $m$ neurons: $n$ proposition neurons $\{\sigma_1, \ldots, \sigma_n\}$ and $k$ rule neurons $\{\overline{\sigma_1}, \ldots, \overline{\sigma_k}\}$ (AND type neurons and OR type neurons), where $m = n + k$.

In what follows, some parameter vectors and matrices are first introduced to be a preparation of FRA.

(1) $U = (u_{ij})_{n \times k}$ is a binary matrix, where $u_{ij} \in \{0, 1\}$. $u_{ij} = 1$ if there is a directed arc (synapse) from proposition neuron $\sigma_i$ to rule neuron $\overline{\sigma_j}$, otherwise, $u_{ij} = 0$.

(2) $V = (v_{ij})_{n \times k}$ is a binary matrix, where $v_{ij} \in \{0, 1\}$. $v_{ij} = 1$ if there is a directed arc (synapse) from rule neuron $\bar{\sigma}_j$ to proposition neuron $\sigma_i$, otherwise, $v_{ij} = 0$.

(3) $\Lambda = \text{diag}(\tau_1, \ldots, \tau_k)$ is a diagonal matrix, where $\tau_j$ represents confidence factor of the $j$th production rule associated with rule neuron $\bar{\sigma}_j$, $1 \le j \le k$.

(4) $H_1 = \text{diag}(h_1, \ldots, h_k)$ is a diagonal matrix. If the $j$th rule neuron is an AND-type neuron, $h_j = 1$; otherwise $h_j = 0$.

(5) $H_2 = \text{diag}(h_1, \ldots, h_k)$ is a diagonal matrix. If the $j$th rule neuron is an OR-type neuron, $h_j = 1$; otherwise $h_j = 0$.

(6) $\alpha_p = (\alpha_{p1}, \ldots, \alpha_{pn})^T$ is a truth value vector, $\alpha_{pi} \in [0, 1]$. $\alpha_{pi}$ represents the truth value of the $i$th proposition neuron. $\alpha_r = (\alpha_{r1}, \ldots, \alpha_{rk})^T$ is also a truth value vector, $\alpha_{rj} \in [0, 1]$. $\alpha_{rj}$ represents the truth value of the $j$th rule neuron.

(7) $a_p = (a_{p1}, \ldots, a_{pn})^T$ is an integer vector, where $a_{pi}$ represents the number of spikes received by the $i$th proposition neuron. $a_r = (a_{r1}, \ldots, a_{rk})^T$ is also an integer vector, where $a_{rj}$ represents the number of spikes received by the $j$th rule neuron.

(8) $\lambda_p = (\lambda_{p1}, \ldots, \lambda_{pn})^T$ is an integer vector, where $\lambda_{pi}$ represents the number of spikes required by firing the $i$th proposition neuron. $\lambda_r = (\lambda_{r1}, \ldots, \lambda_{rk})^T$ is also an integer vector, where $\lambda_{rj}$ represents the number of spikes required by firing the $j$th rule neuron.

(9) $\beta_p = (\beta_{p1}, \ldots, \beta_{pn})^T$ is a truth value vector, $\beta_{pi} \in [0, 1]$. $\beta_{pi}$ represents truth value exported by the $i$th proposition neuron after firing. $\beta_r = (\beta_{r1}, \ldots, \beta_{rk})^T$ is also a truth value vector, $\beta_{rj} \in [0, 1]$. $\beta_{rj}$ represents the truth value exported by the $j$th rule neuron after firing.

(10) $b_p = (b_{p1}, \ldots, b_{pn})^T$ is an integer vector, where $b_{pi} \in \{0, 1\}$ represents the number of spikes exported by the $i$th proposition neuron after firing. $b_r = (b_{r1}, \ldots, b_{rk})^T$ is also an integer vector, where $b_{rj} \in \{0, 1\}$ represents the number of spikes exported by the $j$th rule neuron after firing.

Then we introduce three multiplication operations:

(1) $\oplus$: $C = A \oplus B$, where $A$, $B$ and $C$ are all $r \times s$ matrices, such that $c_{ij} = \max\{a_{ij}, b_{ij}\}$.

(2) $\otimes$: $C = A \otimes B$, where $A$, $B$ and $C$ are $r \times s$, $s \times t$ and $r \times t$ matrices respectively, such that $c_{ij} = \max\limits_{1 \le m \le s} \{a_{im} \cdot b_{mj}\}$.

(3) $\odot$: $C = A \odot B$, where $A$, $B$ and $C$ are $r \times s$, $s \times t$ and $r \times t$ matrices respectively, such that $c_{ij} = \min\limits_{1 \le m \le s} \{a_{im} \cdot b_{mj}\}$.

(4) $\beta = \text{fire}\,(\alpha, a, \lambda)$, where $\beta = (\beta_1, \ldots, \beta_r)^T$, $\alpha = (\alpha_1, \ldots, \alpha_r)^T$, $a = (a_1, \ldots, a_r)^T$, $\lambda = (\lambda_1, \ldots, \lambda_r)^T$. The function is defined as follows:

$$\beta_i = \begin{cases} \alpha_i, & \text{if } a_i = \lambda_i \\ 0 & \text{if } a_i < \lambda_i \end{cases},$$

where $i = 1, 2, \ldots, r$.

(5) $\beta = \text{update}\,(\alpha, a, \lambda)$, where $\beta$, $\alpha$, $a$ and $\lambda$ are vectors described above. The function is defined as follows:

$$\beta_i = \begin{cases} 0 & \text{if } a_i = 0 \\ \beta_i + \alpha_i, & \text{if } 0 < a_i < \lambda_i \\ 0 & \text{if } a_i = \lambda_i \end{cases},$$

where $i = 1, 2, \ldots, r$.

(6) $D = \text{diag}(b)$, where $D = (d_{ij})$ is a $r \times r$ diagonal matrix and $b = (b_1, \ldots, b_r)$. For $1 \le i \le r$, $d_{ii} = b_i$, while $d_{ij} = 0$ for $i \ne j$.

In what follows, an FRA for an rFRSN P system is described.

**FRA**

INPUT: parameter matrices $U$, $V$, $\Lambda$, $H_1$, $H_2$, $\lambda_p$, $\lambda_r$, and initial inputs $\alpha_p^0$, $a_p^0$.

OUTPUT: The fuzzy truth values of propositions associated with the neurons in $O$.

Step (1) Let $\alpha_r^0 = (0, 0, \ldots, 0)^T$, $a_r^0 = (0, 0, \ldots, 0)^T$.

Step (2) Let $t = 0$.

Step (3)

(1) Process the firing of proposition neurons.
$\beta_p^t = \text{fire}(\alpha_p^t, a_p^t, \lambda_p)$, $b_p^t = \text{fire}(1, a_p^t, \lambda_p)$, $\alpha_p^t = \text{update}(\alpha_p^t, a_p^t, \lambda_p)$, $a_p^t = \text{update}(a_p^t, a_p^t, \lambda_p)$, $B_p^t = \text{diag}(b_p^t)$.

(2) Compute the truth values of rule neurons and the number of received spikes.
$\alpha_r^{t+1} = \alpha_r^t \oplus [(H_1 \cdot ((B_p^t \cdot U)^T \odot \beta_p^t)) + (H_2 \cdot ((B_p^t \cdot U)^T \otimes \beta_p^t)]$,
$a_r^{t+1} = a_r^t + [(B_p^t \cdot U)^T \cdot b_p^t]$.

(3) Process the firing of rule neurons.
$\beta_r^{t+1} = \text{fire}(\Lambda \cdot \alpha_r^{t+1}, a_r^{t+1}, \lambda_r)$, $b_r^{t+1} = \text{fire}(1, a_r^{t+1}, \lambda_r)$,
$\alpha_r^{t+1} = \text{update}(\alpha_r^{t+1}, a_p^t, \lambda_p)$, $a_r^{t+1} = \text{update}(a_r^{t+1}, a_p^t, \lambda_p)$, $B_r^{t+1} = \text{diag}(b_r^{t+1})$.

(4) Compute the truth values of proposition neurons and the number of received spikes.
$\alpha_p^{t+1} = \alpha_p^t \oplus [(V \cdot B_r^{t+1}) \otimes \beta_r^{t+1}]$, $a_p^{t+1} = a_p^t + [(V \cdot B_r^{t+1}) \cdot b_r^{t+1}]$.

Step (4) If $a_p^{t+1} = (0, 0, \ldots, 0)^T$ and $a_r^{t+1} = (0, 0, \ldots, 0)^T$ (computation halts), the reasoning results are obtained; otherwise, $t = t + 1$, go to Step (3).

### 5.3.2.2 WMBRA

We first introduce some parameter vectors and matrices before a weighted matrix-based reasoning algorithm (WMBRA) is described.

(1) $\boldsymbol{\theta} = (\theta_1, \ldots, \theta_s)^T$ is a real truth value vector of the $s$ proposition neurons, where $\theta_i$ $(1 \leq i \leq s)$ is a real number in $[0, 1]$ representing the potential value contained in the $i$th proposition neuron. If there is not any spike contained in a proposition neuron, its potential value is 0.

(2) $\boldsymbol{\delta} = (\delta_1, \ldots, \delta_t)^T$ is a real truth value vector of the $t$ rule neurons, where $\delta_j$ $(1 \leq j \leq t)$ is a real number $[0, 1]$ representing the potential value contained in the $j$th rule neuron. If there is not any spike contained in a rule neuron, its potential value is 0.

(3) $\boldsymbol{C} = diag(c_1, \ldots, c_t)$ is a diagonal matrix, where $c_j$ $(1 \leq j \leq t)$ is a real number in $[0, 1]$ representing the confidence factor of the $j$th fuzzy production rule,

(4) $\boldsymbol{W}_{r1} = (\omega_{ij})_{s \times t}$ is a synaptic weight matrix representing the directed connection with weights among *proposition neurons* and *general rule neurons*. If there is a directed arc (synapse) from *proposition neuron* $\sigma_i$ to *general rule neuron* $\sigma_j$, then $\omega_{ij}$ is identical to the output weight of synapse $(i, j)$, otherwise, $\omega_{ij} = 0$.

(5) $\boldsymbol{W}_{r2} = (\omega_{ij})_{s \times t}$ is a synaptic weight matrix representing the directed connection with weights among *proposition neurons* and *and rule neurons*. If there is a directed arc (synapse) from *proposition neuron* $\sigma_i$ to *and rule neuron* $\sigma_j$, then $\omega_{ij}$ is identical to the output weight of synapse $(i, j)$, otherwise, $\omega_{ij} = 0$.

(6) $\boldsymbol{W}_{r3} = (\omega_{ij})_{s \times t}$ is a synaptic weight matrix representing the directed connection with weights among *proposition neurons* and *or rule neurons*. If there is a directed arc (synapse) from *proposition neuron* $\sigma_i$ to *or rule neuron* $\sigma_j$, then $\omega_{ij}$ is identical to the output weight of synapse $(i, j)$, otherwise, $\omega_{ij} = 0$.

(7) $\boldsymbol{W}_p = (\omega_{ji})_{t \times s}$ is a synaptic weight matrix representing the directed connection with weights among rule neurons and proposition neurons. If there is a directed arc (synapse) from rule neuron $\sigma_j$ to proposition neuron $\sigma_i$, then $\omega_{ji}$ is identical to the output weight of synapse $(j, i)$, otherwise, $\omega_{ji} = 0$.

(8) $\boldsymbol{\lambda}_p = (\lambda_{p1}, \ldots, \lambda_{ps})^T$ is a threshold vector of the $s$ proposition neurons, where $\lambda_{pi}$ $(1 \leq i \leq s)$ is a real number in $[0, 1)$ representing the firing threshold of the $i$th proposition neuron.

(9) $\boldsymbol{\lambda}_r = (\lambda_{r1}, \ldots, \lambda_{rt})^T$ is a threshold vector of the $t$ rule neurons, where $\lambda_{rj}$ $(1 \leq j \leq t)$ is a real number in $[0, 1)$ representing the firing threshold of the $j$th rule neuron.

Subsequently, we introduce some multiplication operations as follows.

(1) $\otimes$: $\boldsymbol{W}_{rl}^T \otimes \boldsymbol{\theta} = (\bar{\omega}_1, \ldots, \bar{\omega}_t)^T$, where $\bar{\omega}_j = \omega_{1j} * \theta_1 + \cdots + \omega_{sj} * \theta_s$, $1 \leq j \leq t$, $1 \leq l \leq 3$.

(2) $\oplus$: $\boldsymbol{W}_{rl}^T \oplus \boldsymbol{\theta} = (\bar{\omega}_1, \ldots, \bar{\omega}_t)^T$, where $\bar{\omega}_j = (\omega_{1j} * \theta_1 + \cdots + \omega_{sj} * \theta_s)/(\omega_{1j} + \cdots + \omega_{sj})$, $1 \leq j \leq t$, $1 \leq l \leq 3$.

---

**Algorithm** WMBRA

---

**Input:** $W_{r1}, W_{r2}, W_{r3}, W_p, \lambda_p, \lambda_r, \mathbf{C}, \theta_0, \delta_0$
1: Set the termination condition $\theta = (0, \ldots, 0)_t^T$
2: Let $g = 0$, where $g$ represents the reasoning step
3: **while** $\delta_g \neq \mathbf{0}$ **do**
4:   **for** each input neuron ($g = 0$) or each proposition neuron ($g > 0$) **do**
5:     **if** the firing condition $E = \{a^n, \theta_i \geq \lambda_{pi}, 1 \leq i \leq s\}$ is satisfied **then**
6:       the neuron fires and compute the real truth value vector $\delta_{g+1}$ via $\delta_{g+1} = (\mathbf{W}_{r1}^T \otimes \theta_g) + (\mathbf{W}_{r2}^T \oplus \theta_g) + (\mathbf{W}_{r3}^T \odot \theta_g)$
7:       **if** there is a postsynaptic rule neuron **then**
8:         the neuron transmits a spike to the next rule neuron
9:       **else**
10:          just accumulate the value in the neuron
11:       **end if**
12:     **end if**
13:   **end for**
14:   **for** each rule neuron **do**
15:     **if** the firing condition $E = \{a^n, \delta_j \geq \lambda_{rj}, 1 \leq j \leq t\}$ is satisfied **then**
16:       the rule neuron fires and computes the real truth value vector $\theta_{g+1}$ via $\theta_{g+1} = \mathbf{W}_p^T \odot (\mathbf{C} \otimes \delta_{g+1})$ and transmits a spike to the next proposition neuron
17:     **end if**
18:     $g = g + 1$
19:   **end for**
20: **end while**
**Output:** $\theta_g$, which represents the final states of pulse values contained in proposition neurons.

---

(3) $\odot$: $\mathbf{W}_{rl}^T \odot \theta = (\bar{\omega}_1, \ldots, \bar{\omega}_t)^T$, where $\bar{\omega}_j = max\{\omega_{1j} * \theta_1, \ldots, \omega_{sj} * \theta_s\}$, $1 \leq j \leq t, 1 \leq l \leq 3$. Likewise, $\mathbf{W}_p^T \odot \delta = (\bar{\omega}_1, \ldots, \bar{\omega}_s)^T$, where $\bar{\omega}_i = max \{\omega_{1i} * \delta_1, \ldots, \omega_{ti} * \delta_t\}$, $1 \leq i \leq s$.

Next, we list the pseudocode of WMBRA.

### 5.3.2.3   MBFRA

To adapt tFRSN P systems to solve fault diagnosis problems, we describe MBFRA below.

Given initial truth values of propositions corresponding to all input neurons in a tFRSN P system, MBFRA can perform fuzzy reasoning to obtain the fuzzy truth values of other neurons with unknown pulse values and output reasoning results. Let us assume that the tFRSN P system contains $l$ proposition neurons and $n$ rule neurons, each of which may be *general, and* or *or* rule neurons, with $m = l + n$, where $m$ is the number of all the neurons in this system.

In order to clearly present the reasoning algorithm, we first introduce some parameter vectors and matrices as follows.

(1) $\theta = (\theta_1, \ldots, \theta_l)^T$ is a fuzzy truth value vector of the $l$ proposition neurons, where $\theta_i$ represents the pulse value contained in the $i$th proposition neuron, $1 \leq i \leq l$,

and is expressed by a trapezoidal fuzzy number in [0, 1]. If there is no spike contained in a proposition neuron, its pulse value is *"unknown"* or $(0, 0, 0, 0)$.

(2) $\delta = (\delta_1, \ldots, \delta_n)^T$ is a fuzzy truth value vector of the rule neurons, where $\delta_j$ represents the pulse value contained in the $j$th rule neuron, $1 \leq j \leq n$, and it is expressed by a trapezoidal fuzzy number [0, 1]. If there is no spike contained in a rule neuron, its pulse value is *"unknown"* or $(0, 0, 0, 0)$.

(3) $C = diag(c_1, \ldots, c_n)$ is a diagonal matrix, where $c_j$ is the confidence factor of the $j$th fuzzy production rule, $1 \leq j \leq n$, and it is expressed by a trapezoidal fuzzy number.

(4) $D_1 = (d_{ij})_{l \times n}$ is a synaptic matrix representing the directed connection between proposition neurons and *general* rule neurons. If there is a directed arc (synapse) from the proposition neuron $\sigma_i$ to the *general* rule neuron $\sigma_j$, then $d_{ij} = 1$, otherwise, $d_{ij} = 0$.

(5) $D_2 = (d_{ij})_{l \times n}$ is a synaptic matrix representing the directed connection between proposition neurons and *and* rule neurons. If there is a directed arc (synapse) from the proposition neuron $\sigma_i$ to the *and* rule neuron $\sigma_j$, then $d_{ij} = 1$, otherwise, $d_{ij} = 0$.

(6) $D_3 = (d_{ij})_{l \times n}$ is a synaptic matrix representing the directed connection between proposition neurons and *or* rule neurons. If there is a directed arc (synapse) from the proposition neuron $\sigma_i$ to the *or* rule neuron $\sigma_j$, then $d_{ij} = 1$, otherwise, $d_{ij} = 0$.

(7) $E = (e_{ji})_{n \times l}$ is a synaptic matrix representing the directed connection between rule neurons and proposition rule neurons. If there is a directed arc (synapse) from the rule neuron $\sigma_j$ to the proposition neuron $\sigma_i$, then $e_{ji} = 1$, otherwise, $e_{ji} = 0$.

Subsequently, we introduce some multiplication operations as follows.

(1) $\odot: C \odot \delta = (c_1 \otimes \delta_1, \ldots, c_n \otimes \delta_n)^T; D^T \odot \theta = (\bar{d}_1, \ldots, \bar{d}_n)^T$, where $\bar{d}_j = d_{1j}\theta_1 + \cdots + d_{lj}\theta_l$, $1 \leq j \leq n$.

(2) $\oslash: D^T \oslash \theta = (\bar{d}_1, \ldots, \bar{d}_n)^T$, where $\bar{d}_j = d_{1j}\theta_1 \wedge \cdots \wedge d_{lj}\theta_l$, $1 \leq j \leq n$.

(3) $\circledast: E^T \circledast \delta = (\bar{e}_1, \ldots, \bar{e}_l)^T$, where $\bar{e}_i = e_{1i}\delta_1 \vee \cdots \vee e_{ni}\delta_n$, $1 \leq i \leq l$.

Next, we list the pseudocode of MBFRA.

## 5.4 Fault Diagnosis with Spiking Neural P Systems

This section uses different application examples to detail how to use the three variants of SN P systems discussed in Sect. 5.3 for fault diagnosis. Specifically, fault diagnosis of a transformer, a traction power supply system (TPSS) and a power transmission network, are applied to show the feasibility and effectiveness of rFRSN P systems, WFRSN P systems and tFRSN P systems for solving diagnosis problems, respectively.

---

**Algorithm** MBFRA

---

**Input:** $D_1, D_2, D_3, \mathbf{E}, \mathbf{C}, \theta_0, \delta_0$

1: Set the termination condition $\theta = (unknown, \ldots, unknown)_n^T$
2: Let $t = 0$, where $t$ represents the reasoning step
3: **while** $\delta_t \neq \theta$ **do**
4:     **for** each input neuron ($t = 0$) or each proposition neuron ($t > 0$) **do**
5:         **if** the firing condition $E = a^s$ is satisfied **then**
6:             the neuron fires and computes the fuzzy truth value vector $\delta_{t+1}$ via $\delta_{t+1} = (\mathbf{D}_1^T \odot \theta_t) \oplus$
             $(\mathbf{D}_2^T \odot \theta_t) \oplus (\mathbf{D}_3^T \circledast \theta_t)$
7:             **if** there is a postsynaptic rule neuron **then**
8:                 the neuron transmits a spike to the next rule neuron
9:             **else**
10:                 just accumulate the value in the neuron
11:             **end if**
12:         **end if**
13:     **end for**
14:     **for** each rule neuron **do**
15:         **if** the firing condition $E = a^s$ is satisfied **then**
16:             the rule neuron fires and computes the fuzzy truth value vector $\theta_{t+1}$ via $\theta_{t+1} = \mathbf{E}^T \circledast$
             $(\mathbf{C} \odot \delta_{t+1})$ and transmits a spike to the next proposition neuron
17:         **end if**
18:         $t = t + 1$
19:     **end for**
20: **end while**

**Output:** $\theta_t$, which represents the final states of pulse contained in proposition neurons.

---

### 5.4.1   Transformer Fault Diagnosis with rFRSN P Systems

This subsection uses rFRSN P systems to diagnose the faults in transformers [27]. We first present rFRSN P systems for fuzzy production rules and then discuss the details on the fault diagnosis of transformers.

#### 5.4.1.1   Fuzzy Production Rules

In order to use fuzzy production rules for fault diagnosis of transformers, we need to map them into rFRSN P systems. The basic principle is described as follows. We use a proposition neuron to represent a proposition in the fuzzy diagnosis knowledge base and use a rule neuron (AND-type neuron or OR-type neuron) to a fuzzy production rule. At the beginning, each input proposition neuron has only one spike and its pulse value is assigned to the fuzzy truth value of the proposition associated with it. Then, value $\tau_i$ of each rule neuron is assigned to the confidence factor of the fuzzy production rule associated with it.

According to the above principle, a *simple fuzzy production rule* (5.2) can be modeled by an rFRSN P system $\Pi_1$, as shown in Fig. 5.14.

**Fig. 5.14** An rFRSN P system $\Pi_1$ for simple fuzzy production rules

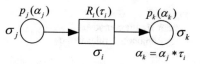

**Fig. 5.15** An rFRSN P system $\Pi_2$ for composite fuzzy conjunctive rules in the antecedent

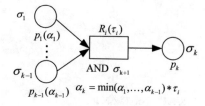

$\Pi_1 = (A, \sigma_i, \sigma_j, \sigma_k, syn, I, O)$, where

(1) $A = \{a\}$
(2) $\sigma_i$ is a rule neuron associated with rule $R_i$ with confidence factor $\tau_i$. Its spiking rule is of the form $E/a^\alpha \rightarrow a^\beta$, where $\beta = \alpha * \tau_i$.
(3) $\sigma_j$ and $\sigma_k$ are two proposition neurons associated with propositions $p_j$ and $p_k$ with truth values $\alpha_j$ and $\alpha_k$ respectively. Their spiking rules are of the form $E/a^\alpha \rightarrow a^\alpha$.
(4) $syn = \{(j, i), (i, k)\}, I = \{\sigma_j\}, O = \{\sigma_k\}$.

A *composite fuzzy conjunctive rule in the antecedent* (5.3) can be modeled by an rFRSN P system $\Pi_2$, as shown in Fig. 5.15.

$\Pi_2 = (A, \sigma_1, \sigma_2, \ldots, \sigma_k, \sigma_{k+1}, syn, I, O)$, where

(1) $A = \{a\}$
(2) $\sigma_j$ ($1 \leq j \leq k$) are proposition neurons associated with propositions $p_j$ ($1 \leq j \leq k$) with truth values $\alpha_j$ ($1 \leq j \leq k$) respectively. Their spiking rules are of the form $E/a^\alpha \rightarrow a^\alpha$.
(3) $\sigma_{k+1}$ is an "AND"-type rule neuron associated with rule $R_i$ with confidence factor $\tau_i$. Its spiking rule is of the form $E/a^\alpha \rightarrow a^\beta$, where $\beta = \alpha * \tau_i$.
(4) $syn = \{(1, k + 1), (2, k + 1), \ldots, (k - 1, k + 1), (k + 1, k)\}$.
(5) $I = \{\sigma_1, \ldots, \sigma_{k-1}\}, O = \{\sigma_k\}$.

A *composite fuzzy conjunctive rule in the consequent* (5.4) can be modeled by an rFRSN P system $\Pi_3$, as shown in Fig. 5.16.

$\Pi_3 = (A, \sigma_1, \ldots, \sigma_{k+1}, syn, I, O)$, where

(1) $A = \{a\}$
(2) $\sigma_j$ ($1 \leq j \leq k$) are proposition neurons associated with propositions $p_j$ ($1 \leq j \leq k$) with truth values $\alpha_j$ ($1 \leq j \leq k$), respectively. Their spiking rules are of the form $E/a^\alpha \rightarrow a^\alpha$.
(3) $\sigma_{k+1}$ is a rule neuron associated with rule $R_i$ with confidence factor $\tau_i$. Its spiking rule is of the form $E/a^\alpha \rightarrow a^\beta$, where $\beta = \alpha * \tau_i$.

**Fig. 5.16** An rFRSN P
system $\Pi_3$ for composite
fuzzy conjunctive rules in the
consequent

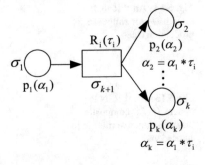

**Fig. 5.17** An rFRSN P
system $\Pi_4$ for composite
fuzzy disjunction rules in the
antecedent

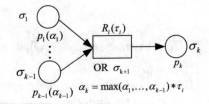

(4) $syn = \{(1, k + 1), (k + 1, 2), (k + 1, 3), \ldots, (k + 1, k)\}$.
(5) $I = \{\sigma_1\}, O = \{\sigma_2, \ldots, \sigma_k\}$.

A *composite fuzzy disjunction rule in the antecedent* (5.5) can be modeled by an
rFRSN P system $\Pi_4$, as shown in Fig. 5.17.
   $\Pi_4 = (A, \sigma_1, \ldots, \sigma_{k+1}\}$, where

(1) $A = \{a\}$
(2) $\sigma_j$ $(1 \leq j \leq k)$ are proposition neurons associated with propositions $p_j$ $(1 \leq j \leq k)$ with truth values $\alpha_j$ $(1 \leq j \leq k)$, respectively. Their spiking rules are of the
   form $E/a^\alpha \rightarrow a^\alpha$.
(3) $\sigma_{k+1}$ is an "OR"-type rule neuron associated with rule $R_i$ with confidence factor
   $\tau_i$. Its spiking rule is of the form $E/a^\alpha \rightarrow a^\beta$, where $\beta = \alpha * \tau_i$.
(4) $syn = \{(1, k + 1), (2, k + 1), \ldots, (k - 1, k + 1), (k + 1, k)\}$.
(5) $I = \{\sigma_1, \ldots, \sigma_{k-1}\}, O = \{\sigma_k\}$.

### 5.4.1.2   Example

In this subsection, an application example in [27] is employed to show how to use
rFRSN P systems and their FRA to diagnose faults in a transformer. The following
fuzzy production rules are obtained from the knowledge base of a transformer fault
diagnosis system.

*Rule 1 (CF = 0.8)*
   *Symptom*:
   (1) Total hydrocarbon is little high ($p_1$);
   (2) $C_2H_2$ is low ($p_2$);
   *Anticipated Fault*: General overheating fault occurs ($p_{11}$).
*Rule 2 (CF = 0.8)*
   (1) Total hydrocarbon is rather high ($p_3$);
   (2) $C_2H_2$ is too high ($p_4$);
   (3) $H_2$ is high ($p_5$);
   (4) $C_2H_2$ in total hydrocarbon occupies a too low proportion ($p_6$);
   *Anticipated Fault*: Serious overheating fault occurs ($p_{11}$).
*Rule 3 (CF = 0.8)*
   (1) Total hydrocarbon is little low ($p_7$);
   (2) $H_2$ is high ($p_5$);
   (3) $CH_4$ in total hydrocarbon occupies a large proportion ($p_8$);
   (4) $CH_4$ in total hydrocarbon occupies a higher proportion than $C_2H_2$ ($p_9$);
   *Anticipated Fault*: The partial discharge occurs ($p_{13}$).
*Rule 4 (CF = 0.8)*
   (1) Total hydrocarbon is rather low ($p_{10}$);
   (2) $C_2H_2$ is too high ($p_4$);
   (3) $H_2$ is high ($p_5$);
   *Anticipated Fault*: The spark discharge occurs ($p_{14}$).

These fuzzy production rules can be modeled by the following rFRSN P system $\Pi_5$, as shown in Fig. 5.18.
   $\Pi_5 = (A, \sigma_1, \ldots, \sigma_{14}, \sigma_{15}, \ldots, \sigma_{18}, syn, I, O)$, where

(1) $A = \{a\}$.
(2) $\sigma_1, \ldots, \sigma_{14}$ are proposition neurons associated with propositions $p_1, \ldots, p_{14}$, respectively.

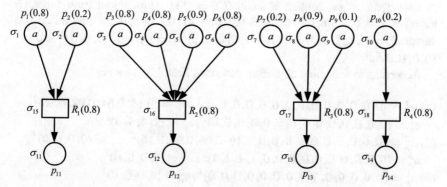

**Fig. 5.18** An example of a transformer fault diagnosis modeled by an rFRSN P system model $\Pi_5$

(3) $\sigma_{15}, \ldots, \sigma_{18}$ are AND-type rule neurons associated with production rules $R_1, \ldots, R_4$, respectively.

(4) $syn = \{(1, 15), (2, 15), (3, 16), (4, 16), (4, 18), (5, 16), (5, 17), (5, 18),$
$(6, 18), (7, 17), (8, 17), (9, 17), (10, 18), (15, 11), (16, 12), (17, 13),$
$(18, 14)\}$.

(5) $I = \{\sigma_1, \sigma_2, \sigma_3, \sigma_4, \sigma_5, \sigma_6, \sigma_7, \sigma_8, \sigma_9, \sigma_{10}\}$, $O = \{\sigma_{11}, \sigma_{12}, \sigma_{13}, \sigma_{14}\}$.

According to the definition of parameter vectors and matrices given in Sect. 1.3.2, $U, V, \Lambda, H_1$ and $H_2$ are follows:

$$U = \begin{bmatrix} 1\,1\,0\,0\,0\,0\,0\,0\,0\,0\,0\,0\,0\,0 \\ 0\,0\,1\,1\,1\,1\,0\,0\,0\,0\,0\,0\,0\,0 \\ 0\,0\,0\,0\,1\,0\,1\,1\,1\,0\,0\,0\,0\,0 \\ 0\,0\,0\,1\,1\,0\,0\,0\,0\,1\,0\,0\,0\,0 \end{bmatrix}^T \qquad H_1 = \begin{bmatrix} 1\,0\,0\,0 \\ 0\,1\,0\,0 \\ 0\,0\,1\,0 \\ 0\,0\,0\,1 \end{bmatrix}$$

$$V = \begin{bmatrix} 0\,0\,0\,0\,0\,0\,0\,0\,0\,0\,1\,0\,0\,0 \\ 0\,0\,0\,0\,0\,0\,0\,0\,0\,0\,0\,1\,0\,0 \\ 0\,0\,0\,0\,0\,0\,0\,0\,0\,0\,0\,0\,1\,0 \\ 0\,0\,0\,0\,0\,0\,0\,0\,0\,0\,0\,0\,0\,1 \end{bmatrix}^T \qquad H_2 = \begin{bmatrix} 0\,0\,0\,0 \\ 0\,0\,0\,0 \\ 0\,0\,0\,0 \\ 0\,0\,0\,0 \end{bmatrix}$$

$$\Lambda = \begin{bmatrix} 0.8 & 0 & 0 & 0 \\ 0 & 0.8 & 0 & 0 \\ 0 & 0 & 0.8 & 0 \\ 0 & 0 & 0 & 0.8 \end{bmatrix}$$

$$\lambda_p = (1, 1, 1, 1, 1, 1, 1, 1, 1, 1, 1, 1, 1, 1)^T \qquad \lambda_r = (2, 4, 4, 3)^T$$

In on-scene information detection of transformer, total hydrocarbon content is high (CF = 0.8), $C_2H_2$ content is high (CF = 0.8), $H_2$ content is high (CF = 0.9), $C_2H_2$ content in total hydrocarbon content is high (CF = 0.8), $CH_4$ content in total hydrocarbon content is small (CF = 0.1). Thus, initial truth value vector $\alpha_p^0 = (0.8, 0.2, 0.8, 0.8, 0.9, 0.8, 0.2, 0.9, 0.1, 0.2, 0, 0, 0, 0)^T$ and initial spike vector $a_p^0 = (1, 1, 1, 1, 1, 1, 1, 1, 1, 1, 0, 0, 0, 0)^T$. Let $\alpha_r^0 = (0, 0, 0, 0)^T$ and $a_r^0 = (0, 0, 0, 0)^T$.

According to reasoning algorithm described Sect. 1.3.2, we get

(1) $\alpha_p^1 = (0, 0, 0, 0, 0, 0, 0, 0, 0, 0, 0, 0, 0, 0)^T$, $\alpha_r^1 = (0.16, 0.64, 0.08, 0.16)^T$,
$a_p^1 = (0, 0, 0, 0, 0, 0, 0, 0, 0, 0, 0, 0, 0, 0)^T$, $a_r^1 = (2, 4, 4, 3)^T$;

(2) $\alpha_p^2 = (0, 0, 0, 0, 0, 0, 0, 0, 0, 0, 0.16, 0.64, 0.08, 0.16)^T$, $\alpha_r^2 = (0, 0, 0, 0)^T$,
$a_p^2 = (0, 0, 0, 0, 0, 0, 0, 0, 0, 0, 1, 1, 1, 1)^T$, $a_r^2 = (0, 0, 0, 0)^T$;

(3) $a_p^3 = (0, 0, 0, 0, 0, 0, 0, 0, 0, 0, 0, 0, 0, 0)^T$, $a_r^3 = (0, 0, 0, 0)^T$.

The computation of this system reaches halting condition ($a_p^3 = (0, 0, 0, 0, 0, 0, 0, 0, 0, 0, 0, 0, 0)^T$ and $a_r^3 = (0, 0, 0, 0)^T$), so the system outputs its reasoning results, namely, the truth values of propositions $p_{11}, p_{12}, p_{13}$ and $p_{14}$ are 0.16, 0.64, 0.08 and 0.16, respectively. According to these reasoning results, there are possible faults: general overheating fault (CF = 0.16), serious overheating fault (CF = 0.64), partial discharge (CF = 0.08) and spark discharge (CF = 0.16). The threshold value of fault occurrence is set to 0.6 in the fault diagnosis system. So the conclusion that the transformer shows a serious overheating fault is consistent with the actual situation.

## 5.4.2 Traction Power Supply Systems Fault Diagnosis with WFRSN P Systems

This subsection first presents WFRSN P system models for fault diagnosis production rules in TPSSs. Then how to build WFRSN P system fault diagnosis models for sections and how to use WFRSN P systems to identify fault sections for feeding sections, are described in detail. Finally, three cases from a local system of a TPSS are considered as application examples to show the effectiveness of WFRSN P systems in fault diagnosis.

### 5.4.2.1 WFRSN P System Models for Fault Diagnosis Production Rules in TPSSs

In the following description, fault diagnosis production rules in TPSSs and their WFRSN P system models in [40], which is shown in Fig. 5.19, are presented.

*Type 1 (Simple Rules)* $R_i$: IF $p_j(\theta_j)$ THEN $p_k(\theta_k)$ (CF = $c_i$), where $p_j$ and $p_k$ are propositions, $c_i$ is a real number in [0, 1] representing the confidence factor of rule $R_i$, $\theta_j$ and $\theta_k$ are real numbers in [0, 1] representing the truth values of $p_j$ and $p_k$, respectively. The weight of proposition $p_j$ is $\omega_j$, where $\omega_j = 1$ because there is only one proposition in the antecedent of this kind of rules. Then the truth values of $p_k$ is $\theta_k = \theta_j * \omega_j * c_i = \theta_j * c_i$.

*Type 2 (Compound And Rules)* $R_i$: IF $p_1(\theta_1)$ and ... and $p_{k-1}(\theta_{k-1})$ THEN $p_k(\theta_k)$ (CF = $c_i$), where $p_1, \ldots, p_k$ are propositions, $c_i$ is a real number in [0, 1] representing the confidence factor of rule $R_i$, $\theta_1, \ldots, \theta_k$ are real numbers in [0, 1] representing the truth values of $p_1, \ldots, p_k$, respectively. Then the weights of propositions $p_1, \ldots, p_{k-1}$ are $\omega_1, \ldots, \omega_{k-1}$, respectively. The truth values of $p_k$ is $\theta_k = [(\theta_1 * \omega_1 + \cdots + \theta_{k-1} * \omega_{k-1})/(\omega_1 + \cdots + \omega_{k-1})] * c_i$.

*Type 3 (Compound Or Rules)* $R_i$: IF $p_1(\theta_1)$ or ... or $p_{k-1}(\theta_{k-1})$ THEN $p_k(\theta_k)$ (CF = $c_i$), where $p_1, \ldots, p_k$ are propositions, $c_i$ is a real number in [0, 1] representing the truth values of $p_1, \ldots, p_k$, respectively. The weights of propositions $p_1, \ldots, p_{k-1}$ are

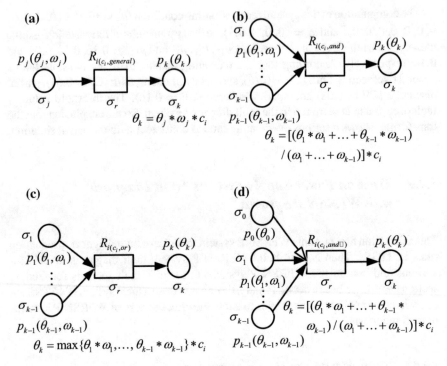

**Fig. 5.19** WFRSN P system models for fault diagnosis production rules in TPSSs. **a** *Type 1*;
**b** *Type 2*; **c** *Type 3*; **d** *Type4*

$\omega_1, \ldots, \omega_{k-1}$, respectively. The truth values of $p_k$ is $\theta_k = max\{\theta_1 * \omega_1, \ldots, \theta_{k-1} * \omega_{k-1}\} * c_i$.

*Type 4* (*Conditional And Rules*) $R_i$: WHEN $p_0(\theta_0)$ is true, IF $p_1(\theta_1)$ and $\ldots$
and $p_{k-1}(\theta_{k-1})$ THEN $p_k(\theta_k)$ (CF $= c_i$), where $p_0, \ldots, p_k$ are propositions, $c_i$ is a
real number in [0, 1] representing the confidence factor of rule $R_i$, $\theta_0, \ldots, \theta_k$ are real
numbers in [0, 1] representing the truth values of $p_0, \ldots, p_k$, respectively. Proposition
$p_0$ is used to judge whether the reasoning condition of rule $R_i$ is satisfied and its truth
value $\theta_0$ is not used in reasoning process. Thus, the weight of $\theta_0$ is not considered in the
model. The weights of propositions $p_1, \ldots, p_{k-1}$ are $\omega_1, \ldots, \omega_{k-1}$, respectively. Then
the truth values of $p_k$ is $\theta_k = [(\theta_1 * \omega_1 + \cdots + \theta_{k-1} * \omega_{k-1})/(\omega_1 + \cdots + \omega_{k-1})] * c_i$.

### 5.4.2.2   WFRSN P System Fault Diagnosis Models for Sections

In what follows, we describe the diagnosis models for sections by using WFRSNP
systems [40]. When a WFRSN P system is used to build a fault diagnosis model,
it is necessary to intuitively describe the causality between a fault and the statuses of its protective devices. Furthermore, the model has to consider all kinds of

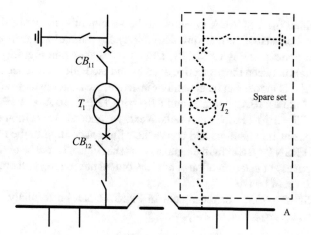

**Fig. 5.20** Single line diagram of bus A in a TSS

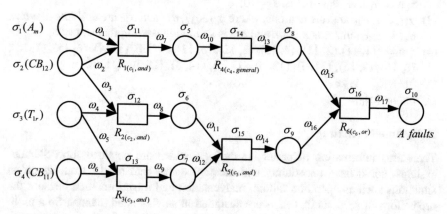

**Fig. 5.21** A WFRSN P system fault diagnosis model for bus A

protective devices including main protective relays, backup protective relays and their corresponding CBs of a faulty section. In the sequel, bus $A$ shown in Fig. 5.20 in a TPSS, and its WFRSN P system fault diagnosis model shown in Fig. 5.21 are used as examples to show the model built process and parameter setting. In Figs. 5.20 and 5.21, $T$ is a transformer; dotted line part is a spare section set; $m, p, r$ are main protection, primary backup protection and remote backup protection, respectively.

### A. Model Built Process

If a fault occurs on a section in a TPSS, the statuses of its protective devices will soon change so as to protect the system. Thus, we can use the observed status information, protective relay operation information and circuit breaker tripping information as the inputs of the WFRSN P system fault diagnosis model of the section. The information can be obtained from SCADA systems. To be specific, the inputs of the diagnosis model of bus $A$ consist of the main protective relay $A_m$, remote backup protective relay $T_{1r}$ and their corresponding CBs, $CB_{11}$ and $CB_{12}$, as shown in Fig. 5.21. According to relationships between the protective devices and the fault occurrence on bus $A$, the other parts of this model can be given. For instance, the relationships with respect to bus $A$ are as follows: IF $A_m$ operates and $CB_{12}$ trips THEN bus $A$ fails; IF $T_{1r}$ operates and $(CB_{11}, CB_{12}$ trip) THEN bus $A$ fails. Next, we choose proposition neurons and different types of rule neurons and create their links according to the relationships. Thus, the WFRSN P system fault diagnosis model in Fig. 5.21 can be obtained. When WMBRA arrives at the termination condition, output neuron $\sigma_{10}$ will output the fault confidence level of bus $A$.

A WFRSN P system for the model in Fig. 5.21 can be formally described as $\Pi_6 = (O, \sigma_1, \ldots, \sigma_{16}, syn, in, out)$, where:

(1) $O = \{a\}$ is the singleton alphabet ($a$ is called spike);
(2) $\sigma_1, \ldots, \sigma_{10}$ are *proposition neurons* corresponding to the propositions with truth values $\theta_1, \ldots, \theta_{10}$; that is, $s = 10$;
(3) $\sigma_{11}, \ldots, \sigma_{16}$ are rule neurons, where $\sigma_{11}, \sigma_{12}, \sigma_{13}$ and $\sigma_{15}$ are *and rule neurons*, $\sigma_{14}$ is a *general rule neuron* and $\sigma_{16}$ is an *or rule neuron*; that is, $t = 6$;
(4) $syn = \{(1, 11), (2, 11), (2, 12), (3, 12), (3, 13), (4, 13), (5, 14), (6, 15), (7, 15), (8, 16), (9, 16), (11, 5), (12, 6), (13, 7), (14, 8), (15, 9), (16, 10)\}$;
(5) $in = \{\sigma_1, \ldots, \sigma_4\}$;
(6) $out = \{\sigma_{10}\}$.

## B. Parameters Setting

The status information of protective devices, which are obtained form SCADA systems, could have uncertainty and incompleteness that results from abnormal situations such as operation failure, malversation and misinformation because the protections of sections in TPSSs are designed in single-ended manner. So a probability value is required to describe the operation confidence level of each section. The operation confidence levels of these protective devices are set the same as those in [34, 44, 46] so that the generality of the reliability of protective relays and CBs in TPSSs and ordinary power systems can be considered. The confidence levels of operated protective devices and non-operate protective devices are shown in Table 5.1, where $FL$, $B$ and $T$ represent the feeder line, bus and transformer, respectively.

At the initial state, each input neuron of a WFRSN P system fault diagnosis model has only one spike with a real number assigned. The number equals the confidence level of the protective device associated with this input neuron. The other neurons in the model do not have spikes and consequently their pulse values are 0. For instance, if bus $A_m$ and $CB_{12}$ operate and $T_{1r}$, $CB_{11}$ do not operate in Fig. 5.21, the spikes contained in $\sigma_1, \ldots, \sigma_4$ are assigned the values of 0.8564, 0.9833, 0.4 and 0.2, respectively; while the pulse values of $\sigma_5, \ldots, \sigma_{16}$ are 0.

A truth value, which represents the confidence factor of the fault diagnosis production rule associated with this rule neuron is assigned to each rule neuron of a

**Table 5.1** Operation and non-operation confidence levels of the protective devices

| Sections | Protective devices (operated) | | | | | | Protective devices (non-operated) | | | | | |
|---|---|---|---|---|---|---|---|---|---|---|---|---|
| | Main | | Primary backup | | Remote backup | | Main | | Primary backup | | Remote backup | |
| | Relays | CBs | Relays | CBs | Relays | CBs | Relays | CBs | Relays | CBs | Relays | CBs |
| $FL$ | 0.9913 | 0.9833 | 0.8 | 0.85 | 0.7 | 0.75 | 0.2 | 0.2 | 0.2 | 0.2 | 0.2 | 0.2 |
| $B$ | 0.8564 | 0.9833 | – | – | 0.7 | 0.75 | 0.4 | 0.2 | – | – | 0.4 | 0.2 |
| $T$ | 0.7756 | 0.9833 | 0.75 | 0.8 | 0.7 | 0.75 | 0.4 | 0.2 | 0.4 | 0.2 | 0.4 | 0.2 |

WFRSN P system fault diagnosis model. As usual, a main protection has a higher reliability than a primary backup protection and a primary backup protection has a higher reliability than a remote backup protection. So we set the truth values of neurons associated with main, primary backup and remote backup protections to 0.975, 0.95 and 0.9, respectively. It is noting that the truth values of *or rule neurons* are set according to their highest protection. In Fig. 5.21, the truth values of $\sigma_{11}, \ldots, \sigma_{16}$ are set to 0.975, 0.9, 0.9, 0.975, 0.9 and 0.975, respectively. The output weights of proposition neurons associated with protective relays and CBs are set as the same value 0.5 because both the protective relay operation information and circuit breaker tripping information is important to a fault diagnosis production rule. If a neuron has only one presynaptic neuron, the output weight of its presynaptic neuron is set to 1. The weight of a protection type is also set to 1. The weights $\omega_1, \ldots, \omega_{17}$ in Fig. 5.21 are set to 0.5, 0.5, 0.5, 0.5, 0.5, 0.5, 1, 1, 1, 1, 0.5, 0.5, 1, 1, 1, 1 and 1, respectively. Since the firing threshold value of each neuron in a WFRSN P system fault diagnosis model is smaller than the minimum pulse value appeared in the neurons in the whole reasoning process, the firing threshold value of each neuron is set to 0.1, according to Table 5.1 and the operation of pulse values in different types of neurons.

### 5.4.2.3 Fault Region Identification for Feeding Sections

Lines in section posts (SPs) are connected in an up and down line paralleling manner in a TSS-ATP-SP feeding section (FS). So when confirmed faults occur in a feeding section, one important task of fault diagnosis for traction power supply systems is to identify fault regions (which parts fail) in FSs. Figure 5.22 shows a single line diagram of a TSS-ATP-SP feeding section and its WFRSN P system fault diagnosis model for fault region identification is shown in Fig. 5.23, where neurons $\sigma_1$ and $\sigma_2$ are associated with the propositions that current directions of $I_{34}$ and $I_{35}$ are positive, respectively; neuron $\sigma_3$ is associated with the proposition that current is detected in SP2; neurons $\sigma_4$ and $\sigma_5$ are associated with the propositions that current directions of $I_{42}$ and $I_{43}$ are negative; a small circle on an arrow tip represents an inverse proposition associated with its presynaptic neuron; a hollow tip represents an assistant synapse, i.e., the proposition associated with its presynaptic neuron is used as a judgement condition; output neuron $\sigma_6$ is associated with the proposition being the first part

**Fig. 5.22** Single line diagram of a TSS-ATP-SP feeding section

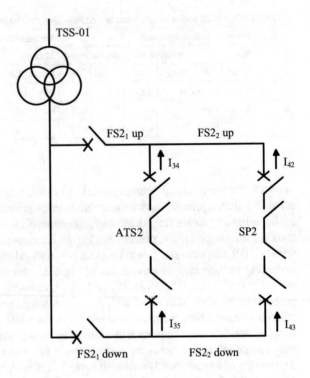

**Fig. 5.23** A WFRSN P system fault diagnosis model for fault region identification of a feeding section

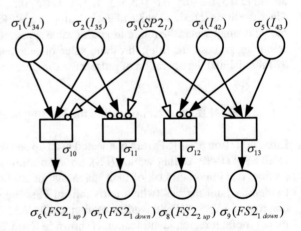

of the up direction feeding section in FS2, i.e., $FS2_{1up}$ has a fault. The meanings of output neurons $\sigma_7$ and $\sigma_9$ are similar. Here, clockwise direction is the positive current direction while counter-clockwise direction is the negative one.

Figures 5.22 and 5.23 show a typical feeding section and its WFRSN P system fault diagnosis model for fault region identification. The models for other feeding sections can be built in a similar way. Causality between currents detected and fault regions is described by a WFRSN P system fault diagnosis model to get the fault regions of feeding sections and no numerical calculation is involved in this identification process. Thus, parameter setting of WFRSN P system fault diagnosis models for fault region identification of feeding sections is not considered.

#### 5.4.2.4   Examples

In this subsection, an example in [40] is described as follows to show the application of WFRSN P systems in fault diagnosis. We consider three cases from the local system of a TPSS in Fig. 5.24 [44]. The first two cases are in normal power supply and the third case is in over zone feeding. The external transmission lines in a power system supplying the TPSS are hypothetical. In Fig. 5.24, $S$ and $R$ are the sending and receiving ends of transmission lines; $L$ is the transmission line. One can note that the complete line connection of FS1, ATP1, SP1, FS3, ATP3 and TPS-02 is the same as that of TSS-01, FS2, SP2 and ATP2 in Fig. 5.24.

*Case 1: normal power supply. $FS2_{1up}$ and AT1 have faults.*
Status information from the SCADA system (in time order): $AT1_m$ operated, $CB_{31}$ tripped, $AT3$ auto switched over; $FS2_m$ operated, $CB_{23}$ and $CB_{24}$ tripped; feeder

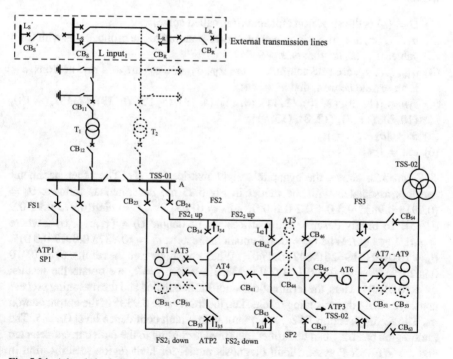

**Fig. 5.24** A local single line sketch map of a TPSS

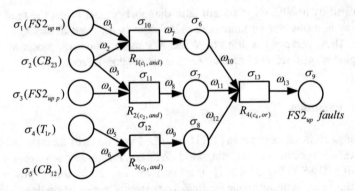

**Fig. 5.25**  A WFRSN P system fault diagnosis model for $FS2_{up}$

lines auto reclosed, $FS2_{up\ m}$ operated quickly, $CB_{23}$ tripped again. When faults occur, current directions of $I_{34}$ and $I_{35}$ are positive, and current is not detected in SP2.

A WFRSN P system $\Pi_2$ (its corresponding WFRSN P system fault diagnosis model is shown in Fig. 5.25) for $FS2_{up}$ is defined as follows:

$\Pi_2 = (O, \sigma_1, \ldots, \sigma_{16}, syn, in, out)$,

where:

(1) $O = \{a\}$ is the singleton alphabet ($a$ is called spike);
(2) $\sigma_1, \ldots, \sigma_9$ are proposition neurons corresponding to the propositions with truth values $\theta_1, \ldots, \theta_9$; that is, $s = 9$;
(3) $\sigma_{10}, \ldots, \sigma_{13}$ are rule neurons, where $\sigma_{10}, \sigma_{11}$ and $\sigma_{12}$ are *and* rule neurons, $\sigma_{14}$ is an *or* rule neuron; that is, $t = 4$;
(4) $syn = \{(1, 10), (2, 10), (2, 11), (3, 11), (4, 12), (5, 12), (6, 13), (7, 13), (8, 13), (10, 6), (11, 7), (12, 8), (13, 9)\}$;
(5) $in = \{\sigma_1, \ldots, \sigma_5\}$;
(6) $out = \{\sigma_9\}$.

Figure 5.26 shows the synaptic weight matrices of $\Pi_2$. The other parameter matrices associated with the model in Fig. 5.25 are described as follows: $\theta_0 = (0.9913\ 0.9833\ 0.8\ 0.4\ 0.2\ 0\ 0\ 0\ 0)^T$, $\delta_0 = (0\ 0\ 0\ 0)^T$, $C = diag(0.975\ 0.95\ 0.9\ 0.975)$. To briefly describe the matrices, let us denote $O_l = (x_1, \ldots, x_l)^T$, where $x_i = 0, 1 \le i \le l$. When $g = 0$, we obtain the results: $\delta_1 = (0.9873\ 0.8917\ 0.3\ 0)^T$, $\theta_1 = (0\ 0\ 0\ 0\ 0\ 0.9626\ 0.8471\ 0.27\ 0)^T$. When $g = 1$, we gain the results: $\delta_2 = (0\ 0\ 0\ 0.9626)^T$, $\theta_2 = (0\ 0\ 0\ 0\ 0\ 0\ 0\ 0\ 0.9385)^T$. When $g = 2$, we obtain the results: $\delta_3 = (0\ 0\ 0\ 0)^T$. Thus, the termination condition is satisfied and the reasoning process ends. We obtain the reasoning results, i.e., the truth value 0.9385 of the output neuron $\sigma_9$. The feeding section $FS2_{up}$ has a fault with a fault confidence level 0.9385. The fault region of $FS2_{up}$ can be further identified according to the fault current detected and the WFRSN P system fault diagnosis model for fault region identification in Fig. 5.23, and then we get the result that $FS21_{up}$ has a fault with a fault confidence level 0.9385.

$$W_{r1} = [\mathbf{O}]_{9\times4}, W_{r2} = \begin{bmatrix} 0.5 & 0 & 0 & 0 \\ 0.5 & 0.5 & 0 & 0 \\ 0 & 0.5 & 0 & 0 \\ 0 & 0 & 0.5 & 0 \\ 0 & 0 & 0.5 & 0 \\ 0 & 0 & 0 & 0 \\ 0 & 0 & 0 & 0 \\ 0 & 0 & 0 & 0 \\ 0 & 0 & 0 & 0 \end{bmatrix}, W_{r3} = \begin{bmatrix} 0 & 0 & 0 & 0 \\ 0 & 0 & 0 & 0 \\ 0 & 0 & 0 & 0 \\ 0 & 0 & 0 & 0 \\ 0 & 0 & 0 & 0 \\ 0 & 0 & 0 & 1 \\ 0 & 0 & 0 & 1 \\ 0 & 0 & 0 & 1 \\ 0 & 0 & 0 & 0 \end{bmatrix}, W_p = \begin{bmatrix} 0 & 0 & 0 & 0 & 0 & 1 & 0 & 0 & 0 \\ 0 & 0 & 0 & 0 & 0 & 0 & 1 & 0 & 0 \\ 0 & 0 & 0 & 0 & 0 & 0 & 0 & 1 & 0 \\ 0 & 0 & 0 & 0 & 0 & 0 & 0 & 0 & 1 \end{bmatrix}.$$

**Fig. 5.26** Synaptic weight matrices of WFRSN P system fault diagnosis model for $FS2_{up}$

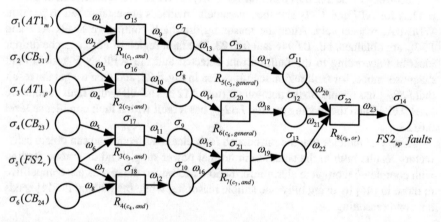

**Fig. 5.27** A WFRSN P system fault diagnosis model for AT1

Similarly, we can design a WFRSN P system for AT1 and its corresponding WFRSN P system fault diagnosis model is shown in Fig. 5.27. The diagnosis process of AT1 is similar to the above description. According to the SCADA data and Table 5.1, the parameter matrices of WFRSN P system fault diagnosis model for AT1 is built to run WMBRA. The fault confidence level 0.8361 of AT1 is finally obtained through the reasoning. Thus, the autotransformer AT1 has a fault with a fault confidence level 0.8361.

*Case 2: normal power supply. $FS2_{1up}$ has faults.*

Status information from the SCADA system (in time order): $FS2_m$ operated, $CB_{24}$ tripped; $T_{1r}$ operated, $CB_{11}$ and $CB_{12}$ tripped. When faults occur, current directions of $I_{34}$ and $I_{35}$ are positive, and the current is not detected in SP2. In this case, $CB_{23}$ refused operation.

According to the SCADA data and Table 5.1, the WFRSN P system fault diagnosis model for $FS2_1$ and its parameter matrices are established to perform WMBRA. Finally, the fault confidence level 0.7439 of $FS2_{up}$ is obtained through reasoning. The fault region of $FS2_{up}$ can be further identified according to the fault current detected and the WFRSN P system fault diagnosis model for fault region identification in Fig. 5.23, and then we get the result that $FS2_{1up}$ has a fault. Thus, the feeding section $FS2_{1up}$ has a fault with a fault confidence level 0.7439.

*Case 3: FS2 is over zone fed by TPS-02. AT7 and $FS2_{2up}$ have faults.*

Status information from the SCADA system (in time order): primary backup protections of feeder lines in SP2 operated, $CB_{42}$ tripped; meanwhile, $CB_{51}$ tripped, AT9 auto switched over; remote backup protection $FS3_s$ of feeder lines in TSS-02 operated, $CB_{63}$ and $CB_{64}$ tripped. When faults occur, current directions of $I_{34}$ and $I_{35}$ are positive, and the current is detected only in SP2 and ATP2. In this case, main protection of feeder lines in SP2, $CB_{43}$ and main protection of AT7 refused operation, and status information of primary backup protection of AT7 lost.

According to the SCADA data and Table 5.1, the WFRSN P system fault diagnosis models for AT7 and $FS2_2$ and their parameter matrices are established to perform WMBRA, respectively. After the reasoning, the fault confidence levels AT7 and $FS2_{up}$ are obtained, i.e., 0.6946 and 0.6123. The fault region of $FS2_{up}$ can be further identified according to the fault current detected and the WFRSN P system fault diagnosis model for fault region identification in Fig. 5.23, and then we get the result that $FS2_{2up}$ has a fault. So the autotransformer AT2 has a fault with a fault confidence level 0.6946 and the feeding section $FS2_{2up}$ has a fault with a fault confidence level 0.6123.

*Cases 1–3* indicate that the introduced fault diagnosis technique can obtain satisfactory results both in the situation in normal power supply and over zone feeding with complete/incomplete alarm information. Moreover, the results are competitive to those in [44] by using only one simple reasoning while the method in [44] needs a second reasoning.

## 5.4.3   Power Transmission Networks Fault Diagnosis with tFRSN P Systems

This subsection first presents fault fuzzy production rule sets for main sections in transmission networks. Second, fault fuzzy productions rules based on tFRSN P systems are described. Third, an algorithmic elaboration of fault diagnosis method based on tFRSN P systems (FDSNP, for short) for power transmission networks is discussed. Finally, seven cases of a local system in an EPS are considered as application examples to show the effectiveness of tFRSN P systems in fault diagnosis.

### 5.4.3.1   Fault Fuzzy Production Rule Sets

In this subsection, fault fuzzy production rule sets for main sections including transmission lines $(L)$, buses $(B)$ and transformers $(T)$ in transmission networks are described to obtain the causality between a fault and the statuses of its protective devices [42]. The fault fuzzy production rules are composed of propositions and confidence factors. A confidence factor represents the confidence degree that a fault appears. Each rule has one confidence factor. According to the characteristic of the uncertainty of the knowledge of experts and senior dispatchers, linguistic terms are used to describe confidence factors.

In what follows, the fault fuzzy production rule sets for three main sections, lines, buses and transformers, are presented. A line has six types of protections: sending end main protections, sending end first backup protections, sending end second backup protections, receiving end main protections, receiving end first backup protections and receiving end second backup protections. The fault fuzzy production rule set for lines consist of nine rules. Table 5.2 shows the meaning of each proposition.

$$R_1(c_1 = AH) : (p_1 \text{ and } p_2 \text{ operate}) \otimes (CB_1 \text{ and } CB_2 \text{ trip}) \rightarrow L \text{ fails}$$
$$R_2(c_2 = AH) : (p_1 \text{ and } p_4 \text{ operate}) \otimes (CB_1 \text{ and } CB_4 \text{ trip}) \rightarrow L \text{ fails}$$
$$R_5(c_3 = AH) : (p_1 \text{ and } p_6 \text{ operate}) \otimes (CB_1 \text{ and } CB_6 \text{ trip}) \rightarrow L \text{ fails}$$
$$R_3(c_4 = AH) : (p_3 \text{ and } p_2 \text{ operate}) \otimes (CB_3 \text{ and } CB_2 \text{ trip} \rightarrow L \text{ fails}$$
$$R_4(c_5 = AH) : (p_3 \text{ and } p_4 \text{ operate}) \otimes (CB_3 \text{ and } CB_4 \text{ trip}) \rightarrow L \text{ fails}$$
$$R_7(c_6 = AH) : (p_3 \text{ and } p_6 \text{ operate}) \otimes (CB_3 \text{ and } CB_6 \text{ trip}) \rightarrow L \text{ fails}$$
$$R_6(c_7 = AH) : (p_5 \text{ and } p_2 \text{ operate}) \otimes (CB_5 \text{ and } CB_2 \text{ trip}) \rightarrow L \text{ fails}$$
$$R_8(c_8 = AH) : (p_5 \text{ and } p_4 \text{ operate}) \otimes (CB_5 \text{ and } CB_4 \text{ trip}) \rightarrow L \text{ fails}$$
$$R_9(c_9 = VH) : (p_5 \text{ and } p_6 \text{ operate}) \otimes (CB_5 \text{ and } CB_6 \text{ trip}) \rightarrow L \text{ fails}$$

The fault fuzzy production rule set for buses consist of two rules and the meaning for each proposition is shown in Table 5.3.

$$R_1(c_1 = AH) : (p_1 \text{ operates}) \otimes (\text{all or partial } CB_1 \text{ trips}) \rightarrow B \text{ fails}$$
$$R_2(c_2 = VH) : (p_2 \text{ operates}) \otimes (CB_2 \text{ trips}) \rightarrow B \text{ fails}$$

**Table 5.2** Meaning of each proposition in rule set of L

| Protections | Protective devices | CB | Relationship |
|---|---|---|---|
| $P_1$ | Sending end main protections of L | $CB_1$ | $CB_s$ related to $P_1$ |
| $P_2$ | Receiving end main protections of L | $CB_2$ | $CB_s$ related to $P_2$ |
| $P_3$ | Sending end first backup protections of L | $CB_3$ | $CB_s$ related to $P_3$ |
| $P_4$ | Receiving end first backup protections of L | $CB_4$ | $CB_s$ related to $P_4$ |
| $P_5$ | Sending end second backup protections of L | $CB_5$ | $CB_s$ related to $P_5$ |
| $P_6$ | Receiving end second backup protections of L | $CB_6$ | $CB_s$ related to $P_6$ |

**Table 5.3** Meaning of each proposition in rule set of B

| Protections | Protective devices | Circuit breaker | Relationship |
|---|---|---|---|
| $P_1$ | Main protections of B | $CB_1$ | $CB_s$ related to $P_1$ |
| $P_2$ | Backup protections of B | $CB_2$ | $CB_s$ related to $P_2$ |

**Table 5.4** Meaning of each proposition in rule set of T

| Protections | Protective devices | Circuit breaker | Relationship |
|---|---|---|---|
| $P_1$ | Main protections of T | $CB_1$ | $CB_s$ related to $P_1$ |
| $P_2$ | First backup protections of T | $CB_2$ | $CB_s$ related to $P_2$ |
| $P_3$ | Second backup protections of T | $CB_3$ | $CB_s$ related to $P_3$ |

The fault fuzzy production rule set for transformers consist of three rules. Table 5.4 shows the meaning for each proposition.

$$R_1(c_1 = AH) : (p_1 \ operates) \wedge (all \ or \ partial \ CB_1 \ trips) \rightarrow T \ fails$$
$$R_2(c_2 = AH) : (p_2 \ operates) \wedge (all \ or \ partial \ CB_2 \ trips) \rightarrow T \ fails$$
$$R_3(c_3 = VH) : (p_3 \ operates) \wedge (CB_3 \ trips) \rightarrow T \ fails$$

As usual, the data from the SCADA system may have operation failure, maloperation and misinformation, so it is necessary to use a confidence level to describe the operation accuracy of each section. Here, an empirical confidence level is assigned to each protective device including the protective relay of each line, bus or transformer, or each of its corresponding CBs. The confidence levels of the operated protective devices and the non-operated protective devices are shown in Tables 5.5 and 5.6, respectively.

### 5.4.3.2   Fault Fuzzy Productions Rules Based on tFRSN P Systems

In what follows, fault fuzzy productions rules based on tFRSN P systems in [42] are described. *Fuzzy production rules* are modeled by using tFRSN P systems and further they are used to model fault diagnosis in power transmission networks. Four types

**Table 5.5** Confidence levels of the operated protective devices

| Sections | Protective devices | | | | | |
|---|---|---|---|---|---|---|
| | Main | | First backup | | Second backup | |
| | Relays | CBs | Relays | CBs | Relays | CBs |
| $L$ | VH | VH | H | H | MH | MH |
| $B$ | VH | VH | - | - | MH | MH |
| $T$ | VH | VH | H | H | MH | MH |

**Table 5.6** Confidence levels of the non-operated protective devices

| Sections | Protective devices | | | | | |
|---|---|---|---|---|---|---|
| | Main | | First backup | | Second backup | |
| | Relays | CBs | Relays | CBs | Relays | CBs |
| $L$ | L | L | L | L | L | L |
| $B$ | ML | L | - | - | ML | L |
| $T$ | ML | L | ML | L | L | L |

**Fig. 5.28** Modeling process of *Type 1* using one tFRSN P system

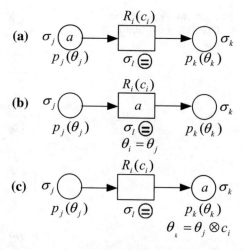

of fuzzy production rules are considered. A tFRSN P system can be used to model one or more fuzzy production rules. In the following description, $R_i$ $(i = 1, \ldots, N_r)$ is the $i$th fuzzy production rule, $N_r$ represents the number of fuzzy production rules, $c_i$ is a trapezoidal fuzzy number in $[0, 1]$ representing the confidence factor of $R_i$, $p_j$ $(1 \leq j \leq N_p)$ is the $j$th proposition appearing in the antecedent or consequent part of $R_i$, $N_p$ represents the number of propositions, and $\theta_j$ is a trapezoidal fuzzy number in $[0, 1]$ representing the fuzzy truth value of proposition $p_j$.

*Type 1*: $R_i(c_i) : p_j(\theta_j) \rightarrow p_k(\theta_k)$ $(1 \leq j, k \leq N_p)$. The modeling process of this rule type by using one tFRSN P system is shown in Fig. 5.28, where (a), (b) and (c) represent spike $a$ being transmitted from input neuron $\sigma_j$ to output neuron $\sigma_k$. The fuzzy truth value of the proposition $p_k$ is $\theta_k = \theta_j \otimes c_i$.

*Type 2*: $R_i(c_i) : p_1(\theta_1) \otimes \cdots \otimes p_{k-1} (\theta_{k-1}) \rightarrow p_k (\theta_k)$. The process of this rule type modeled by using one tFRSN P system is shown in Fig. 5.29, where (a), (b) and (c) represent spike $a$ being transmitted from input neurons $\sigma_1, \ldots, \sigma_{k-1}$ to output neuron $\sigma_k$. The fuzzy truth value of the proposition $p_k$ is $\theta_k = (\theta_1 \otimes \cdots \otimes \theta_{k-1}) \otimes c_i$.

*Type 3*: $R_i(c_i) : p_1(\theta_1) \rightarrow p_2(\theta_2) \otimes \cdots \otimes p_k (\theta_k)$. The process of this rule type modeled by using one tFRSN P system is shown in Fig. 5.30, where (a), (b) and (c) represent spike $a$ being transmitted from input neuron $\sigma_1$ to output neurons $\sigma_2$, $\ldots$, $\sigma_k$. The fuzzy truth values of the propositions $p_2, \ldots, p_k$ are identical, i.e., $\theta_2 = \cdots = \theta_k = \theta_1 \otimes c_i$.

*Type 4*: $R_i(c_i) : p_1(\theta_1) \oslash \cdots \oslash p_{k-1} (\theta_{k-1}) \rightarrow p_k(\theta_k)$. The process of this rule type modeled by using one tFRSN P system is shown in Fig. 5.31, where (a), (b) and (c) represent spike $a$ being transmitted from input neurons $\sigma_1, \ldots, \sigma_{k-1}$ into output neuron $\sigma_k$. The fuzzy truth value of the proposition $p_k$ is $\theta_k = (\theta_1 \oslash \cdots \oslash \theta_{k-1}) \otimes c_i$.

It is worth pointing out that linguistic terms can be also used to describe the fuzzy truth values of the propositions in the fuzzy production rules and the confidence factor of each fuzzy production rule. The linguistic terms are represented by the trapezoidal fuzzy numbers shown in Table 5.7.

**Fig. 5.29** Modeling process of *Type 2* using one tFRSN P system

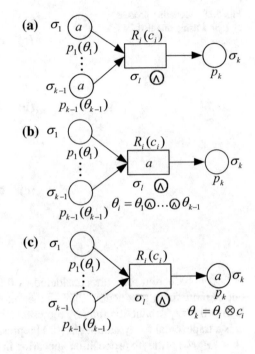

### 5.4.3.3  Algorithmic Elaboration of FDSNP

In this subsection, we elaborate the algorithm for implementing fault diagnosis method based on tFRSN P systems (FDSNP for short) [42]. Figure 5.32 shows the FDSNP flowchart, where each step is described as follows:

*Step 1*: Read data. The operation messages about protective relays and/or CBs in a power transmission network are read from the SCADA system.

*Step 2*: Search for outage areas. The network topology analysis is used to search the outage areas. This method can decrease the number of candidate diagnosing areas and reduce the subsequent computational workload [46]. The searching process is as follows:

(1) The search iteration $t = 1$;
(2) *Construct a set $Q_t$ of section numbers*: a number to each section is assigned in the power transmission network. The numbers of all sections constitute the set $Q_t$;
(3) *Construct a subset $M_t$ of section numbers*: the number of a randomly chosen section from $Q_t$ is put into the subset $M_t$. If there is a closed CB connecting this chosen section, all the closed CBs connecting it can be found. Otherwise, the algorithm goes to *Step 2* (6). Find all the other sections linking with each of the closed CBs and put their numbers from $Q_t$ into $M_t$. Continue to find the closed CBs and sections according to those in $M_t$;
(4) $t$ is increased by one;

**Fig. 5.30** Modeling process of *Type 3* using one tFRSN P system

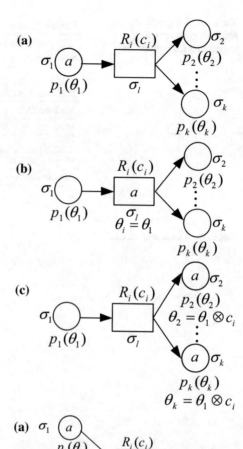

**Fig. 5.31** Modeling process of *Type 4* using one tFRSN P system

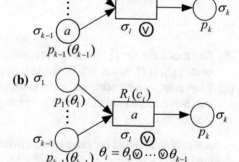

**Table 5.7** Linguistic terms and their corresponding trapezoidal fuzzy numbers

| Linguistic terms | Trapezoidal fuzzy numbers |
|---|---|
| Absolutely-false (AF) | (0, 0, 0, 0) |
| Very-low (VL) | (0, 0, 0.02, 0.07) |
| Low (L) | (0.04, 0.1, 0.18, 0.23) |
| Medium-low (ML) | (0.17, 0.22, 0.36, 0.42) |
| Medium (M) | (0.32, 0.41, 0.58, 0.65) |
| Medium-high (MH) | (0.58, 0.63, 0.80, 0.86) |
| High (H) | (0.72, 0.78, 0.92, 0.97) |
| Very-high (VH) | (0.975, 0.98, 1, 1) |
| Absolutely-high (AH) | (1, 1, 1, 1) |

**Fig. 5.32** The flowchart of FDSNP

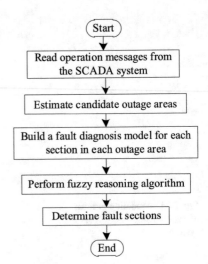

(5) *Construct the set* $Q_t$: remove the numbers of the sections in $M_t$ from $Q_{t-1}$ and obtain $Q_t$. If $Q_t$ is not empty, the search process goes to *Step 2* (3);

(6) Find passive networks, i.e., outage areas, from $M_1, \ldots, M_{N_s}$, where $N_s$ is the maximum number of all numbers referring to section subsets. The search process stops.

*Step 3*: If there exists only one section in the passive networks found in *Step 2*, this section is the faulty one and the algorithm stops, otherwise, a fault diagnosis model based on a tFRSN P system is built for each section. The model-building process is described as follows. A section in the passive network is randomly chosen. Fault fuzzy production rules are designed according to the relay protections of the section. Proposition and rule neurons are decided and their linking relationship is created to obtain the tFRSN P system. We set the confidence factor of each rule empirically. The confidence levels for main protections, first backup protections, second backup protections and their CBs can be assigned according to Tables 5.5 and 5.6. Then, we construct a one-to-one relationship between the fuzzy truth value of each input neuron and the confidence level of each protection to obtain the initial values of the model.

*Step 4*: To obtain the fault confidence level of each section, the algebraic fuzzy reasoning algorithm is used.

*Step 5*: If the confidence level $\theta$ of a section satisfies the condition $\theta \geq (0.58, 0.63, 0.80, 0.86)$, the section is faulty. If $\theta$ satisfies the condition $\theta \leq (0.17, 0.22, 0.36, 0.42)$, the section is not faulty. Otherwise, the section may be faulty.

### 5.4.3.4 Examples

To show the application of FDSNP, we use seven cases with single and multiple fault situations of the local system in an EPS shown in Fig. 5.3 to conduct experiments [42]. Table 5.8 shows the status information (with/without incompleteness and uncertainty) about protective relays and CBs. The symbol "*" in Table 5.8 indicates that a case includes incomplete or uncertain status information from the SCADA system. We consider four diagnosis methods, fuzzy logic (FL) [3], fuzzy Petri nets (FPN) [28], genetic algorithm-tabu search (GATS) [19] and genetic algorithm (GA) [43], as benchmarks to draw comparisons. After we use FDSNP to diagnose faults in the seven cases, the results including faulty sections and their fault confidence levels are obtained and shown in Table 5.9. Table 5.10 shows the comparisons of five methods. The symbol "–" means that the case was not considered in the corresponding reference. It is noting that only the information about CBs was used in FL [3] and GA may have multiple solutions such as in *Cases 5–7* [43].

Table 5.9 indicates that the fault confidence levels represented by trapezoidal fuzzy numbers can provide a quantitative description for the faulty sections, which makes the results more reliable. In addition, the linguistic terms related to the trapezoidal fuzzy numbers are more intuitive and flexible for experts and dispatchers than

**Table 5.8** Status information about protective relays and CBs

| Cases | Status information | |
|-------|-------------------|---|
| | Operated relays | Tripped CBs |
| 1 | $B_{1m}, L_{2Rs}, L_{4Rs}$ | $CB_4, CB_5, CB_7, CB_9, CB_{12}, CB_{27}$ |
| 2* | $L_{2Rs}, L_{4Rs}$ | $CB_4, CB_5, CB_7, CB_9, CB_{12}, CB_{27}$ |
| 3 | $B_{1m}, L_{1Sp}, L_{1Rm}$ | $CB_4, CB_5, CB_6, CB_7, CB_9, CB_{11}$ |
| 4 | $B_{1m}, L_{1Sm}, L_{1Rp}$ | $CB_4, CB_5, CB_6, CB_7, CB_8$ |
| | $B_{2m}, L_{2Sp}, L_{2Rm}$ | $CB_9, CB_{10}, CB_{11}, CB_{12}$ |
| 5 | $T_{3p}, L_{7Sp}, L_{7Rp}$ | $CB_{14}, CB_{16}, CB_{29}, CB_{39}$ |
| 6 | $L_{1Sm}, L_{1Rp}, L_{2Sp}, L_{2Rp}$ | $CB_7, CB_8, CB_{11}, CB_{12}$ |
| | $L_{7Sp}, L_{7Rm}, L_{8Sm}, L_{8Rm}$ | $CB_{29}, CB_{30}, CB_{39}, CB_{40}$ |
| 7* | $T_{7m}, T_{8P}, B_{7m}, B_{8m}, L_{5Sm}$ | $CB_{19}, CB_{20}, CB_{29}, CB_{30}, CB_{32}$ |
| | $L_{5Rp}, L_{6Ss}, L_{7Sp}, L_{7Rm}, L_{8Ss}$ | $CB_{33}, CB_{34}, CB_{35}, CB_{36}, CB_{37}, CB_{39}$ |

**Table 5.9** Fault sections and their fault confidence levels obtained by using FDSNP

| Cases | Diagnosis results of FDSNP | | |
|---|---|---|---|
| | Fault sections | Fault confidence levels | |
| | | Trapezoidal fuzzy numbers | Linguistic terms |
| 1 | $B_1$ | (0.975, 0.98, 1, 1) | VH |
| 2 | $B_1$ | (0.5655, 0.6174, 0.80, 0.86) | [M, MH] |
| 3 | $B_1$ | (0.975, 0.98, 1, 1) | VH |
| | $L_1$ | (0.9506, 0.9604, 1, 1) | [H, VH] |
| 4 | $B_1$ | (0.975, 0.98, 1, 1) | VH |
| | $B_2$ | (0.975, 0.98, 1, 1) | VH |
| | $L_1$ | (0.9506, 0.9604, 1, 1) | [H, VH] |
| | $L_2$ | (0.9506, 0.9604, 1, 1) | [H, VH] |
| 5 | $T_3$ | (0.72, 0.78, 0.92, 0.97) | H |
| | $L_7$ | (0.9506, 0.9604, 1, 1) | [H, VH] |
| 6 | $L_1$ | (0.702, 0.7644, 0.92, 0.97) | [H, VH] |
| | $L_2$ | (0.702, 0.7644, 0.92, 0.97) | [H, VH] |
| | $L_7$ | (0.702, 0.7644, 0.92, 0.97) | [H, VH] |
| | $L_8$ | (0.9506, 0.9604, 1, 1) | [H, VH] |
| 7 | $L_5$ | (0.702, 0.7644, 0.92, 0.97) | [H, VH] |
| | $L_7$ | (0.702, 0.7644, 0.92, 0.97) | [H, VH] |
| | $B_7$ | (0.975, 0.98, 1, 1) | [H, VH] |
| | $B_8$ | (0.975, 0.98, 1, 1) | [H, VH] |
| | $T_7$ | (0.975, 0.98, 1, 1) | [H, VH] |
| | $T_8$ | (0.72, 0.78, 0.92, 0.97) | H |

probability values because they often express their knowledge by using linguistic terms with a certain degree of uncertainty.

Table 5.10 shows that FDSNP, in *Case 1* and *Cases 3–6*, obtains the same diagnosis results as the methods in [3, 28]. So FDSNP is feasible and effective for diagnosing faults in a power transmission network. Furthermore, it is indicated in Table 5.10 that FDSNP has an advantage over FL, FPN and GA with respect to diagnosis correctness in some cases. To be specific, FDSNP obtains different results in *Case 7* from those in [3, 28, 43]. Actually, for section $L_8$ in this case, only its second backup protective relay $S_{L8Ss}$ operated and $S_{L8Ss}$ operated as the second backup protective relay of section $B_8$, so $L_8$ is not a faulty section; in addition, for section $B_7$, its main protective relay $B_{7m}$ operated and tripped its corresponding CBs, $CB_{33}$, $CB_{34}$ and $CB_{35}$, so $B_7$ is a faulty section. Therefore, FDSNP and GATS are better than those in [3, 28, 43] in *Case 7*. The comparisons of diagnosis results between FDSNP and the methods in [43] in *Cases 5–7* imply that FDSNP can solve the non-uniqueness problem of the diagnosis solution. Moreover, *Cases 2* and *7* indicate that FDSNP can obtain satisfactory results to the case with incomplete or uncertain alarm information.

To show more details on how to use FDSNP to diagnose faults, *Case 1* and *Case 2* are used as examples to elaborate the diagnosis process.

**Table 5.10** Comparisons between FDSNP and four fault diagnosis methods

| Cases | Diagnosis results | | | | |
|---|---|---|---|---|---|
| | FDSNP | FL [3] | FPN [28] | GATS [19] | GA [43] |
| 1 | $B_1$ | $B_1$ | $B_1$ | – | $B_1$ |
| 2 | $B_1$ | – | – | – | – |
| 3 | $B_1$ | $B_1$ | $B_1$ | | $B_1$ |
| | $L_1$ | $L_1$ | $L_1$ | – | $L_1$ |
| 4 | $B_1$ | $B_1$ | $B_1$ | | $B_1$ |
| | $B_2$ | $B_2$ | $B_2$ | – | $B_2$ |
| | $L_1$ | $L_1$ | $L_1$ | | $L_1$ |
| | $L_2$ | $L_2$ | $L_2$ | | $L_2$ |
| 5 | | | | | $(1)T_3, L_7$ |
| | $T_3$ | $T_3$ | $T_3$ | $T_3$ | $(2)T_3$ |
| | $L_7$ | $L_7$ | $L_7$ | $L_7$ | $(3)L_7$ |
| | | | | | $(4)No$ |
| 6 | $L_1$ | $L_1$ | $L_1$ | $L_1$ | $(1)L_1, L_2$ |
| | $L_2$ | $L_2$ | $L_2$ | $L_2$ | $L_7, L_8$ |
| | $L_7$ | $L_7$ | $L_7$ | $L_7$ | $(2)L_1, L_7$ |
| | $L_8$ | $L_8$ | $L_8$ | $L_8$ | $L_8$ |
| 7 | $L_5$ | $L_5$ | $L_5$ | $L_5$ | |
| | $L_7$ | $L_7$ | $L_7$ | $L_7$ | $(1)L_5, L_7$ |
| | $B_7$ | $B_8$ | $B_7$ | $B_7$ | $B_7, B_8$ |
| | $B_8$ | $T_7$ | $B_8$ | $B_8$ | $T_7, T_8$ |
| | $T_7$ | $T_8$ | $T_7$ | $T_7$ | $(2)L_5, L_7$ |
| | $T_8$ | | $T_8$ | $T_8$ | $T_7, B_8$ |
| | | | $L_8$ | | |

*Case 1*: The SCADA system provides complete information. Operated relays: $B_{1m}$, $L_{2Rs}$ and $L_{4Rs}$. Tripped CBs: $CB_4$, $CB_5$, $CB_7$, $CB_9$, $CB_{12}$ and $CB_{27}$.

The search process of outage areas is detailed as follows:

(1) *Construct the set $Q_1$ of section numbers*: $Q_1 = \{01, 02, 03, 04, 1, 2, 3, 4, 5, 6, 7, 8, 9, 10, 11, 12, 13, 14, 15, 16, 17, 18, 19, 20, 21, 22, 23, 24, 25, 26, 27, 28\}$, where numbers $01 \sim 04$ denote four joint nodes considered as active nodes, between the local power system in Fig. 5.3 and other parts of the power system.

(2) *Construct the subset $M_1$ of section numbers*: add number *1* into $M_1$ and find all the closed CBs, i.e., $CB_1$, $CB_2$ and $CB_3$, connecting the section *1*. Find all the other sections, i.e., *01*, *13*, and *14*, linking with $CB_1$, $CB_2$ and $CB_3$ and add them into $M_1$. No other closed CB is found according to the sections in $M_1$. Thus, $M_1 = \{01, 1, 13, 14\}$.

(3) *Construct the set $Q_2$*: remove the numbers of the sections in $M_1$ from $Q_1$ and obtain $Q_2 = \{02, 03, 04, 2, 3, 4, 5, 6, 7, 8, 9, 10, 11, 12, 15, 16, 17, 18, 19, 20, 21, 22, 23, 24, 25, 26, 27, 28\}$.

(4) *Construct the subset $M_2$*: add number *2* into $M_2$ and execute *Step 2* (3) in Sect. 5.4.3.3. We get $M_2 = \{2, 3, 16, 20\}$.
(5) *Construct the set $Q_3$*: remove the numbers of the sections in $M_2$ from $Q_2$ and obtain $Q_3 = \{02, 03, 04, 4, 5, 6, 7, 8, 9, 10, 11, 12, 15, 17, 18, 19, 21, 22, 23, 24, 25, 26, 27, 28\}$.
(6) *Construct the subset $M_3$*: add number *4* into $M_3$ and execute *Step 2* (3) in Sect. 5.4.3.3. We get $M_3 = \{02, 03, 04, 4, 5, 6, 7, 8, 9, 10, 11, 12, 15, 17, 18, 19, 21, 22, 23, 24, 25, 26, 27, 28\}$.
(7) *Construct the set $Q_4$*: remove the numbers of the sections in $M_3$ from $Q_3$ and obtain $Q_4 = \emptyset$.
(8) Find one passive network $M_2 = \{2, 3, 16, 20\}$ in $M_1$, $M_2$, $M_3$, where the numbers 2, 3, 16, 20 represent bus $B_1$, bus $B_2$, line $L_2$ and line $L_4$, respectively. The search process stops.

To show how to use *Steps 3–5* in Sect. 5.4.3.3, bus $B_1$ is used as an example. Figure 5.33 shows the fault diagnosis model of $B_1$ based on a tFRSN P system

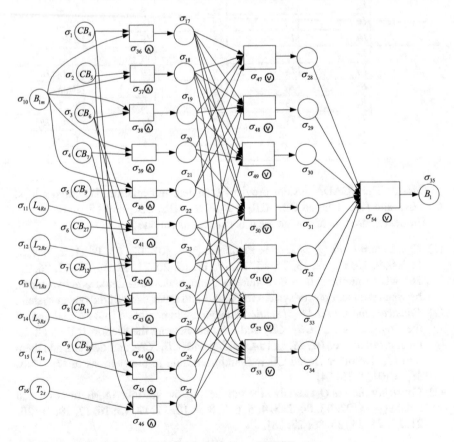

**Fig. 5.33** Fault diagnosis model of bus $B_1$ based on a tFRSN P system

constructed. The model consists of 35 proposition neurons and 19 rule neurons. There are four assistant arcs (synapses), i.e., $(3, 41)$, $(3, 42)$, $(4, 43)$ and $(22, 52)$, marked by hollow tips. The arc $(3, 41)$ means that if $CB_6$ opens, the operation of $L_{4Rs}$ and $CB_{27}$ is invalid and then the values of $L_{4Rs}$ and $CB_{27}$ are set to $(0, 0, 0, 0)$; otherwise, the operation of $L_{4R_s}$ and $CB_{27}$ is valid.

The fuzzy reasoning process is described as follows:

First of all, the trapezoidal fuzzy numbers $\theta_0$ and $\delta_0$ are obtained according to the alarm information in *Case 1* and Tables 5.5, 5.6 and 5.7. Numbers $\theta$ and $\delta$ are 35-dimension and 19-dimension vectors, respectively.

$$
\theta_0 = \begin{bmatrix}
(0.975, 0.98, 1, 1) \\
(0.975, 0.98, 1, 1) \\
(0.04, 0.1, 0.18, 0.23) \\
(0.975, 0.98, 1, 1) \\
(0.975, 0.98, 1, 1) \\
(0.58, 0.63, 0.80, 0.86) \\
(0.58, 0.63, 0.80, 0.86) \\
(0.04, 0.1, 0.18, 0.23) \\
(0.04, 0.1, 0.18, 0.23) \\
(0.975, 0.98, 1, 1) \\
(0.58, 0.63, 0.80, 0.86) \\
(0.58, 0.63, 0.80, 0.86) \\
(0.17, 0.22, 0.36, 0.42) \\
(0.17, 0.22, 0.36, 0.42) \\
(0.17, 0.22, 0.36, 0.42) \\
(0.17, 0.22, 0.36, 0.42) \\
\mathbf{O}
\end{bmatrix}, \quad \delta_0 = \begin{bmatrix} \mathbf{O} \end{bmatrix}.
$$

When $g = 1$, the results are

$$
\delta_1 = \begin{bmatrix}
(0.975, 0.98, 1, 1) \\
(0.975, 0.98, 1, 1) \\
(0.04, 0.1, 0.18, 0.23) \\
(0.975, 0.98, 1, 1) \\
(0.975, 0.98, 1, 1) \\
(0.58, 0.63, 0.80, 0.86) \\
(0.58, 0.63, 0.80, 0.86) \\
(0.04, 0.1, 0.18, 0.23) \\
(0.04, 0.1, 0.18, 0.23) \\
(0.17, 0.22, 0.36, 0.42) \\
(0.17, 0.22, 0.36, 0.42) \\
\mathbf{O}
\end{bmatrix},
$$

$$\theta_1 = \begin{bmatrix} \mathbf{O} \\ (0.975, 0.98, 1, 1)_{i=17} \\ (0.975, 0.98, 1, 1) \\ (0.04, 0.1, 0.18, 0.23) \\ (0.975, 0.98, 1, 1) \\ (0.975, 0.98, 1, 1) \\ (0.5655, 0.6174, 0.80, 0.86) \\ (0.5655, 0.6174, 0.80, 0.86) \\ (0.039, 0.098, 0.18, 0.23) \\ (0.039, 0.098, 0.18, 0.23) \\ (0.0156, 0.2156, 0.36, 0.42) \\ (0.0156, 0.2156, 0.36, 0.42) \\ \mathbf{O} \end{bmatrix}.$$

When $g = 2$, the results are

$$\delta_2 = \begin{bmatrix} \mathbf{O} \\ (0.975, 0.98, 1, 1)_{j=12} \\ (0.975, 0.98, 1, 1) \\ (0.975, 0.98, 1, 1) \\ (0.975, 0.98, 1, 1) \\ (0.5655, 0.6174, 0.80, 0.86) \\ (0.5655, 0.6174, 0.80, 0.86) \\ (0.975, 0.98, 1, 1) \\ (0, 0, 0, 0) \end{bmatrix},$$

$$\theta_2 = \begin{bmatrix} \mathbf{O} \\ (0.975, 0.98, 1, 1)_{i=28} \\ (0.975, 0.98, 1, 1) \\ (0.975, 0.98, 1, 1) \\ (0.975, 0.98, 1, 1) \\ (0.5514, 0.6051, 0.80, 0.86) \\ (0.5514, 0.6051, 0.80, 0.86) \\ (0.975, 0.98, 1, 1) \\ (0, 0, 0, 0) \end{bmatrix}.$$

When $g = 3$, the results are

$$\delta_3 = \begin{bmatrix} \mathbf{O} \\ (0.975, 0.98, 1, 1) \end{bmatrix}, \quad \theta_3 = \begin{bmatrix} \mathbf{O} \\ (0.975, 0.98, 1, 1) \end{bmatrix}.$$

When $g = 4$, the results are

$$\delta_4 = \begin{bmatrix} \mathbf{O} \end{bmatrix}.$$

To date, the termination criterion is met and the reasoning process stops. So the reasoning results, the fuzzy truth values (0.975, 0.98, 1, 1) from output neuron $\sigma_{35}$, are obtained. According to the condition in *Step 5* in Sect. 5.4.3.3, it is obvious that $B_1$ is a faulty section with a confidence level *AH*.

The rest may be deduced by analogy. $B_2$, line $L_2$ and line $L_4$ gain the same confidence level (0.04, 0.1, 0.18, 0.23). According to the condition in *Step 5* in Sect. 5.4.3.3, $B_2$, line $L_2$ and line $L_4$ are not faulty sections.

*Case 2*: The SCADA system provides incomplete information. Operated relays: $L_{2Rs}$ and $L_{4Rs}$. Tripped CBs: $CB_4$, $CB_5$, $CB_7$, $CB_9$, $CB_{12}$ and $CB_{27}$. It is noting that the status information of $B_{1m}$ is missing.

Similar to the process of *Case 1*, we obtain one passive network {2, 3, 16, 20}. The following description shows how to use *Steps 3–5* in Sect. 5.4.3.3 by considering bus $B_1$ as an example.

According to the alarm information in *Case 2* and Tables 5.5, 5.6 and 5.7, we obtain the trapezoidal fuzzy numbers $\theta_0$ and $\delta_0$.

$$
\theta_0 = \begin{bmatrix}
(0.975, 0.98, 1, 1) \\
(0.975, 0.98, 1, 1) \\
(0.04, 0.1, 0.18, 0.23) \\
(0.975, 0.98, 1, 1) \\
(0.975, 0.98, 1, 1) \\
(0.58, 0.63, 0.80, 0.86) \\
(0.58, 0.63, 0.80, 0.86) \\
(0.04, 0.1, 0.18, 0.23) \\
(0.04, 0.1, 0.18, 0.23) \\
(0.17, 0.22, 0.36, 0.42) \\
(0.58, 0.63, 0.80, 0.86) \\
(0.58, 0.63, 0.80, 0.86) \\
(0.17, 0.22, 0.36, 0.42) \\
(0.17, 0.22, 0.36, 0.42) \\
(0.17, 0.22, 0.36, 0.42) \\
(0.17, 0.22, 0.36, 0.42) \\
\mathbf{O}
\end{bmatrix}, \quad \delta_0 = [\mathbf{O}]
$$

Finally, the reasoning results of *Case 2*, the fuzzy truth values (0.5655, 0.6174, 0.80, 0.86) of the output neuron $\sigma_{35}$, are obtained.

Thus, $B_1$ may be a faulty section with a confidence level (0.5655, 0.6174, 0.80, 0.86), according to the condition in *Step 5* in Sect. 5.4.3.3. The rest may be deduced by analogy. $B_2$, $L_2$ and $L_4$ obtain the same confidence level (0.04, 0.1, 0.18, 0.23). So bus $B_2$, line $L_2$ and line $L_4$ are not faulty sections, in terms of the condition in *Step 5* in Sect. 5.4.3.3. So the final result is achieved, i.e., $B_1$ is the faulty section with a confidence level (0.5655, 0.6174, 0.80, 0.86) ($M \leq (0.5655, 0.6174, 0.80, 0.86) \leq MH$).

## 5.5  Conclusions

The rFRSN P systems, WFRSN P systems and tFRSN P systems are collectively called FRSN P systems in this chapter. The FRSN P systems are novel graphical models for representing fuzzy knowledge and information. This chapter employs them to deal with fault diagnose problems. The fault diagnosis ability of a method is usually associated with the knowledge availability and the reasoning process. A comparison between FRSN P systems and several fault diagnosis approaches, such as expert systems (ESs), fuzzy set theory (FST), artificial neural networks (ANNs) and fuzzy Petri nets (FPNs), regarding knowledge representation and inference process can be referred to [9, 35, 41, 45].

This chapter shows the effectiveness and correctness of fault diagnosis with fuzzy reasoning spiking neural P systems and the results of application examples are obtained by manual computation. To test the speed, convergence and accuracy of FRA, WMBRA and MBFRA, and to explore automatic generation of FRSN P systems in fault diagnosis, our future work will simulate them on software platforms, MATLAB, P-Lingua or MeCoSim [25–27, 29–31]. Meanwhile, how to verify and realize the parallelism of tFRSN P systems and MBFRA on hardware such as FPGA and CUDA is also our further research task.

Moreover, valuable research interests refer to the extension of models, algorithms and application areas. For models and algorithms, one promising direction is to design new variants of SN P systems and their reasoning algorithms according to requirements of different fault diagnosis problems, such as on-line diagnosis, fast fault diagnosis, high-precision diagnosis. Another promising research avenue is to propose FRSN P systems with learning ability. For application areas, FRSN P systems can be used for more different systems, such as power supply systems for urban rail transit, mechanical fault diagnosis and power systems with new energies.

## References

1. Cardoso, G., J.G. Rolim, and H.H. Zurn. 2008. Identifying the primary fault section after contingencies in bulk power systems. *IEEE Transition on Power Delivery* 23 (3): 1335–1342.
2. Cavaliere, M., O.H. Ibarra, O. Gh Păun, M.Ionescu Egecioglu, and S. Woodworth. 2009. Asynchronous spiking neural P systems. *Theoretical Computer Science* 410 (24–25): 2352–2364.
3. Chang, C.S., J.M. Chen, D. Srinivasan, F.S. Wen, and A.C. Liew. 1997. Fuzzy logic approach in power system fault section identification. *IEE Proceedings of Generation, Transmission and Distribution* 144 (5): 406–414.
4. Chen, S.M. 1996. A fuzzy reasoning approach for rule-based systems based on fuzzy logics. *IEEE Transactions on Systems, Man and Cybernetics, Part B* 26 (5): 769–778.
5. Chen, W.H. 2011. Fault section estimation using fuzzy matrix-based reasoning methods. *IEEE Transactions on Power Delivery* 26 (1): 205–213.
6. Chen, W.H. 2012. Online fault diagnosis for power transmission networks using fuzzy digraph models. *IEEE Transactions on Power Delivery* 27 (2): 688–698.

7. Chen, H., T.O. Ishdorj, Gh. Păun, and M.J. Pérez-Jiménez. 2006. Handling languages with spiking neural P systems with extended rules. *Romanian Journal of Information Science and Technology* 9 (3): 151–162.

8. Chen, S.M., J.S. Ke, and J.F. Chang. 1990. Knowledge representation using fuzzy Petri nets. *IEEE Transactions on Knowledge and Data Engineering* 2 (3): 311–319.

9. Chen, W.H., S.H. Tsai, and H.I. Lin. 2011. Fault section estimation for power networks using logic cause-effect models. *IEEE Transactions on Power Delivery* 26 (2): 963–971.

10. Chien, C.F., S.L. Chen, and Y.S. Lin. 2002. Using Bayesian network for fault location on distribution feeder. *IEEE Transactions on Power Delivery* 17 (13): 785–793.

11. Davidson, E.M., S.D.J. McArthur, J.R. McDonald, and I. Watt. 2006. Applying multi-agent system technology in practice: automated management and analysis of SCADA and digital fault recorder data. *IEEE Transactions on Power Systems* 21 (2): 559–567.

12. Freund, R., M. Ionescu, and M. Oswald. 2008. Extended spiking neural P systems with decaying spikes and/or total spiking. *International Journal of Foundations of Computer Science* 19 (5): 1223–1234.

13. Grzegorzewski, P., and E. Mrowka. 2005. Trapezoidal approximations of fuzzy numbers. *Fuzzy Sets and Systems* 2005 (153): 115–135.

14. Guo, W.X., F.S. Wen, G. Ledwich, Z.W. Liao, X.Z. He, and J.H. Liang. 2010. An analytic model for fault diagnosis in power systems considering malfunctions of protective relays and circuit breakers. *IEEE Transactions on Power Delivery* 25 (3): 1393–1401.

15. He, Z.Y., H.D. Chiang, C.W. Li, and Q.F. Zeng. 2009. Fault-section estimation in power systems based on improved optimization model and binary particle swarm optimization. In *Proceedings of IEEE Power and Energy Society General Meeting*, 1–8.

16. Hossack, J.A., J. Menal, S.D.J. McArthur, and J.R. McDonald. 2003. A multiagent architecture for protection engineering diagnostic assistance. *IEEE Transactions on Power Systems* 18 (2): 639–647.

17. Ionescu, M., Gh. Păun, and T. Yokomori. 2006. Spiking neural P systems. *Fundamenta Informaticae* 71 (2–3): 279–308.

18. Lee, H.J., B.S. Ahn, and Y.M. Park. 2000. A fault diagnosis expert system for distribution substations. *IEEE Transactions on Power Delivery* 15 (1): 92–97.

19. Lin, X.N., S.H. Ke, Z.T. Li, H.L. Weng, and X.H. Han. 2010. A fault diagnosis method of power systems based on improved objective function and genetic algorithm-tabu search. *IEEE Transactions on Power Delivery* 25 (3): 1268–1274.

20. Lin, S., Z.Y. He, and Q.Q. Qian. 2010. Review and development on fault diagnosis in power grid. *Power System Protection and Control* 38 (4): 140–150.

21. Liu, H.C., L. Liu, Q.L. Lin, and N. Liu. 2013. Knowledge acquisition and representation using fuzzy evidential reasoning and dynamic adaptive fuzzy Petri nets. *IEEE Transactions on Cybernetics* 43 (3): 1059–1072.

22. Luo, X., and M. Kezunovic. 2008. Implementing fuzzy reasoning Petri-nets for fault section estimation. *IEEE Transactions on Power Systems* 23 (2): 676–685.

23. Ma, D.Y., Y.C. Liang, X.S. Zhao, R.C. Guan, and X.H. Shi. 2013. Multi-BP expert system for fault diagnosis of power system. *Engineering Applications of Artificial Intelligence* 26 (3): 937–944.

24. Marks II, R.J. (ed.). 1994. *Fuzzy logic technology and their applications*. IEEE: IEEE Technology Update Series.

25. Pan, L.Q., and Gh Păun. 2010. Spiking neural P systems: an improved normal form. *Theoretical Computer Science* 411 (6): 906–918.

26. Păun, Gh, M.J. Pérez-Jiménez, and G. Rozenberg. 2006. Spike trains in spiking neural P systems. *International Journal of Foundations of Computer Science* 17 (4): 975–1002.

27. Peng, H., J. Wang, M.J. Pérez-Jiménez, H. Wang, J. Shao, and T. Wang. 2013. Fuzzy reasoning spiking neural P system for fault diagnosis. *Information Sciences* 235: 106–116.

28. Sun, J., S.Y. Qin, and Y.H. Song. 2004. Fault diagnosis of electric power systems based on fuzzy Petri nets. *IEEE Transactions on Power Systems* 19 (4): 2053–2059.

29. The Matlab Website. http://www.mathworks.es/products/matlab/.

30. The P-Lingua Website. http://www.p-lingua.org. Research Group on Natural Computing, University of Seville.
31. The MeCoSim Website. http://www.p-lingua.org/mecosim. Research Group on Natural Computing, University of Seville.
32. Tang, L., H.B. Sun, B.M. Zhang, and F. Gao. 2003. Online fault diagnosis for power system based on information theory. *Proceedings of the Chinese Society for Electrical Engineering* 23 (7): 5–11.
33. Thukaram, D., H.P. Khincha, and H.P. Vijaynarasimha. 2005. Artificial neural network and support vector machine approach for locating faults in radial distribution systems. *IEEE Transactions on Power Delivery* 20 (2): 710–721.
34. Tu, M., J. Wang, H. Peng, and P. Shi. 2014. Application of adaptive fuzzy spiking neural P systems in fault diagnosis of power systems. *Chinese Journal of Electronics* 23 (1): 87–92.
35. Wang, J., P. Shi, H. Peng, M.J. Pérez-Jiménez, and T. Wang. 2013. Weighted fuzzy spiking neural P system. *IEEE Transactions on Fuzzy Systems* 21 (2): 209–220.
36. Wang, T., J. Wang, H. Peng, and Y. Deng. 2010. Knowledge representation using fuzzy spiking neural P systems. In *Proceedings of IEEE Fifth International Conference on Bio-inspired Computing: Theories and Applications*, 586–590.
37. Wang, T., J. Wang, H. Peng, and H. Wang. 2011. Knowledge representation and reasoning based on FRSN P systems. In *Proceedings of the 9th World Congress on Intelligent Control and Automation*, 255–259.
38. Wang, T., G.X. Zhang, and M.J. Pérez-Jiménez. 2014. Fault diagnosis models for electric locomotive systems based on Fuzzy Reasoning Spiking Neural P Systems. Lecture Notes in Computer Science. In *Membrane Computing (CMC 2014)*, ed. M. Gheorghe, G. Rozenberg, A. Salomaa, P. Sosík, and C. Zandron, 385–395. Berlin Heidelberg: Springer.
39. Wang, T., G.X. Zhang, and M.J. Pérez-Jiménez. 2015. Fuzzy membrane computing: theory and applications. *International Journal of Computer, Communications and Control* 10 (6): 904–935.
40. Wang, T., G.X. Zhang, M.J. Pérez-Jiménez, and J.X. Cheng. 2015. Weighted fuzzy reasoning spiking neural P systems: application to fault diagnosis in traction power supply systems of high-speed railways. *Journal of Computer and Theoretical Nanoscience* 12 (7): 1103–1114.
41. Wang, T., G.X. Zhang, H.N. Rong, and M.J. Pérez-Jiménez. 2014. Application of fuzzy reasoning spiking neural P systems to fault diagnosis. *International Journal of Computer, Communications and Control* 9 (6): 786–799.
42. Wang, T., G.X. Zhang, J.B. Zhao, Z.Y. He, J. Wang, and M.J. Pérez-Jiménez. 2015. Fault diagnosis of electric power systems based on fuzzy reasoning spiking neural P systems. *IEEE Transactions on Power Systems* 30 (3): 1182–1194.
43. Wen, F.S., and Z.X. Han. 1995. Fault section estimation in power systems using a genetic algorithm. *Electric Power Systems Research* 34 (3): 165–172.
44. Wu, S., Z.Y. He, C.H. Qian, and T.L. Zang. 2011. Application of fuzzy Petri net in fault diagnosis of traction power supply system for high-speed railway. *Power System Technology* 35 (9): 79–85.
45. Xiong, G.J., D.Y. Shi, L. Zhu, and X.Z. Duan. 2013. A new approach to fault diagnosis of power systems using fuzzy reasoning spiking neural P systems, *Mathematical Problems in Engineering*, 2013, Article ID 815352, 13 p.
46. Yang, J.W., Z.Y. He, and T.L. Zang. 2010. Power system fault-diagnosis method based on directional weighted fuzzy Petri nets. *Proceedings of the Chinese Society for Electrical Engineering* 30 (34): 42–49.
47. Zhang, G.X., H.N. Rong, F. Neri, and M.J. Pérez-Jiménez. 2014. An optimization spiking neural P system for approximately solving combinatorial optimization problems, *International Journal of Neural Systems*, 24, 5, Article No. 1440006, 16 p.
48. Zhu, Y.L., L.M. Huo, and J.L. Liu. 2006. Bayesian networks based approach for Power Systems Fault Diagnosis. *IEEE Transactions on Power Delivery* 21 (2): 634–639.

# Chapter 6
# Robot Control with Membrane Systems

**Abstract** Numerical and Enzymatic Numerical P systems are used to design mobile robot controllers and for implementing simulators running on webots platform.

## 6.1 Introduction

P systems were first investigated as a computational model offering the possibility of solving NP-complete problems at a polynomial time, but much attention has also been paid to their applications in recent years, especially to real-life applications, in parallel with richly theoretical results. For instance, probabilistic membrane systems were applied to model ecological systems [12]; spiking neural P systems were used for fault diagnosis [34]; the hybrid methods of P systems and meta-heuristics have been developed to solve broadcasting problems [44], image processing [43] and constrained engineering optimization problems [40]. This chapter discusses another promising real-life application of P systems, i.e., the use of Numerical P Systems (NPS) to model the behaviors of autonomous robots.

Although most of the P systems have been inspired by biological phenomena [27], there are some variants of P systems that are based on other processes. For instance, NPS model, introduced in [26], was inspired by processes and phenomena of the economic reality and can also be represented as a tree-like structure. Compared to the widely investigated classes of P systems [27], numerical P systems were also proved to be Turing universal [32] and are suitable for designing robot controllers [8] due to their flexible evolution rules and parallel distributed framework.

Differing from most of the variants of P systems [27], numerical P systems use numerical variables to evolve inside the compartments by means of programs. A program (or rule) is composed of a production function and a repartition protocol. The variables have a given initial value and the production function is a multivariate polynomial. The value of the production function for the current values of the variables is distributed among variables in certain compartments according to a repartition protocol. NPS were designed both as deterministic and non-deterministic systems [26]. Non-deterministic NPS allow the existence of more rules inside a single membrane and the best rule is selected by a special mechanism, while the deterministic

NPS can have only one rule inside each membrane. The robot controller requires a deterministic mechanism and consequently is designed by using the deterministic NPS. But there is a drawback that a large number of membranes are needed in the design of NPS controllers, which usually affects the computing efficiency. Thus, inspired by the behavior of biological systems with enzymatic reactions, an extension of NPS, called Enzymatic Numerical P Systems (ENPS), in which enzyme-like variables allow the existence of more than one program (rule) in each membrane while keeping the deterministic nature of the system, was proposed in [29, 30] for robot controllers. ENPS are a kind of more powerful modeling tool for controlling robot behaviors than classical NPS because there are more possibilities of selecting and executing more rules inside each membrane in ENPS.

This chapter will discuss how (enzymatic) numerical P systems can be used to design the mobile robot controllers for some behaviors [8, 29, 30, 33]. We also present the simulator of (enzymatic) numerical P systems and its implementation on webots platform. So we will first introduce NPS and ENPS in Sect. 6.2. In Sect. 6.3, the preliminaries of mobile robot control are discussed. Section 6.4 presents the design principle of membrane controllers. The simulators for NPS, ENPS and Webots (a software simulator for robots) are introduced in Sect. 6.5. The membrane controller design for several robotic behaviors is detailed in Sect. 6.6. Experiments and results are shown in Sect. 6.7. Finally, we conclude this chapter by listing some future work in Sect. 6.8.

## 6.2   Numerical P Systems

The focus of this section is on the introductions of *numerical P systems* (NPS, for short) and *enzymatic numerical P systems* (ENPS, for short). We begin with the basic concepts and principles of the simple numerical P systems, which only include a single rule inside each membrane. Then, we introduce the extension of numerical P systems, ENPS, which may include more than one production function inside each membrane inspired by biologically enzyme reactions.

### 6.2.1   NPS

Inspired by economic and business processes, NPS model uses numerical variables to fulfill its computation from initial values by means of *programs* [26]. An NPS model of degree $m$ ($m \geq 1$) is formally expressed as a tuple

$$\Pi = (m, \mu, (Var_1, Pr_1, Var_1(0)), \dots, (Var_m, Pr_m, Var_m(0)), x_{j_0, i_0})$$

where

(1) $m$ is the number of membranes used in the system (the degree of $\Pi$);
(2) $\mu$ is a membrane structure (a rooted tree) consisting of $m$ membranes, with the membranes (and hence the *regions* or *compartments*: space between a membrane and the immediately inner membranes in the case of non-elementary membranes) injectively labeled with $1, \ldots, m$;
(3) $Var_i$ $(1 \leq i \leq m)$ is a finite set of variables $x_{j,i}$ from compartment $i$ whose values must be natural numbers (they can also be zero), the value of $x_{j,i}$ at time $t \in \mathbf{N}$ is denoted by $x_{j,i}(t)$, and the initial values for the variables in compartment $i$ is denoted by the vector $Var_i(0)$ (if the initial value of variable $x_{j,i}$ is $k$, then will denote it by $x_{j,i}[k]$, that is $x_{j,i}[k]$ means that $x_{j,i}(0) = k$);
(4) $Pr_i$ $(1 \leq i \leq m)$ is a finite set of programs from compartment $i$. Programs process variables and have two components: a production function and a repartition protocol. The $l$-th program in $Pr_i$ has the following form:

$$Pr_{l,i} = (F_{l,i}(x_{1,i}, \ldots, x_{k_l,i}), c_{l,1}|v_1, \ldots, c_{l,n_l}|v_{n_l}) \qquad (6.1)$$

which can be described as $F_{l,i}(x_{1,i}, \ldots, x_{k_l,i}) \rightarrow c_{l,1}|v_1, \ldots, c_{l,n_l}|v_{n_l}$, where

(i) $F_{l,i}(x_{1,i}, \ldots, x_{k_l,i})$ is the production function, being $x_{1,i}, \ldots, x_{k_l,i}$ variables in $Var_i$ (usually, only polynomials with integer coefficients are considered);
(ii) $c_{l,1}|v_1, \ldots, c_{l,n_l}|v_{n_l}$ is the repartition protocol associated with the program, being $c_{l,1}, \ldots, c_{l,n_l}$ natural numbers (the idea is that the coefficients specify the proportion of the current production distributed to each variable $v_1, \ldots, v_{n_l}$ in $\bigcup_{i=1}^{m} Var_i$);

(5) $x_{j_0,i_0}$ is a distinguished variable (from a distinguished compartment $i_0$) which provides the result of a computation.

Next we describe how a program processes variables. Consider a program

$$(F_{l,i}(x_{1,i}, \ldots, x_{k_i,i}), c_{l,1}|v_1, \ldots, c_{l,n_l}|v_{n_l})$$

At a time instant $t \geq 0$ we compute $(F_{l,i}(x_{1,i}(t), \ldots, x_{k_i,i}(t))$. The value

$$q_{l,i}(t) = \frac{F_{l,i}(x_{1,i}(t), \ldots, x_{k_i,i}(t))}{c_{l,1} + \cdots + c_{l,n_l}}$$

represents the *unary portion* at instant $t$ to be distributed according to the repartition expression to variables $v_1, \ldots, v_{n_l}$. Then $q_{l,i}(t) \cdot c_{l,s}$ is the contribution added to the current value from $v_s$ $(1 \leq s \leq n_l)$, at step $t + 1$.

A production function may use only part of the variables from a compartment (a variable is *productive* if it appears in a production rule). The values of productive variables are "consumed" (become zero) when the production function is used and the other variables retain their values. To these values one adds all "contributions" received from the neighboring compartments, according to the repartition protocols, forming the new values. Each program in each membrane can only be used once in every computation step. All the programs are executed in a parallel manner.

### 6.2.2   An Example for NPS

We illustrate the previous definition with the numerical P systems $\Pi_1$ of degree 2 given in Fig. 6.1:

$$\Pi_1 = (2, \mu, (Var_1, Pr_1, Var_1(0)), (Var_2, Pr_2, Var_2(0)), x_{j_0, i_0}),$$

where

- the hierachical membrane structure $\mu$ consists of two membranes: the skin membrane labeled by 1 and the inner membrane labeled by 2;
- $Var_1 = \{x_{1,1}, x_{2,1}\}$, $Var_2 = \{x_{1,2}, x_{2,2}\}$, $Var_1(0) = (1, 2)$ and $Var_2(0) = (3, 1)$;
- $Pr_1 = \{Pr_{1,1}\}$ and $Pr_2 = \{Pr_{1,2}\}$, being

$$
\begin{aligned}
Pr_{1,1} &= (F_{1,1}(x_{1,1}, x_{2,1}), c_{1,1}|v_1 + c_{1,2}|v_2 + c_{1,3}|v_3) \\
&= (4x_{1,1} + x_{2,1}, 1|x_{1,1} + 1|x_{2,1} + 1|x_{2,2}) \\
Pr_{1,2} &= (F_{1,2}(x_{1,2}), c'_{1,1}|v'_1 + c'_{1,2}|v'_2 + c'_{1,3}|v'_3) \\
&= (3x_{1,2} - 3, 1|x_{1,1} + 1|x_{1,2} + 1|x_{2,2})
\end{aligned}
$$

that is,

$$
\begin{aligned}
&Pr_{1,1} \text{ is the program } 4x_{1,1} + x_{2,1} \to 1|x_{1,1} + 1|x_{2,1} + 1|x_{2,2} \\
&Pr_{1,2} \text{ is the program } 3x_{1,2} - 3 \to 1|x_{1,1} + 1|x_{1,2} + 1|x_{2,2}.
\end{aligned}
$$

The production functions are:

($\star$) Associated with program $Pr_{1,1}$ : $F_{1,1} = 4x_{1,1} + x_{2,1}$;
($\star$) Associated with program $Pr_{1,2}$ : $F_{1,2} = 3x_{1,2} - 3$.

The repartition protocols are:

($\star$) Associated with program $Pr_{1,1}$ :
$$c_{1,1}|v_1 + c_{1,2}|v_2 + c_{1,3}|v_3 = 1|x_{1,1} + 1|x_{2,1} + 1|x_{2,2};$$
($\star$) Associated with program $Pr_{1,2}$ :
$$c'_{1,1}|v'_1 + c'_{1,2}|v'_2 + c'_{1,3}|v'_3 = 1|x_{1,1} + 1|x_{1,2} + 1|x_{2,2}.$$

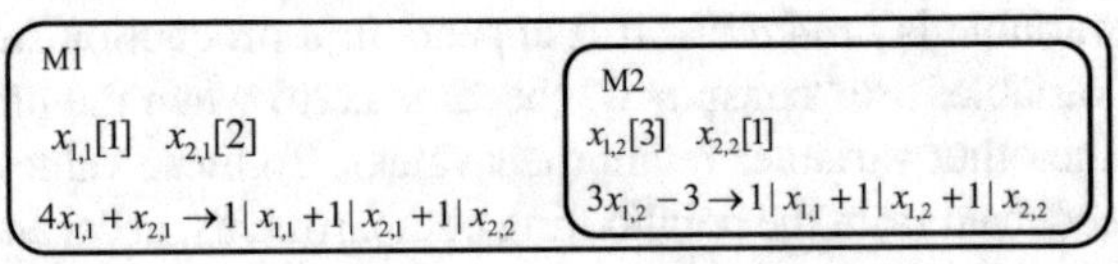

**Fig. 6.1** The numerical P system $\Pi_1$

| M1 | M2 |
|---|---|
| $x_{1,1}[1]$   $x_{2,1}[2]$ | $x_{1,2}[3]$   $x_{2,2}[1]$ |
| $4x_{1,1} + x_{2,1} \to 1\mid x_{1,1} +1\mid x_{2,1} +1\mid x_{2,2}$ | $3x_{1,2} -3 \to 1\mid x_{1,1} +1\mid x_{1,2} +1\mid x_{2,2}$ |

- $x_{j_0,i_0}$ is, for instance, $x_{1,1}$ (the first variable in compartment 1).

Next we summarize the composition of each membrane:

- Membrane labeled by 1 (the skin membrane):

  - Variables: $x_{1,1}, x_{2,1}$. The initial values are $x_{1,1}(0) = 1, x_{2,1}(0) = 2$.
  - Programs: only one program, $Pr_{1,1}$.
  - Production function from $Pr_{1,1}$: $F_{1,1} = 4x_{1,1} + x_{2,1}$.
  - Repartition protocol from $Pr_{1,1}$: $1|x_{1,1} + 1|x_{2,1} + 1|x_{2,2}$.

- Membrane labeled by 2 (the inner membrane):

  - Variables: $x_{2,1}, x_{2,2}$. The initial values are $x_{2,1}(0) = 3, x_{2,2}(0) = 1$.
  - Programs: only one program, $Pr_{2,1}$.
  - Production function from $Pr_{1,2}$: $F_{1,2} = 3x_{1,2} - 3$.
  - Repartition protocol from $Pr_{1,2}$: $1|x_{1,1} + 1|x_{1,2} + 1|x_{2,2}$.

Next we describe how the system $\Pi_1$ evolves from step $t$ to step $t + 1$, for each $t \geq 0$:

- We compute the unary portions associated with each program $Pr_{1,1}$ and $Pr_{1,2}$ at instant $t$:

  - $q_{1,1}(t) = \frac{F_{1,1}(x_{1,1}(t), x_{2,1}(t))}{c_{1,1} + c_{1,2} + c_{1,3}} = \frac{4x_{1,1}(t) + x_{2,1}(t)}{1+1+1}$;
  - $q_{1,2}(t) = \frac{F_{1,2}(x_{1,2}(t))}{c'_{1,1} + c'_{1,2} + c'_{1,3}} = \frac{3x_{1,2}(t) - 3}{1+1+1}$.

- We distribute the corresponding portions to variables to the repartition protocols:

  - Variable $x_{1,1}$ appears in the production function $F_{1,1}$ and the repartition protocols of membrane 1 and membrane 2, hence

  $$x_{1,1}(t + 1) = 0 + c_{1,1} \cdot q_{1,1}(t) + c'_{1,1} \cdot q_{1,2}(t).$$

  - Variable $x_{2,1}$ appears in the production function $F_{1,1}$, but only in the repartition protocol from membrane 1, so

  $$x_{2,1}(t + 1) = 0 + c_{1,2} \cdot q_{1,1}(t).$$

  - Variable $x_{1,2}$ appears in the production function $F_{1,2}$, but only in the repartition protocol of membrane 2, so

  $$x_{1,2}(t + 1) = 0 + c'_{1,2} \cdot q_{1,2}(t).$$

  - Variable $x_{2,2}$ appears only in the repartition protocols of membrane 1 and membrane 2, consequently

  $$x_{2,2}(t + 1) = x_{2,2}(t) + c_{1,3} \cdot q_{1,1}(t) + c'_{1,3} \cdot q_{1,2}(t).$$

We illustrate the previous process in four computation steps:

**Step 1.**

- We compute the unary portions associated with each program $Pr_{1,1}$ and $Pr_{1,2}$ at instant $t = 0$:

  - $q_{1,1}(0) = \frac{4x_{1,1}(0)+x_{2,1}(0)}{c_{1,1}+c_{1,2}+c_{1,3}} = \frac{4\cdot 1+2}{1+1+1} = 2$;
  - $q_{1,2}(0) = \frac{3x_{1,2}(0)-3}{c'_{1,1}+c'_{1,2}+c'_{1,3}} = \frac{3\cdot 3-3}{1+1+1} = 2$.

- We distribute the corresponding portions to variables to the repartition protocols obtaining the new variables' values for membranes 1 and 2:

  - $x_{1,1}(1) = 0 + c_{1,1} \cdot q_{1,1}(0) + c'_{1,1} \cdot q_{1,2}(0) = 0 + 1\cdot 2 + 1\cdot 2 = 4$;
  - $x_{2,1}(1) = 0 + c_{1,2} \cdot q_{1,1}(0) = 0 + 1\cdot 2 = 2$;
  - $x_{1,2}(1) = 0 + c'_{1,2} \cdot q_{1,2}(0) = 0 + 1\cdot 2 = 2$;
  - $x_{2,2}(1) = x_{2,2}(0) + c_{1,3} \cdot q_{1,1}(0) + c'_{1,3} \cdot q_{1,2}(0) = 1 + 1\cdot 2 + 1\cdot 2 = 5$.

**Step 2.**

- We compute the unary portions associated with each program $Pr_{1,1}$ and $Pr_{1,2}$ at instant $t = 1$:

  - $q_{1,1}(1) = \frac{4x_{1,1}(1)+x_{2,1}(1)}{c_{1,1}+c_{1,3}} = \frac{4\cdot 4+2}{1+1+1} = 6$;
  - $q_{1,2}(1) = \frac{3x_{1,2}(1)-3}{c'_{1,1}+c'_{1,2}+c'_{1,3}} = \frac{3\cdot 2-3}{1+1+1} = 1$.

- We distribute the corresponding portions to variables of the repartition protocols obtaining the new variables' values for membranes 1 and 2:

  - $x_{1,1}(2) = 0 + c_{1,1} \cdot q_{1,1}(1) + c'_{1,1} \cdot q_{1,2}(1) = 0 + 1\cdot 6 + 1\cdot 1 = 7$;
  - $x_{2,1}(2) = 0 + c_{1,2} \cdot q_{1,1}(1) = 0 + 1\cdot 6 = 6$;
  - $x_{1,2}(2) = 0 + c'_{1,2} \cdot q_{1,2}(1) = 0 + 1\cdot 1 = 1$;
  - $x_{2,2}(2) = x_{2,2}(1) + c_{1,3} \cdot q_{1,1}(1) + c'_{1,3} \cdot q_{1,2}(1) = 5 + 1\cdot 6 + 1\cdot 1 = 12$.

**Step 3.**

- We compute the unary portions associated with each program $Pr_{1,1}$ and $Pr_{1,2}$ at instant $t = 1$:

  - $q_{1,1}(2) = \frac{4x_{1,1}(2)+x_{2,1}(2)}{c_{1,1}+c_{1,2}+c_{1,3}} = \frac{4\cdot 7+6}{1+1+1} = \frac{34}{3}$;
  - $q_{1,2}(2) = \frac{3x_{1,2}(2)-3}{c'_{1,1}+c'_{1,2}+c'_{1,3}} = \frac{3\cdot 1-3}{1+1+1} = 0$;

- We distribute the corresponding portions to variables of the repartition protocols obtaining the new variables' values for membranes 1 and 2:

  - $x_{1,1}(3) = 0 + c_{1,1} \cdot q_{1,1}(2) + c'_{1,1} \cdot q_{1,2}(2) = 0 + 1\cdot \frac{34}{3} + 1\cdot 0 = \frac{34}{3}$;
  - $x_{2,1}(3) = 0 + c_{1,2} \cdot q_{1,1}(2) = 0 + 1\cdot \frac{34}{3} = \frac{34}{3}$;
  - $x_{1,2}(3) = 0 + c'_{1,2} \cdot q_{1,2}(2) = 0 + 1\cdot 0 = 0$;
  - $x_{2,2}(3) = x_{2,2}(2) + c_{1,3} \cdot q_{1,1}(2) + c'_{1,3} \cdot q_{1,2}(2) = 12 + 1\cdot \frac{34}{3} + 1\cdot 0 = \frac{70}{3}$.

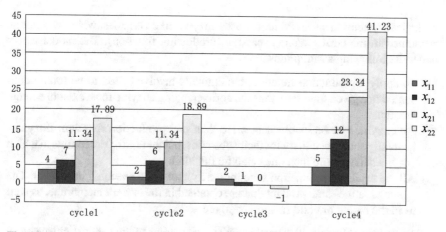

**Fig. 6.2** Evolution of variables during four computation steps of NPS

## Step 4.

- We compute the unary portions associated with each program $Pr_{1,1}$ and $Pr_{1,2}$ at instant $t = 1$:

  - $q_{1,1}(3) = \frac{4x_{1,1}(3)+x_{2,1}(3)}{c_{1,1}+c_{1,2}+c_{1,3}} = \frac{4 \cdot \frac{34}{3}+\frac{34}{3}}{1+1+1} = \frac{170}{9}$;
  - $q_{1,2}(3) = \frac{3x_{1,2}(3)-3}{c'_{1,1}+c'_{1,2}+c'_{1,3}} = \frac{3 \cdot 0 - 3}{1+1+1} = -1$.

- We distribute the corresponding portions to variables of the repartition protocols obtaining the new variables' values for membranes 1 and 2:

  - $x_{1,1}(4) = 0 + c_{1,1} \cdot q_{1,1}(3) + c'_{1,1} \cdot q_{1,2}(3) = 0 + 1 \cdot \frac{170}{9} + 1 \cdot (-1) = \frac{161}{9}$;
  - $x_{2,1}(4) = 0 + c_{1,2} \cdot q_{1,1}(3) = 0 + 1 \cdot \frac{170}{9} = \frac{160}{9}$;
  - $x_{1,2}(4) = 0 + c'_{1,2} \cdot q_{1,2}(3) = 0 + 1 \cdot (-1) = -1$;
  - $x_{2,2}(4) = x_{2,2}(3) + c_{1,3} \cdot q_{1,1}(3) + c'_{1,3} \cdot q_{1,2}(3) = \frac{70}{3} + 1 \cdot \frac{170}{9} + 1 \cdot (-1) = \frac{371}{9}$.

The evolution of the variables during the four computation steps can be presented in a graphical way, as shown in Fig. 6.2.

## 6.2.3 ENPS

Deterministic numerical P systems have only one production function inside each membrane. Even if there are multiple production functions inside each membrane, only one is still selected in a random/non-deterministic way. To make NPS model suitable for more complex applications and improve the computational efficiency, Pavel et al. [28] proposed an extended version of NPS, called enzymatic numerical P systems model and its computational power was shown by Vasile et al. in [32]. In an ENPS model, an enzyme is a protein, which catalyzes reactions occurring in biological systems. The ENPS model uses enzyme-like variables, which will not be consumed in the corresponding reaction (production function). The model works under the following assumptions:

(1) To apply a production function, the amounts involved need to be totally consumed in a given time step, thus the corresponding enzyme must exist in enough amounts.
(2) An enzyme can be involved in more than one production function. Enzymes generally show specificity for their substrates, but there exist exceptions in which enzymes have more than one catalytic function.
(3) All the available production functions are chosen and executed in parallel. The amounts of the same product that are obtained in the current membrane, as well as in the parent or child membrane(s) are added up.

The enzyme-like variables used in ENPS are similar to the catalyst objects used by other classes of P systems [25]. Catalysts are objects that are needed for the reaction to take place, but are not modified in it. Although enzyme-like variables behave similarly to catalysts (they are needed in the reaction, but are not consumed in it), there are two differences: on the one hand, an enzyme can be produced or consumed in other reactions, except for the ones that are catalyzed by it, just like non-enzymatic variables; on the other hand, the amount of the enzyme is also important in establishing if a rule is active. Like biological enzymes, a number of enzyme molecules can catalyze only a certain amount of a specific substrate (compound).

An ENPS model of degree $m$ $(m \geq 1)$ is formally expressed as a tuple

$$\Pi = (m, \mu, (Var_1, Pr_1, Var_1(0)), \ldots, (Var_m, Pr_m, Var_m(0)), x_{j_0,i_0}) \qquad (6.2)$$

where

(1) $m$ is the number of membranes used in the system (the degree of $\Pi$);
(2) $\mu$ is a membrane structure (a rooted tree) consisting of $m$ membranes, with the membranes (and hence the *regions* or *compartments*: space between a membrane and the immediately inner membranes in the case of non-elementary membranes) injectively labeled by $1, \ldots, m$;
(3) $Var_i$ $(1 \leq i \leq m)$ is a finite set of variables $x_{j,i}$ from compartment $i$ whose values must be natural numbers (they can also be zero), the value of $x_{j,i}$ at time $t \in \mathbf{N}$ is denoted by $x_{j,i}(t)$, and the initial values for the variables in compartment $i$ are denoted by the vector $Var_i(0)$ (if the initial value of variable $x_{j,i}$ is $k$, then will denote it by $x_{j,i}[k]$, that is $x_{j,i}[k]$ means that $x_{j,i}(0) = k$);
(4) $E_i$ $(1 \leq i \leq m)$ is a finite set of enzyme variables from compartment $i$, that is, $E_i \subseteq Var_i$;
(5) $Pr_i$ $(1 \leq i \leq m)$ is a finite set of programs from compartment $i$. Programs process variables and have two components: a production function and a repartition protocol. The $l$-th program in $Pr_i$ has one of the following forms:

(a) Non-enzymatic form, which is exactly like the form in NPS:

$$Pr_{l,i} = (F_{l,i}(x_{1,i}, \ldots, x_{k_l,i}), c_{l,1}|v_1, \ldots, c_{l,n_l}|v_{n_l}) \qquad (6.3)$$

which can be described as $F_{l,i}(x_{1,i}, \ldots, x_{k_l,i}) \to c_{l,1}|v_1, \ldots, c_{l,n_l}|v_{n_l}$, where

- $F_{l,i}(x_{1,i}, \ldots, x_{k_l,i})$ is the production function and $x_{1,i}, \ldots, x_{k_l,i}$ variables in $Var_i$ (usually, only polynomials with integer coefficients are considered);
- $c_{l,1}|v_1, \ldots, c_{l,n_l}|v_{n_l}$ is the repartition protocol associated with the program and $c_{l,1}, \ldots, c_{l,n_l}$ natural numbers (the idea is that the coefficients specify the proportion of the current production distributed to each variable $v_1, \ldots, v_{n_l}$ in $\bigcup_{i=1}^{m} Var_i$);

(b) Enzymatic form:

$$Pr_{l,i} = (F_{l,i}(x_{1,i}, \ldots, x_{k_l,i}), e_{j,i}, c_{l,1}|v_1, \ldots, c_{l,n_l}|v_{n_l}) \qquad (6.4)$$

which can be described as $F_{l,i}(x_{1,i}, \ldots, x_{k_l,i})(e_{j,i} \to)c_{l,1}|v_1, \ldots, c_{l,n_l}|v_{n_l}$, where

- $F_{l,i}(x_{1,i}, \ldots, x_{k_l,i})$ is the production function and $x_{1,i}, \ldots, x_{k_l,i}$ variables in $Var_i$ (usually, only polynomials with integer coefficients are considered);
- $c_{l,1}|v_1, \ldots, c_{l,n_l}|v_{n_l}$ is the repartition protocol associated with the program and $c_{l,1}, \ldots, c_{l,n_i}$ natural numbers (the idea is that the coefficients specify the proportion of the current production distributed to each variable $v_1, \ldots, v_{n_l}$ in $\bigcup_{i=1}^{m} Var_i$);
- $e_{j,i} \in E_i$.

The computing process for an ENPS is shown in Fig. 6.3 [28]. The process is divided into four main steps:

- selection of the active rules in each membrane;
- combination of the active production functions (optional);
- computation of the active production functions within all membranes;
- distribution of the computed results to the variables of the membranes.

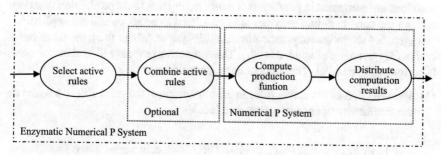

**Fig. 6.3** Computing process of ENPS [28]

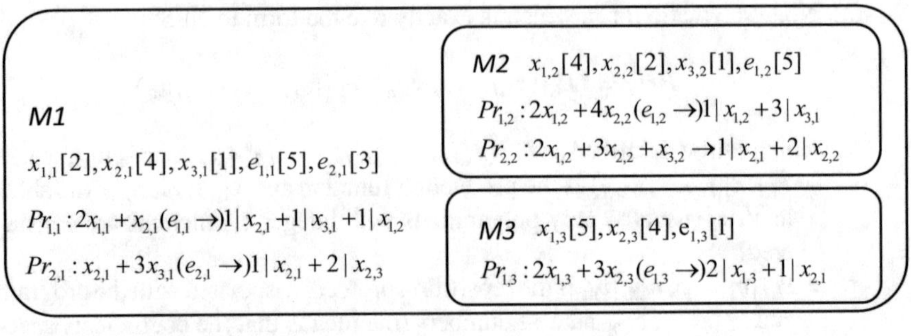

$M1$

$x_{1,1}[2], x_{2,1}[4], x_{3,1}[1], e_{1,1}[5], e_{2,1}[3]$

$Pr_{1,1} : 2x_{1,1} + x_{2,1}(e_{1,1} \rightarrow)1 \mid x_{2,1} + 1 \mid x_{3,1} + 1 \mid x_{1,2}$

$Pr_{2,1} : x_{2,1} + 3x_{3,1}(e_{2,1} \rightarrow)1 \mid x_{2,1} + 2 \mid x_{2,3}$

$M2$   $x_{1,2}[4], x_{2,2}[2], x_{3,2}[1], e_{1,2}[5]$

$Pr_{1,2} : 2x_{1,2} + 4x_{2,2}(e_{1,2} \rightarrow)1 \mid x_{1,2} + 3 \mid x_{3,1}$

$Pr_{2,2} : 2x_{1,2} + 3x_{2,2} + x_{3,2} \rightarrow 1 \mid x_{2,1} + 2 \mid x_{2,2}$

$M3$   $x_{1,3}[5], x_{2,3}[4], e_{1,3}[1]$

$Pr_{1,3} : 2x_{1,3} + 3x_{2,3}(e_{1,3} \rightarrow)2 \mid x_{1,3} + 1 \mid x_{2,1}$

**Fig. 6.4**  ENPS with three membranes

To illustrate the ENPS computing process in detail, a simple ENPS with three membranes will be presented. Before the description, it is worth pointing out that the enzymatic mechanism is inspired by biological processes, but P systems themselves do not aim at modeling chemical reactions and they are usually studied for their computational power. So, the enzymatic mechanism will not constrain the computational model by considering all the real biological facts.

Let us consider one membrane M with the following variables: two molecule variables $x_{11}, x_{21}$ whose initial values are 2 and 4, respectively; one enzymatic variable $e_{11}$ whose initial value is 5, and one production function $2x_{11} + x_{21}(e_{11} \rightarrow)$. This production function represents a biochemical reaction taking place in a cell: two molecule variables $x_{11}(2)$ will react with one molecule $x_{21}$ catalyzed by one enzymatic variable $e_{11}$. Assuming that enough enzymatic molecule are available, then the biochemical reactions can be executed. The initial values of the variables are the basis of the initial concentration of molecules. So the rule can be applied only if the concentration of the enzyme molecules is greater than the minimum between the concentrations of the reactants. In this example, $e_{11} > \min(x_{11}/2, x_{21})$ is considered as the condition. The simplified and more general condition is $e_{11} > \min(x_{11}, x_{21})$. So considering the initial amounts of the reactants and the enzyme in this case: $5 > \min(2, 4)$, this rule is active.

If more than one production function is active in a membrane, all of them should be combined in some way. A simple way to apply the selected production functions is to compute them in parallel. In the following example, all the active production functions are computed in parallel. It is worth noting that those production functions without the corresponding enzymes are considered to be active and are applied.

Figure 6.4 shows an enzymatic numerical P system, where there are three membranes ($M_2$ and $M_3$ are included in $M_1$). The computing process within all membranes is performed in parallel. The process has four main steps:

(1) The selection of the active rules in each membrane, the rules $Pr_{j,i}$, is based on the availability of enough enzyme molecules

 – $M_1$:
 – if $Pr_{1,1} : e_{1,1} > \min(x_{1,1}/2, x_{2,1}) = TRUE \Rightarrow$ active production function;
 – if $Pr_{2,1} : e_{2,1} > \min(x_{2,1}, x_{3,1}/3) = TRUE \Rightarrow$ active production function;

- $M_2$:
- if $Pr_{1,2} : e_{1,2} > \min(x_{1,2}/2, x_{2,2}/4) = TRUE \Rightarrow$ active production function;
- $Pr_{2,2} : has\ no\ enzyme \Rightarrow$ active production function;
- $M_3$:
- if $Pr_{1,3} : e_{1,3} > \min(x_{1,3}/2, x_{2,3}/3) = FALSE \Rightarrow$ inactive production function.

(2) The combination of the active production functions is based on the fact that all the rules which are active will be applied independently and in parallel

(3) The computation of the active production functions within all membranes is given by

- $M_1$:
- $Pr_{1,1} : 2x_{1,1} + x_{2,1} = 2 \cdot 2 + 4 = 8;$
- $Pr_{2,1} : x_{2,1} + 3x_{3,1} = 4 + 3 \cdot 1 = 7;$
- $M_2$:
- $Pr_{1,2} : 2x_{1,2} + 4x_{2,2} = 2 \cdot 4 + 4 \cdot 2 = 16;$
- $Pr_{2,2} : 2x_{1,2} + 3x_{2,2} + x_{3,2} = 2 \cdot 4 + 3 \cdot 2 + 1 = 15;$
- $M_3$:
- no rule is computed.

(4) The distribution of the computed results to the membranes' variables based on the distribution protocol is as follows

- $q_{1,1} = \frac{2x_{1,1}+x_{2,1}}{c_{1,1}+c_{2,1}+c_{3,1}} = \frac{2 \cdot 2+4}{1+1+1} = \frac{8}{3};$

- $q_{2,1} = \frac{x_{2,1}+3x_{3,1}}{c'_{1,1}+c'_{2,1}} = \frac{4+3 \cdot 1}{1+2} = \frac{7}{3};$

- $q_{1,2} = \frac{2x_{1,2}+4x_{2,2}}{c_{1,2}+c_{2,2}} = \frac{2 \cdot 4+4 \cdot 2}{1+3} = \frac{16}{4} = 4;$

- $q_{2,2} = \frac{2x_{1,2}+3x_{2,2}+x_{3,2}}{c'_{1,2}+c'_{2,2}} = \frac{2 \cdot 4+3 \cdot 2+1}{1+2} = \frac{15}{3} = 5.$

We distribute the corresponding portions to variables of the repartition protocols obtaining the new variables' values for membranes, i.e., the amounts of the same variable which are produced in the native membrane, or in the parent membrane or in a child membrane are added:

- $x_{2,1} = c_{1,1} \cdot q_{1,1} + c'_{1,1} \cdot q_{2,1} + c'_{1,2} \cdot q_{2,2} = 1 \cdot \frac{8}{3} + 1 \cdot \frac{7}{3} + 1 \cdot 5 = 10;$

- $x_{3,1} = c_{2,1} \cdot q_{1,1} + c_{2,2} \cdot q_{1,2} = 1 \cdot \frac{8}{3} + 3 \cdot 4 = \frac{44}{3};$

- $x_{1,2} = c_{3,1} \cdot q_{1,1} + c_{1,2} \cdot q_{1,2} = 1 \cdot \frac{8}{3} + 1 \cdot 4 = \frac{20}{3};$

- $x_{2,2} = c'_{2,2} \cdot q_{2,2} = 2 \cdot 5 = 10;$

- $x_{2,3} = c'_{2,1} \cdot q_{2,1} = 2 \cdot \frac{7}{3} = \frac{14}{3}.$

ENPS is a distributed and parallel computing model with flexible process control: enzyme variables can be used for conditional transmembrane transport and the program flow control; active rules are computed in parallel in their membrane and unnecessary rules are not executed; the calculated results are distributed in a globally uniform way; thus the computing power of ENPS is optimized and the membrane structure representation is very efficient for designing robotic behaviors.

## 6.3  Preliminaries of Mobile Robot Control

A mobile robot is an automatic machine that is capable of moving around in its environment and is not fixed to one physical location. It has sensors allowing it to perceive the environment, and has actuators allowing them to modify its environment, and also has a micro-processor allowing them to process the sensory information and control its actuators accordingly. So, mobile robots can be "autonomous" (usually called *autonomous mobile robot* (AMR)), which means they are capable of navigating an uncontrolled environment without the direct guidance from physical or electro-mechanical guidance devices. Alternatively, mobile robots can rely on guiding devices that allow them to travel a pre-defined navigation route in relatively controlled space (also called *autonomous guided vehicle* (AGV)). Mobile robots have become more commonplace in commercial, military, security, industrial settings, etc. The mobile robot is also a major focus of the current research and almost every major university has one or more labs that focus on mobile robot research [24].

To adapt the environment and to complete the task, from the mechanical and electronic points of view, robots ought to act as artificial animals. They are equipped with many sensors (distance sensors, cameras, touch sensors, position sensors, temperature sensors, battery level sensors, accelerometers, microphones, wireless communication, etc.) and actuators (motors, speakers, LEDs, etc.). According to today's technology, the hardware technology for intelligent robots is currently abundant and available, however, it is still necessary to develop a better software technology (control algorithm) to drive the robots to act as intelligent ones. In other words, we currently have the bodies of our intelligent robots, but we lack their minds. Robotics research trends will not focus on robot hardware, but on robot softwares because this is the most challenging topic in the design of more and more intelligent robots.

There are many types of mobile robots for different tasks, such as insect robots, humanoid robots, unmanned aerial vehicles (UAV), underwater robots and wheeled robots. The wheeled robots can also be classified into two-wheel, four-wheel and multi-wheel robots. Accordingly, the design of robot softwares (controller design) must aim at the special robots, further the special controller design for special robots with special sensors and actuators. This chapter mainly discusses the controller design of a commonly useful wheeled robot, especially the mini wheel mobile robot (e-puck and khepera) [16, 23].

In order to understand how to design an e-puck controller, first we need to know its physical characteristics. E-puck was designed by Francesco Mondada and Michael

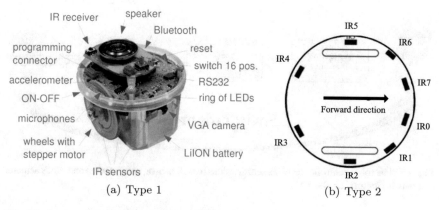

**Fig. 6.5**  Sensors and actuators of e-puck robots [8, 13, 16, 23]

**Table 6.1**  Features of the e-puck robot [8, 13, 16, 23]

| Features | Technical information |
|---|---|
| Size, weight | 70 mm diameter, 55 mm height, 150 g |
| Battery autonomy | 5 Wh LiION rechargeable and removable battery providing about 3 h autonomy |
| Processor | dsPIC 30F6014A @ 60 Mhz (15 MIPS) 16 bit microcontroller with DSP core |
| Memory | RAM: 8 KB; FLASH: 144 KB |
| Motors speed | 2 stepper motors with a 50:1 reduction gear, resolution: 0.13 mm, Max: 15 cm/s |
| IR sensors | 8 infra-red sensors measuring ambient light and proximity of objects up to 6 cm |
| Camera | VGA color camera with resolution 480x640 (typical use: 52x39 or 480x1) |
| Microphones | 3 omni-directional microphones for sound localization |
| Accelerometer | 3D accelerometer along the X, Y and Z axis |
| LEDs | 8 independent red LEDs on the ring, green LEDs in the body, 1 strong red LED in front |
| PC connection | Standard serial port up to 115 kbps |
| Wireless | Bluetooth for robot-computer and robot-robot wireless communication |
| Remote control | Infra-red receiver for standard remote control commands |
| Expansion bus | Large expansion bus designed to add new capabilities |
| Simulation | Webots facilitates the use of the e-puck robot: powerful simulation, remote control, graphical and C programming systems |

Bonani in 2006 at EPFL, the Swiss Federal Institute of Technology in Lausanne. It has become a tool for research and university education. The e-puck robot is powered by a dsPIC (Digital Signal Programmable Integrated Circuit) processor. This is a micro-controller processor produced by the Microchip company and is able

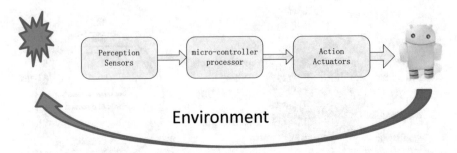

**Fig. 6.6** The robot controller receives sensor values (ex: IR sensor, camera, etc.) and sends actuator commands (motors, LEDs, etc.)

to efficiently perform digital signal processing and works like the robot's brain. The e-puck robot also features a large number of sensors and actuators as depicted in Fig. 6.5 and devices as described by Table 6.1 [8, 13, 16, 23]. The electronic layout is described in [17]. Some of the sensors and actuators will be discussed in detail later in this chapter when the design of some controllers will be described.

A robot micro-controller processor perceives and handles only the values measured by the robot sensors and sends commands to the robot actuators as depicted in Fig. 6.6. This model is also known as the Observe-Decide-Act loop (ODA), which is also used to represent self-adaptive computing systems [19]. The sensor values are modified by the environment and processed by some behavior controllers in processors. A robot can adapt to the environment with its actuators for avoiding motion or stationary obstacles, tracking some motion targets or recording its own motion trajectories.

## 6.4   Membrane Controllers

The architecture of robot controllers usually uses a cognitive architecture, which is a domain-generic cognitive model exhibiting a wide range of cognitive abilities, such as high level cognitive architectures (logical reasoning, organization, learning) and low level cognitive architectures (perception, action, etc.) [5, 15]. Cognitive modeling is the process of developing computational models of cognitive processes with the aim of understanding human cognition. In the community of designing robots controller, there is a long-term tendency to mimic the organization and functional complexity of the human brain. The cognitive architectures of intelligent robots are designed by using various bio-inspired computing methods, such as neural networks, evolutionary computation, swarm intelligence, fuzzy logics and expert knowledge [1, 10, 35, 41].

Membrane controllers are a sort of controllers with cognitive architectures and designed by using the syntax (such as the membrane structure, initial multisets and evolution rules) and semantics of membrane systems. Here we use Numerical P Systems (NPS) and Enzymatic Numerical P Systems (ENPS) to devise membrane controllers to control wheeled mobile robots, to be specific, to solve the mobile robot environmental adaptability problems, such as obstacle avoidance, follower, wall following and localization. The use of NPS or ENPS in the membrane controllers is based on the following considerations: the values of the variables received from various sensors are real-valued numbers and the computation is performed on real-valued variables, instead of symbols like in usual P systems.

The algorithm design for robot controllers is fundamentally different from the algorithm design for solving the problems with well-defined and completely known data. The robot world is only partially known and imperfectly modeled, so we usually use a world model to approximate the movement law of a robot. Thus, the robot controllers have to be designed to overcome uncertainty, errors in modeling and measurement, and noise. In [5, 7], Buiu et al. started to discuss the possibility of directly using NPS as a new computational paradigm and cognitive architecture for implementing such robot controllers. In [8], Buiu et al. presented several examples of robot behavior controller design with NPS for obstacle avoidance, wall following, following another robot, and used two differential robot E-puck to verify the feasibility of membrane controllers. The controller performance was measured by the mean execution time of a cycle. The cycle represents the computation of a loop in which the robotic system reads the information from the sensors, computes the speeds of the motors based on the information received from the sensors, and sets the new values of the motors' speeds.

The generic structure of membrane controllers is summarized as shown in Algorithm 1 [5, 7, 8], where there are four main steps. The first step is the initialization phase of the robot parameters with respect to the specific robot and to the selected behavior. The second step reads the sensors. At the third step, the NPS corresponding to the current membrane computing process is executed (this is done as an http request). The last step is the setting of the motors' speeds. Then the execution of the controller goes back to the second step.

---

**Algorithm 1** Generic structure of membrane controllers

---

**Require:** parameters = initializeBehaviourParameters(robot, behavior)
    While (True)
        sensors = readSensors(robot)
        # simulate Numerical P System on the network server
        query = constructQuery(robot, behaviour, sensors, parameters)
        response = queryWebApp(address, query)
        speed = extractContent(response)
        setSpeed(robot, speed)
    End While

---

Suppose that $u$ and $y$ are the inputs and outputs of a membrane controller and $r$ is the predefined setpoints. Then, a standard membrane controller (MeC) is defined [8] as a tuple:

$$MeC = (m, \mu, (Var_1, Pr_1, Var_1(0)), \dots, (Var_m, Pr_m, Var_m(0)), y, u) \quad (6.5)$$

where

- $m$ ($m \geq 1$) is the number of membranes used in the system (the degree of $MeC$);
- $\mu$ is a membrane structure;
- $Var_i$ is a set of variables from compartment $i$, and the initial values for these variables are denoted by vector $Var_i(0)$;
- $Pr_i$ is a set of programs from compartment $i$, which are composed of a polynomial production function and a repartition protocol, and chosen such that the process outputs $y$ match the predefined setpoints $r$ as "good" as possible;
- $u$ is a set of input variables of the MeC (from a subset of compartments);
- $y$ is a set of output variables (from a subset of compartments), which provide the result of the computation.

The membrane controller can be understood as a hierarchical cognitive architecture [8]. It may receive higher level commands from the upper levels, which are also membranes (using the communication rules coming from the membrane systems). By using the membrane-based structure, the membrane controller can be easily interfaced with the other modules and integrated into the global cognitive architecture. This is one of the main advantages of the use of numerical P systems as building blocks for cognitive/control architectures. On the other hand, the membrane controller inherits numerical variables and parallel distributed structure of numerical P systems as a controller modeling tool. Membranes in an NPS can be distributed over a grid or over a network of micro controllers in a robot. The computation in each membrane region (the execution of the membranes programs) can also be performed in parallel.

To further improve the design and operation efficiency of the membrane controller, the research team at Politehnica University of Bucharest proposed an extension of the NPS model, Enzymatic Numerical P Systems (ENPS) in the context of modeling robot behaviors. The ENPS model allows the parallel execution of more rules (programs) inside each membrane while keeping the determinism (the design and implementation of robot controllers require deterministic computational models). The possibility of selecting and executing more production functions inside each membrane makes ENPS be a more flexible modeling tool than NPS. The ENPS robot controller in [29, 30] has a less complex structure than the NPS robot controller in [8], and therefore needs less step computations and has better performance.

Due to the parallelism and distributed structure of membrane computing models and the experimental verification by using a single task membrane controller in [8, 29, 30], the membrane controller has a great potential in the area of controller design for robot applications. Taking advantage of the scalability, modularity, various types of flexible structures and multiple communication rules between membranes, we can combine membrane computing models with various advanced control algorithms to obtain more complex and efficient robot membrane controller. Furthermore, if a variety of membrane computing models are combined with the methodology of artificial intelligence, swarm intelligence or evolutionary computation, we may develop multiple behavioral collaborative work membrane controller for an intelligent robot. For example, we have developed a mobile robot trajectory tracking controller with a good balance between efficiency and effectiveness by combining the efficiency and flexible ENPS with the classic proportional-integral-derivative (PID) control algorithms and neural networks [33].

## 6.5 Software Platforms

This section introduces the software platforms for conducting virtual experiments on the use of membrane controllers associated with mobile robots. The software platform consists of the simulator of NPS/ENPS and the robot simulator. The former is the simulator for numerical P systems (SNUPS) and enzymatic numerical P systems (SimP, for short) and the latter uses Webots, a specialized simulator for mobile robots. SNUPS was developed by the research team at the University of Politehnica [2, 6]. SimP is a parallelized GPU-based simulator for ENPS and was jointly developed by the Research Group on Natural Computing at the University of Seville and the research team at the Politehnica University of Bucharest [18]. Webots is a professional simulation software for mobile robots [14].

A schematic framework for testing membrane controllers on real and simulated robots is shown in Fig. 6.7 [31], where XML files, an NPS/ENPS simulator and Webots are integrated into this framework. The XML files store robot behaviors in a platform-independent way. The NPS/ENPS simulator simulates the computation of NPS/ENPS. The Webots provide a virtual robot platform and realize the connection between the simulated and real robots.

In what follows, we will separately describe SNUPS and Webots simulators in detail.

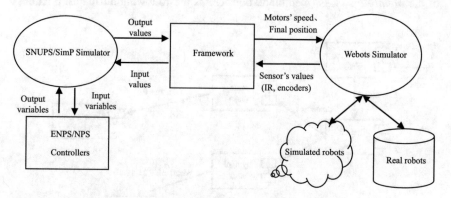

**Fig. 6.7** Schematic framework for testing membrane controllers of robots [31]

### 6.5.1  SNUPS: Numerical P System Simulator

SNUPS, a simulator for numerical P systems, was developed on the standard Java platform [2, 6]. So, it can work on multiple operating systems, such as Linux, UNIX, MAC OS, or Microsoft, equipped with a Java Virtual Machine (JDK 1.6.0 or later). SNUPS has a friendly interface environment. In SNUPS, the rules in different membranes can be performed in a parallel way by using multi-threading software development technology so that they can save resources and minimize computing time. In [18], the research team at the University of Politehnica and the Research Group on Natural Computing have also launched a hardware platform based on GPU for SNUPS/ENPS.

The software can be downloaded from the web site [36]. The online resources include the SNUPS manual and routines. SNUPS is distributed as a Java archive with two mandatory libraries (xml.jar and xmlparserv2.jar) for XML operations. There are two applications: a batch application, where it is supposed that the membranes have already been encoded as an XML file and the aim is to get the results by simulating the system, and a graphical user interface (GUI) application, where a new membrane system, which includes the membrane structure, symbols, evolution rules and contribution table, is created from scratch by using visual components and is saved as an XML file.

Here we first explain the concepts of basic elements in SNUPS in the form of classes. The basic concepts of a membrane system wrapped into corresponding classes are shown in Fig. 6.8 [2]. In the following description, we introduce several main classes, *Membrane*, *Region*, *Symbol*, *Contribution* and *Rule*, step by step.

The *Membrane* class is used to formalize the membranes of the NPS. This class has two components: region and membranes (sub membranes). The instantiation of the *Membrane* class (membrane object) has the following attributes: a name, a

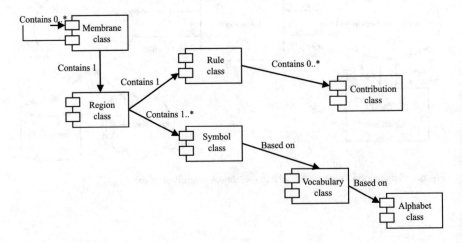

**Fig. 6.8**  Class diagram in SNUPS [2]

membrane-object children list and one region-object. A membrane can have 0 or at least 1 membrane child.

The *Region* class is applied to formalize the membrane's region and represents the container for rules and symbols. The instantiation of the region class (region object) has the following attributes: a name, a rule object list and a symbol object list.

The *Symbol* class is employed to formalize the membrane's symbol class. The instantiation of the symbol class (symbol object) has the following attributes: contribution list symbol, rule's symbol and region's symbol. The contribution list and rule concepts are explained in the following paragraphs. Each symbol object has the following attributes: a name, an initial value (floating numbers), a current value (floating numbers) and an available boolean value. The initial value of the symbol is used for starting computation and depends on the output of the rule and the corresponding contribution list to change the current value of symbols after each rule's computing has finished. If the rules computing restarts and all the current values will be reset, the initial values will be preserved. As mentioned in the last section, each rule's computing step is based on some production functions. The computing output will be distributed among the symbols which belong to the other region or current regions.

The *Contribution* class is used to formalize the membrane's contribution class. The instantiation of the contribution class (contribution object) has the following attributes: a distribution coefficient (contribution value) and a symbol object. The distribution coefficient attribute holds the numerical value from the contributions table region.

The *Rule* class is for formalizing the membrane's rule class. The instantiation of the rule class (rule object) has the following attributes: a contribution-object list and a production function. The contribution object list can be reset: add new symbols, remove symbols, or change symbols (name and/or values).

SNUPS GUI has three main panels: membranes tree, symbols assignment and rules definitions. We take the example shown in Fig. 6.9 to describe how to build a membrane system like this case with SNUPS in detail.

The GUI of this membrane system is shown in Fig. 6.10. The left hand side panel shows the membrane structures (e.g., M1 is the skin membrane), shown in Fig. 6.10a. It can be noticed that the membrane tree structure in the panel containing the regions (e.g., R-2 is the region defined by the second membrane), symbols, rules and contribution tables belong to this membrane system (e.g., $x_{12}[3]$, $x_{22}[1]$ and $x_{32}[0]$ are the symbols; $(x_{12}^2 - x_{22} - 3x_{32})$ is the evolution rule; $x_{22}, x_{32}, x_{23}$ are the contribution table symbols, the variables that receive part of the production function through the coefficients of the current repartition protocol). The symbols allocation panel is shown on the right hand side in Fig. 6.10a, where the symbols of the current membrane are in the white rows, while the symbols belonging to the other membranes are in the gray rows. We can set new symbols and initial values under the same panel.

It is necessary to define the evolution rule and the contribution table for each membrane after symbols have been allocated. The variables of the production function (mathematical function) are the symbols allocated to the corresponding membrane. Figure 6.10b shows an evolution rule like $2x_{11}^2 + 7$ allocated to the first membrane,

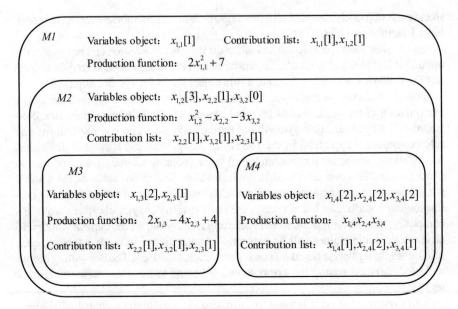

**Fig. 6.9** A sample membrane structure with four membranes

which is edited in the rule editor placed on the bottom right corner panel, and this rule must be validated before assigning it to the membrane system. The validation of each rule consists of the following verifications: (1) if the symbols of the rule belong to the current membrane; (2) if the syntax of the mathematical function is correct. The symbols are added in the contribution table. The rule editor panel from rules tab on the bottom left corner of rules panel allows editing the function of the rule and its syntax validation. The following math operations are allowed in the rule editor:

- addition: +;
- multiplication: ∗;
- subtraction: −;
- division: /;
- powers: ∧;

More details are now described according to [6]. The evolution rules inside each membrane are considered as threads depicting one computing cycle to be performed in a parallel way. The entire P system will terminate after a predefined number of cycles. The values of symbols are updated after each computing cycle by using the sum of the contributions received from the membranes. The updating process of symbols is synchronous, where a barrier point is set so that each thread waits for the arrival of all other threads. Therefore, all the codes are ensured for each thread before the barrier has been completed across all other threads. Finally, a file is used to store the computing results (in the current cycle number and all symbols values). In this file, a comma is applied to separate values to be easily accessed, understood, and further used.

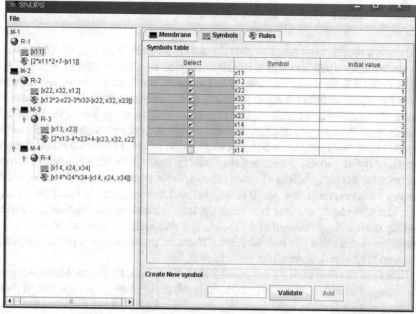

(a) Symbols panel

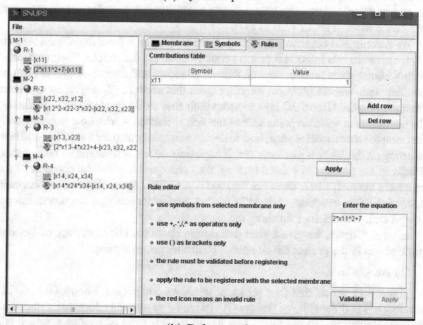

(b) Rules panel

**Fig. 6.10** SNUPS GUI

## 6.5.2 Webots: Mobile Robot Simulator

Webots is a professional mobile robot simulation software package, which has been developed since 1996 and was originally designed by Olivier Michel at EPFL, the Swiss Federal Institute of Technology in Lausanne, Switzerland, in the lab of Jean-Daniel Nicoud. It offers a rapid prototyping environment that allows the user to create 3D virtual worlds with physics properties such as mass, joints, friction coefficients, etc. The user can add simple passive objects or active objects called mobile robots. These robots can have different locomotion schemes (wheeled robots, legged robots, or flying robots). Moreover, they may be equipped with a number of sensor and actuator devices, such as distance sensors, drive wheels, cameras, servos, touch sensors, emitters, receivers, etc. It actually allows the designers to visualize rapidly their ideas, to check whether they meet the requirements of the application, and to develop the intelligent control of the robots, and eventually, to transfer the simulation results into a real robot like two-wheeled differential mobile robot (e-puck, Khepera), humanoids (Asimo), Sony's dog robot (Aibo), etc.

Users can develop their own controller in C, C++, Java, Python or MATLAB programming, and in VRML97 (Virtual Reality Modeling Language) and run Webots on the operating systems like Linux, Windows (Windows 7, Windows Vista and Windows XP, but they are not supported on Windows 98, ME, 2000 or NT4), Macintosh. Users can download more technical details, such as software codes, user's guide and Reference Manual about Webots [37].

We recommend beginners to start with the user's guide from the official website [37]. There are two important points about Webots: (1) the chapter 'Getting Started with Webots' gives an overview of Webots windows and menus. Beginners can be familiar with Webots platform interface from this section; (2) beginners can learn from the chapter 'Tutorials' how to setup their first world (*mybot.wbt*), how to set the environment variable parameter of the first simulated world, how to create their first simple robot model *mybot*, how to design a simple controller for *mybot* and how to debug on the Webots platform, etc. If beginners do not understand some technical details in the process of guided leaning, they can quickly view the contents of the reference manual, where there are detailed descriptions about Webots, such as each node of the data structure, each API function, parameter settings. Beginners can get started with Webots after finishing the User's Guide.

In what follows, we give a brief description about the User Interface of Webots, basic stages in the project development, controller programming.

(1) Webots GUI

Figure 6.11 shows that four principal windows composing Webots GUI: (i) 3D window lies in the middle of Webots GUI, which displays and allows to interact with the 3D simulation; (ii) the Scene tree lies in the left hand of windows, which is a hierarchical representation of the current world; (iii) the Text editor is in the right hand of windows, which allows to edit source codes; (iv) the Console is in the bottom of windows, which displays both controller outputs and compilation. The GUI has eight menus: File, Edit, View, Simulation, Build, Tools, Wizard and Help.

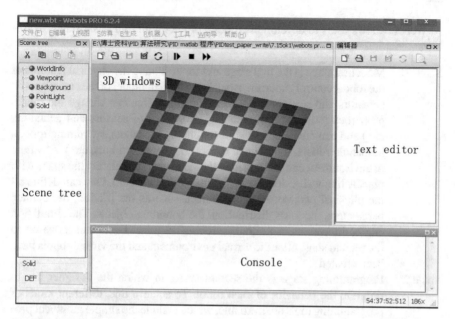

**Fig. 6.11** Webots GUI

- 3D window
  Users can select a solid object in the 3D window by single-clicking. Double-clicking on the object can open the Scene Tree or Robot Window. Dragging the mouse while pressing a mouse button can move the camera of the 3D window. Selecting an object while holding down the shift key and dragging the mouse can move an object.
- The Scene Tree
  The scene tree contains the information that describes a simulated world, such as robots and environment, and its graphical representation. The scene tree is structured like a VRML97 file and composed of a list of nodes, each of which contains fields. Fields can contain values (text strings, numerical values) or other nodes.
- Text editor
  Users can edit their own controller codes in the text editor window. One can run his own controller on a special robot by changing the robot controller attribution, which lies in the scene tree window into the custom controller (edit in the text editor window). One can also change the behavior control of a robot by changing the codes in the text editor.
- Console window
  Console window of Webots GUI shows the debugging information or output parameters of the robot controller just like Visual Studio console.

(2) Basic stages in the project development

Webots allows users to develop his robotic project with four basic stages as follows:

Stage 1:   Modeling stage is the first stage and for designing the physical features of the robots control experiment, such as physical model of the environment, actuators and sensors for the special robot. One can change the robots properties (color, shape, technical properties of sensors and actuators, etc.) and can create any kind of robots (flying robots, swimming robots, humanoid robots, four legged robots and wheeled robots, etc.). Any type of environment can be created in the same way (populating the space with objects like walls, doors, steps, balls, obstacles, etc.). One can define all the physical parameters of the object, such as the friction, the bounce parameters, the mass distribution, the bounding objects. The simulation engine in Webots can simulate robots' physics. Users can move on to the second stage after the virtual environment and the virtual robots have been created

Stage 2:   Programming stage is the second stage, in which the designer has to program the behavior of each robot. To achieve this, different kinds of programming tools are available, which include the simple graphical programming tools for beginners and programming languages (like C, C++ or Java), which are more powerful and enable the development of more complex behaviors. The program controlling a robot is generally an endless loop, which is divided into three sections: (1) read the values measured by the sensors of the robot; (2) compute what should be the next action(s) of the robot and (3) send actuators commands to perform these actions. The easiest section are parts (1) and (3). The most difficult one is part (2), which can be divided into sub-parts like sensor data processing, learning, motor pattern generation, etc.

Stage 3:   Simulation stage is the third stage allowing the designers to test the correctness of their programs. By running the simulation, users can see the robot executing the program. One is able to play interactively with his robot by moving obstacles using the mouse, moving the robot itself. One is also able to visualize the values measured by the sensors, the results of the processing of his program. It is likely that one has to return several times back to the second stage to fix or improve his program and test it again in the simulation stage.

Stage 4:   The final stage is the transfer to a real robot. The control program is transferred into the real robot running in the real world. One could then see whether his control program behaves the same as in simulation or not. If the simulation model of the robot is carefully performed and is calibrated against its real counterpart, the real robot should behave roughly the same as the simulated robot, otherwise, it is necessary to go back to stage 1 and refine the model of the robot so that the simulated robot will behave like the real one. In this case, one has to go through the second and third stages again, but mostly for slightly tuning, rather than redesigning the program.

## 6.6 Design of Membrane Controller for Robots

In [8, 29, 30], NPS and ENPS controllers for several behaviors of mobile robots, such as obstacle avoidance, wall following and following a leader, were designed to show good performance of membrane controllers. This section uses an example, a membrane controller for solving a complex mobile robot trajectory tracking problem (MR2TP), to introduce the design procedure of membrane controllers for wheeled mobile robots' behaviors.

### 6.6.1 Mobile Robot Trajectory Tracking Problem

The mobile robot trajectory tracking problem were briefly described in [33]. In what follows, we redescribe it in detail. A wheeled mobile robot (WMR) is a complex system having a large delay, nonholonomic constraints and highly nonlinear characteristic, so it is quite difficult to build a precisely mathematical model for trajectory tracking control. The aim of solving MR2TP is to design control rules to make WMR move along a desired path quickly and accurately. The WMR mechanical system that are considered is shown in Fig. 6.12, where there are two differential driving wheels and a back unpowered universal wheel. The passive wheel does not have any effect on the freedom degree of the kinematic model [9] and works with the nonholonomic

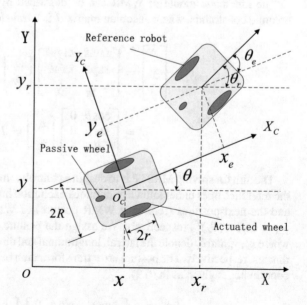

**Fig. 6.12** The mobile robot structure and motion

constraints, which are represented by the following formula:

$$\dot{y} \cdot \cos\theta + \dot{x} \cdot \sin\theta = 0. \tag{6.6}$$

The Cartesian coordinate vectors with three freedom degrees $p = (x, y, \theta)^T$ is used to denote the posture of WMR in two dimensional coordinates $XOY$. Thus, the reference posture of the reference WMR is denoted as $p_R = (x_r, y_r, \theta_r)^T$, where $(x_r, y_r)$ is the centroid of the WMR. The positive direction of $\theta$ for guiding the angle of a robot is anticlockwise. A linear velocity $v_c$ and an angular velocity $\omega_c$ forming a vector $V = (v_c, \omega_c)^T$ (also called the steering system of WMR) decide the motion posture of WMR. Correspondingly, the velocity of the reference WMR is represented as the vector $V_R = (v_R, \omega_R)^T$. It is noting that the two wheels are driven by independent torques from two direct current (DC) motors, where the radius of two wheels and the tread of WMR are denoted by $r$ and $2R$, respectively. If the speeds of two wheels are different, then WMR steers to the corresponding direction. Suppose that the WMR mass center is located at $O_c$ and mounted with non-deformable wheels. Let us denote $v_l$ and $v_r$ (resp., $\omega_l$ and $\omega_r$) the linear velocities (resp., angular velocities) of the left and right driving wheels. We describe the relationship between the WMR body and the wheels velocity as follows:

$$\begin{aligned} v_l = v_c - (\omega_c \cdot 2R)/2\,, \quad \omega_l = v_l/r \\ v_r = v_c + (\omega_c \cdot 2R)/2\,, \quad \omega_r = v_r/r \end{aligned} \tag{6.7}$$

The kinematic model of WMR can be described by using (6.8) with the non-holomic constraints, where Jacobian matrix $J$ is a transformation matrix.

$$\begin{aligned} \begin{bmatrix} \dot{x} \\ \dot{y} \\ \dot{\theta} \end{bmatrix} &= \begin{bmatrix} \frac{r \cdot \cos\theta}{2} & \frac{r \cdot \cos\theta}{2} \\ \frac{r \cdot \sin\theta}{2} & \frac{r \cdot \sin\theta}{2} \\ -\frac{r}{2R} & \frac{r}{2R} \end{bmatrix} \begin{bmatrix} \omega_l \\ \omega_r \end{bmatrix} \\ &= \begin{bmatrix} \cos\theta & 0 \\ \sin\theta & 0 \\ 0 & 1 \end{bmatrix} \begin{bmatrix} v_c \\ \omega_c \end{bmatrix} = J \cdot V \end{aligned} \tag{6.8}$$

The aim for solving MR2TP is to design a controller making WMR real-time track the reference ones under known continuous excitation inputs $v_R$ and $\omega_R$. The current and the next postures (reference WMR posture) of WMR are $p = (x, y, \theta)^T$ and $p_R = (x_r, y_r, \theta_r)^T$, respectively. We assign the posture error as $p_e = (x_e, y_e, \theta_e)^T$, where $x_e$, $y_e$ and $\theta_e$ denote the lateral, longitudinal and direction errors in mobile coordinates, respectively. The posture error transformation between mobile and Cartesian coordinates is given as in (6.9).

$$\begin{bmatrix} x_e \\ y_e \\ \theta_e \end{bmatrix} = \begin{bmatrix} \cos\theta & \sin\theta & 0 \\ -\sin\theta & \cos\theta & 0 \\ 0 & 0 & 1 \end{bmatrix} \begin{bmatrix} x_r - x \\ y_r - y \\ \theta_r - \theta \end{bmatrix} \tag{6.9}$$

Thus, we derive the differential equations of WMR posture errors from (6.8) and (6.9) as follows:

$$\begin{bmatrix} \dot{x}_e \\ \dot{y}_e \\ \dot{\theta}_e \end{bmatrix} = \begin{bmatrix} y_e \cdot \omega_c - v_c + v_R \cdot \cos\theta_e \\ -x_e \cdot \omega_c + v_R \cdot \sin\theta_e \\ \omega_R - \omega_c \end{bmatrix} \tag{6.10}$$

By further analysis and matrix transformation, we can get

$$\begin{bmatrix} \dot{x}_e \\ \dot{y}_e \\ \dot{\theta}_e \end{bmatrix} = \begin{bmatrix} -1 & y_e \\ 0 & -x_e \\ 0 & -1 \end{bmatrix} \begin{bmatrix} v_c \\ \omega_c \end{bmatrix}$$

$$+ \begin{bmatrix} \cos\theta_e & 0 \\ \sin\theta_e & 0 \\ 0 & 1 \end{bmatrix} \begin{bmatrix} v_R \\ \omega_R \end{bmatrix} \tag{6.11}$$

where $(v_c, \omega_c)^T$ and $(v_R, \omega_R)^T$ are the input velocity vector of the real WMR and the reference velocity vector of the reference WMR, respectively. $v_R$ can be represented by using (6.12).

$$v_R = \sqrt{\dot{x}_r^2 + \dot{y}_r^2}, \quad \omega_R = \frac{\dot{x}_r \cdot \ddot{y}_r - \dot{y}_r \cdot \ddot{x}_r}{\dot{x}_r^2 + \dot{y}_r^2} \tag{6.12}$$

In order to obtain $(v_c, \omega_c)^T$ making posture error $p_e$ converge to zero as $t \to \infty$, under the effect of $(v_R, \omega_R)^T$, the classic speed controller based on Lyapunov theory [21] is applied to attain (6.13):

$$V_c = \begin{bmatrix} v_c \\ \omega_c \end{bmatrix} = \begin{bmatrix} v_R \cdot \cos\theta_e + k_1 \cdot x_e \\ \omega_R + k_2 \cdot v_R \cdot y_e + k_3 \cdot v_R \cdot \sin\theta_e \end{bmatrix} \tag{6.13}$$

where $v_R > 0$ and $k_1, k_2$ and $k_3$ are three positive constants. In the design of the kinematic controller, the backstepping method in [10, 11] is used. The aim of kinematic controller is to produce the input velocities of wheels $(v_c, \omega_c)$ for dynamic controller, which is illustrated in Fig. 6.13 (the part with dashdot line).

Actually, it is necessary to consider the maximal velocity and acceleration constraints of the WMR wheels during inner dynamic control. That is, we bound the maximally linear velocity and angular velocity of WMR as $v_{max} = \omega_m \cdot r$ and $\omega_{max} = \omega_M \cdot r/R$, where $\omega_M, r$ and $R$ are the maximally rotational velocity of WMR wheels, the radius of wheels and the half of tread, respectively. The acceleration of left and right wheels are set to $a_{max}$ so as to avoid wheel skid. Thus, WMR will not slip and the controller designed approaches to the actual working condition.

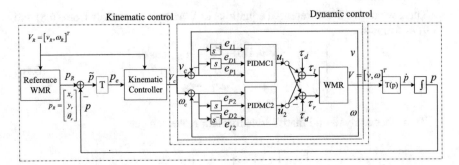

**Fig. 6.13** Kinematic and dynamic controllers for WMR

## 6.6.2   Kinematic Controller Design for Wheeled Mobile Robots

The kinematic controller of WMR is shown in Fig. 6.13 (inside the dash dot lines). The generic control law in (6.13) for the WMR kinematic model actually consists of two parts: feed-forward and feedback. They can be described as $u_f = (v_f, \omega_f)^T$ and $u_b = (v_b, \omega_b)^T$, respectively. Thus, (6.13) can be rewritten as

$$u = \begin{bmatrix} v \\ \omega \end{bmatrix} = \begin{bmatrix} v_f + v_b \\ \omega_f + \omega_b \end{bmatrix} \tag{6.14}$$

and (6.14) is substituted into (6.10) to obtain

$$\begin{bmatrix} \dot{x}_e \\ \dot{y}_e \\ \dot{\theta}_e \end{bmatrix} = \begin{bmatrix} y_e \cdot (\omega_f + \omega_b) - (v_f + v_b) + v_R \cdot \cos\theta_e \\ -x_e \cdot (\omega_f + \omega_b) + v_R \cdot \sin\theta_e \\ \omega_R - (\omega_f + \omega_b) \end{bmatrix}. \tag{6.15}$$

### 6.6.2.1   Design of Feed-Forward Control

The design of feed-forward part in the kinematic model is mainly based on sliding-mode control (SMC) method [11]. The use of SMC for the nonholonomic system with many constraints has many benefits such as fast system response, robustness to un-modeled dynamic factors, insensitivity to disturbance and parameter changes. The switching function and approaching law are improved to enhance the tracking accuracy. It is worth noting that the inherent jitter of SMC has little direct impact on the WMR DC motor because the output velocity of kinematic model is only treated as the input velocity of the WMR dynamic model,

**Lemma 1** ([39]) *For any $x \in R$ and $|x| < \infty$, $\varphi(x) = x \cdot \sin(\arctan x) \geq 0$, the equality occurs only $x = 0$.*

Here SMC is designed by using the backstepping method in [21]. We can deduce that $\theta_e = -\arctan(v_R y_e)$ can lead to the convergence of $y_e$, as $x_e = 0$, based on the partial Lyapunov function $V_y = \frac{1}{2} y_e^2$ [39], the second equation of posture error differential equation in (6.10) and Lemma 1. That is, as $x_e \to 0$ and $\theta_e \to -\arctan(v_R y_e)$, $y_e$ will converge to zero. If the system enters the slide mode, to guarantee $x_e = 0$ and $\theta_e = -\arctan(v_R y_e)$, a switching function is introduced as follows:

$$s(p_e) = \begin{bmatrix} s_1(p_e) \\ s_2(p_e) \end{bmatrix} = \begin{bmatrix} x_e \cdot y_e \\ \theta_e + arctg(v_R \cdot y_e) \end{bmatrix} \tag{6.16}$$

Within a limited time the system will go to the surface of slide mode $s = 0$ under varying structure control, as $x_e = 0$ and $\theta_e = -\arctan(v_R y_e)$, on the basis of the SMC principle. The posture error will converge to zero $p_e = \{x_e, y_e, \theta_e\}^T \to 0$ as $t \to 0$, due to the asymptotic stability of this sliding mode. If the system states tend to stay on the surface of sliding mode, the condition $\dot{s}^T(p_e) \cdot s(p_e) < 0$ must be satisfied, where $s(p_e)$ is

$$\dot{s}(p_e) = -\eta \cdot \mathrm{sgn}(s(p_e)) - \kappa \cdot s(p_e) \tag{6.17}$$

where $\mu$ is an adjustable parameter having effect on the shape of the sigmoid function and the boundary layer around the surface of switching mode; $\eta$ and $\kappa$ are positive constants. To weaken the jitter effect of SMC, the sigmoid function $\frac{2(1-e^{-s\cdot\mu})}{\mu(1+e^{-s\cdot\mu})}$ is used to replace the sign function $sgn(s)$. Thus, (6.17) can be changed to (6.18)

$$\dot{s}_i(p_e) = -\eta_i \frac{2(1 - e^{-s_i(pe)\cdot\mu_i})}{\mu_i(1 + e^{-s_i(pe)\cdot\mu_i})} - \kappa_i \cdot s_i(p_e), i = 1, 2 \tag{6.18}$$

As $\phi = \arctan(v_R \cdot y_e) \geq 0$, $v_b = 0$ and $\omega = 0$, (6.18) and (6.15) are substituted into (6.16) to obtain

$$s(p_e) = \begin{bmatrix} \dot{s}_1(p_e) \\ \dot{s}_2(p_e) \end{bmatrix} = \begin{bmatrix} -\eta_1 \frac{2(1-e^{-s_1(pe)\cdot\mu_1})}{\mu_1(1+e^{-s_1(pe)\cdot\mu_1})} - \kappa_1 s_1(p_e) \\ -\eta_2 \frac{2(1-e^{-s_2(pe)\cdot\mu_2})}{\mu_2(1+e^{-s_2(pe)\cdot\mu_2})} - \kappa_2 s_1(p_e) \end{bmatrix}$$

$$= \begin{bmatrix} \dot{x}_e \cdot y_e + \dot{y}_e \cdot x_e \\ \dot{\theta}_e + \frac{\partial\phi}{\partial v_R}\dot{v}_R + \frac{\partial\phi}{\partial y_e}\dot{y}_e \end{bmatrix}$$

$$= \begin{bmatrix} (y_e \cdot \omega_f - v_f + v_R \cdot \cos\theta_e) \cdot y_e + (-x_e \cdot \omega_f + v_R \cdot \sin\theta_e) \cdot x_e \\ \omega_R - \omega_f + \frac{\partial\phi}{\partial v_R}\dot{v}_R + \frac{\partial\phi}{\partial y_e}(-x_e \cdot \omega_f + v_R \cdot \sin\theta_e) \end{bmatrix} \tag{6.19}$$

The feed-forward control law can be derived from (6.19):

$$u_f = \begin{bmatrix} v_f \\ \omega_f \end{bmatrix}$$

$$= \begin{bmatrix} y_e \cdot \omega_f + v_R \cdot \cos\theta_e + \dfrac{x_e \cdot v_R \cdot \sin\theta_e - x_e^2 \cdot \omega_f + \eta_1 \frac{2(1-e^{-s_1(pe)\cdot\mu_1})}{\mu_1(1+e^{-s_1(pe)\cdot\mu_1})} + \kappa_1 s_1(p_e)}{y_e} \\[4mm] \dfrac{\omega_R + \frac{\partial\phi}{\partial v_R}\dot{v}_R + \frac{\partial\phi}{\partial y_e}(v_R \cdot \sin\theta_e) + \eta_2 \frac{2(1-e^{-s_2(pe)\cdot\mu_2})}{\mu_2(1+e^{-s_2(pe)\cdot\mu_2})} - \kappa_2 s_2(p_e)}{1 + \frac{\partial\phi}{\partial y_e} x_e} \end{bmatrix} \tag{6.20}$$

where,

$$\frac{\partial\phi}{\partial y_e} = \frac{v_R}{1 + (v_R \cdot y_e)^2} \tag{6.21}$$

$$\frac{\partial\phi}{\partial v_R} = \frac{y_e}{1 + (v_R \cdot y_e)^2}. \tag{6.22}$$

### 6.6.2.2   Design of Feedback Control

In this subsection, Lyapunov function is considered to design feedback control including the reference path and real posture of WMR under known conditions. On the basis of (6.13), the feedback part of this trajectory tracking controller can be obtained from (6.23) as follows:

$$u_b = \begin{bmatrix} v_b \\ \omega_b \end{bmatrix} = \begin{bmatrix} k_1 \cdot x_e \\ v_R(k_2 \cdot y_e + k_3 \cdot \sin\theta_e) \end{bmatrix} \tag{6.23}$$

Actually, a simple control law based on (6.13) was used in [4] as the feedback part of the kinematic controller. The law is

$$u_b = \begin{bmatrix} v_b \\ \omega_b \end{bmatrix} = \begin{bmatrix} k_1 \cdot x_e \\ k_2 \cdot v_R \cdot y_e \frac{\sin\theta_e}{\theta_e} + k_3 \cdot \theta_e \end{bmatrix} \tag{6.24}$$

By integrating (6.20) with (6.24), the kinematic controller including feed-forward and feedback is described as follows:

$$u = \begin{bmatrix} v \\ \omega \end{bmatrix} = \begin{bmatrix} v_f + v_b \\ \omega_f + \omega_b \end{bmatrix}$$

$$= \begin{bmatrix} y_e \cdot \omega_f + v_R \cdot \cos\theta_e + \dfrac{x_e \cdot v_R \cdot \sin\theta_e - x_e^2 \cdot \omega_f + \eta_1 \frac{2(1-e^{-s_1(pe)\cdot\mu_1})}{\mu_1(1+e^{-s_1(pe)\cdot\mu_1})} + \kappa_1 s_1(p_e)}{y_e} + k_1 \cdot x_e \\[4mm] \dfrac{\omega_R + \frac{\partial\phi}{\partial v_R}\dot{v}_R + \frac{\partial\phi}{\partial y_e}(v_R \cdot \sin\theta_e) + \eta_2 \frac{2(1-e^{-s_2(pe)\cdot\mu_2})}{\mu_2(1+e^{-s_2(pe)\cdot\mu_2})} - \kappa_2 s_2(p_e)}{1 + \frac{\partial\phi}{\partial y_e} x_e} + k_2 \cdot v_R \cdot y_e \frac{\sin\theta_e}{\theta_e} + k_3 \cdot \theta_e \end{bmatrix}.$$

$$\tag{6.25}$$

### 6.6.3 Dynamic Controller Design for Wheeled Mobile Robots

The use of the kinematic model control is to make sure that the input velocity vector $(v_c, \omega_c)^T$ of the WMR movement can follow the reference velocity vector $(v_R, \omega_R)^T$. In fact, the trajectory tracking of the WMR kinematic model requires to ultimately reflect on the motor driving of the wheels. So it is not enough to obtain the WMR control performance in the kinematic model control and further the dynamic characteristics of the WMR movement is required to be considered. The proportional-integral-derivative (PID) controller is quite popular in the traditional robot motion control, but there is an extreme difficulty to properly tune PID controller parameters because the WMR movement is a nonlinear system with large external disturbances and system uncertainties. In what follows, a PID based membrane controller (PIDMC) is presented to control the WMR dynamic model.

#### 6.6.3.1 Dynamic Model of Wheeled Mobile Robots

The classical well-known Euler Lagrange dynamic equation [22] is used to model the considered WMR with an $n$-dimensional configuration space $p \in R^{n \times 1}$ and $m$ constraints.

$$M(p) \cdot \ddot{p} + V_m(p, \dot{p}) \cdot \dot{p} + F_G(\dot{p}) + G(p) + \tau_d = B(p) \cdot \tau - J^T(p) \cdot \lambda \tag{6.26}$$

$$M(p) = \begin{bmatrix} m_R & 0 & 0 \\ 0 & m_R & 0 \\ 0 & 0 & I \end{bmatrix}, B(p) = \frac{1}{r} \begin{bmatrix} \cos\theta & \cos\theta \\ \sin\theta & \sin\theta \\ R & -R \end{bmatrix} \tag{6.27}$$

where $M(p) \in R^{n \times n}$ is a positive definite and symmetric inertia matrix; $\ddot{p}$ and $\dot{p}$ are acceleration and velocity vectors, respectively; $V_m(p, \dot{p}) \in R^{n \times n}$ is a coriolis and centripetal matrix; $F_G(\dot{p}) \in R^{n \times 1}$ is the surface friction; $G(p) \in R^{n \times 1}$ is the gravitational vector; $\tau_d \in R^{n \times 1}$ represents bounded unknown disturbances including unstructured un-modeled dynamics; $\tau = [\tau_r \quad \tau_l]^T$, where $\tau_r$ and $\tau_l$ are right and left torque inputs generated by DC motors of the two wheels, respectively; $B(p) \in R^{n \times (n-m)}$ is the input transformation matrix; $I$ is the moment of inertia of the WMR about the mass center; $m_R$ is the mass of WMR.

Here we can omit the effect of gravity, that is, let $G(p) = 0$, because the WMR motion considered is in the horizontal plane. If the surface friction $F_G$ and disturbance torque $\tau_d$ are considered as the disturbance and un-modeled uncertainties items (such as free wheel), a simpler and more appropriate dynamic model can be described as follows [10, 11]:

$$\bar{M}(p) \cdot \dot{v} + \bar{\tau}_d = \bar{B}(p) \cdot \tau \tag{6.28}$$

where,

- $\bar{M}(p) = J^T M(p)$,

- $J = \begin{bmatrix} m_R & 0 \\ 0 & I \end{bmatrix} \in R^{2 \times 2}$,
- $\bar{B} = J^T B = \frac{1}{r} \begin{bmatrix} 1 & 1 \\ R & -R \end{bmatrix}$,
- $\bar{\tau}_d = J^T \tau_d$.

We consider the disturbance torque $\tau_d$ as an external additional item, which is similar to that in Fig. 6.13. Thus, we can derive from (6.28) the relation between the WMR velocity and the torque of DC motor, i.e.,

$$\dot{v}(t) = E \cdot \tau(t) \tag{6.29}$$

where the transformation matrix $E$ is

$$E = \bar{M}^{-1}(p) \cdot \bar{B}(p) = \begin{bmatrix} m_R & 0 \\ 0 & I \end{bmatrix}^{-1} \cdot \frac{1}{r} \begin{bmatrix} 1 & 1 \\ R & -R \end{bmatrix}$$
$$= \frac{1}{m_R \cdot r \cdot I} \begin{bmatrix} I & I \\ R m_R & -R m_R \end{bmatrix}. \tag{6.30}$$

### 6.6.3.2   Proportional-Integral-Derivative Based Membrane Controller Design

In an enzymatic numerical P system, the enzymes enable it to perform several rules (protocols or programs) inside a single membrane, while the deterministic behavior can be kept, as compared with NPS, where a single membrane contains only one rule. The rules or protocols can be derived from the process of finding the most suitable proportional-integral-derivative (PID) parameters. These rules can be executed in a distributed and parallel way and can flexibly alter the form of PID algorithm according to the behavior of the biological enzymes. Benefited from the enzymatic numerical P system, an adaptive controller PIDMC with a powerful capability of continuously learning is introduced. In what follows, we first briefly describe PID.

PID controller in discrete time can be represented by the standard formula (6.31) or improved formula (6.32):

$$u(k) = K_P \cdot e(k) + K_I \sum_{j=0}^{k} e(j) + K_D \cdot [e(k) - e(k-1)] \tag{6.31}$$

$$\Delta u(k) = u(k) - u(k-1)$$
$$= K_P \cdot \Delta e(k) + K_I \cdot e(k) + K_D \cdot [\Delta e(k) - \Delta e(k-1)] \tag{6.32}$$
$$u(k) = u(k-1) + \Delta u(k)$$

where $k = 0, 1, \ldots$ is the sample number; $u(k)$ is the output value at $k$th sample; $e(k)$ is the error input value at $k$th sample; $\Delta e(k) = e(k) - e(k-1)$; $K_P$, $K_I$ and $K_D$ are proportional, integral and derivative coefficients [3], respectively. The incremental

PID $\Delta u(k)$ in (6.32) has some benefits, such as little switching impact, little impact on malfunction and accumulation avoidance. These advantages make it suitable for the stepper motor applications like WMR DC motor. In what follows, we will describe the design procedure of PIDMC in detail. Figure 6.13 shows the structure of PIDMC. The motion of WMR is controlled by linear velocity $v_c$ and angular velocity $\omega_c$, where $v_c$ and $\omega_c$ are the output velocities of kinematic controller compared with the reference velocities. The aim of using double PIDMC is to force the linear velocity error $e_1 = v_c - v$ and the angular velocity error $e_2 = \omega_c - \omega$ to converge to zero. The WMR outputs of PIDMC are

$$\begin{bmatrix} \tau_l \\ \tau_r \end{bmatrix} = \begin{bmatrix} u_1 + u_2 \\ u_1 - u_2 \end{bmatrix} \tag{6.33}$$

where $u_1$ and $u_2$ are the outputs of two PIDMC, PIDMC1 and PIDMC2, respectively.

To obtain an auto-tuning strategy for PID parameters based on the deterministic ENPS, PIDMC has the following three general steps:

(1) the empirical knowledge of experts is transformed into a series of rules which are allocated to the corresponding membranes;
(2) to enhance the adaptivity of the PID controller, the continuous learning capacity is endowed to some rules inspired from the idea of neural networks;
(3) the rules are performed according to the action of enzymes.

The inputs of the PIDMC controller in Fig. 6.13 are the reference velocity $V_c(k)$ and the WMR real velocity $V(k)$. In controller online learning, $V(k)$ is preprocessed as state variables $x_1(k)$, $x_2(k)$, $x_3(k)$, which are error, error change rate and change rate of error change rate, respectively. They are

$$\begin{aligned} x_1(k) &= V_c(k+1) - V(k+1) = e(k) \\ x_2(k) &= \Delta e(k) = e(k) - e(k-1) \\ x_3(k) &= \Delta e(k) - \Delta e(k-1) \\ &= e(k) - 2e(k-1) + e(k-2) \end{aligned} \tag{6.34}$$

To trade off different kinds of errors at each time sample $k$, parameters $K_I$, $K_P$, $K_D$ in (6.32) are regarded as time-varying weighting factors $w_1(k)$, $w_2(k)$, $w_3(k)$:

$$u(k) = u(k-1) + K \sum_{i=1}^{3} w_i(k) \cdot x_i(k) \tag{6.35}$$

where $K$ is a scale factor. In [1, 42], the quadratic error was used as the performance index function. To avoid the overshoot, PIDMC considers the quadratic of the control incremental value $\Delta u(k)$ as performance index function $I$:

$$I = \tfrac{1}{2} P \cdot E^2(k) + \tfrac{1}{2} Q \cdot \Delta u^2(k) \tag{6.36}$$

where $P$ and $Q$ are the proportional coefficients for output error $E(k) = V_c(k) - V(k)$, and control increment $\Delta u(k)$ in (6.36), respectively. The adjustment of weighting factors $w_i(k)$ should be along the decreasing direction of performance index $I$, i.e., along the negative gradient direction of $E(k)$ for $w_i(k)$. By performing the partial differential operator on $w_i(k)$, we obtain

$$
\begin{aligned}
\frac{\partial I}{\partial w_i(k)} &= -P \cdot E(k) \cdot \frac{\partial V(k)}{\partial u(k)} \cdot \frac{\partial u(k)}{\partial w_i(k)} \\
&\quad + Q \cdot K \cdot \left[ \sum_{i=1}^{3} w_i(k) x_i(k) \right] \cdot x_i(k)
\end{aligned}
\tag{6.37}
$$

So the adjustment $w_i(k)$ is

$$
\Delta w_i(k) = w_i(k+1) - w_i(k) = -\zeta_i \cdot \frac{\partial I}{\partial w_i(k)}
\tag{6.38}
$$

where $\zeta_i$ is the convergence rate. According to (6.33), (6.35) and (6.37), $w_i(k)$ can be expanded as:

$$
\begin{aligned}
\Delta w_1(k) &= \zeta_1 \cdot P \cdot K \cdot E(k) \cdot x_1(k) \cdot \frac{\partial V(k)}{\partial u(k)} \\
&\quad - \zeta_1 \cdot Q \cdot K \cdot \left[ \sum_{i=1}^{3} w_i(k) x_i(k) \right] \cdot x_1(k) \\
\Delta w_2(k) &= \zeta_2 \cdot P \cdot K \cdot E(k) \cdot x_2(k) \cdot \frac{\partial V(k)}{\partial u(k)} \\
&\quad - \zeta_2 \cdot Q \cdot K \cdot \left[ \sum_{i=1}^{3} w_i(k) x_i(k) \right] \cdot x_2(k) \\
\Delta w_3(k) &= \zeta_3 \cdot P \cdot K \cdot E(k) \cdot x_3(k) \cdot \frac{\partial V(k)}{\partial u(k)} \\
&\quad - \zeta_3 \cdot Q \cdot K \cdot \left[ \sum_{i=1}^{3} w_i(k) x_i(k) \right] \cdot x_3(k)
\end{aligned}
\tag{6.39}
$$

As usual, $\frac{\partial V(k)}{\partial u(k)}$ cannot be directly obtained in a PID algorithm, but can be approximately denoted by the sigmoid function $S(x)$. PIDMC uses sign function $sgn(u(k))$ as the approximate part. The convergence rate $\zeta_i$ can compensate the inaccurate calculation results. By combining (6.38) with (6.39), the gain tuning formula can be obtained as follows

$$
\begin{aligned}
w_1(k+1) &= w_1(k) + \zeta_{pk1} \cdot x_1(k) \cdot x_1(k) \cdot S_{gn}(k) \\
&\quad - \zeta_{qk1} \cdot x_1(k) \cdot S_{um}(k) \\
w_2(k+1) &= w_2(k) + \zeta_{pk2} \cdot x_1(k) \cdot x_2(k) \cdot S_{gn}(k) \\
&\quad - \zeta_{qk2} \cdot x_2(k) \cdot S_{um}(k) \\
w_3(k+1) &= w_3(k) + \zeta_{pk3} \cdot x_1(k) \cdot x_3(k) \cdot S_{gn}(k) \\
&\quad - \zeta_{qk3} \cdot x_3(k) \cdot S_{um}(k)
\end{aligned}
\tag{6.40}
$$

where $\zeta_{pk1} = \zeta_1 \cdot P \cdot K$, $\zeta_{pk2} = \zeta_2 \cdot P \cdot K$, $\zeta_{pk3} = \zeta_3 \cdot P \cdot K$, $\zeta_{qk1} = \zeta_1 \cdot Q \cdot K$, $\zeta_{qk2} = \zeta_2 \cdot Q \cdot K$, $\zeta_{qk3} = \zeta_3 \cdot Q \cdot K$, $S_{gn}(k) = \text{sgn}(u(k))$, $S_{um}(k) = \sum_{i=1}^{3} w_i(k) \cdot x_i(k)$.

The substitution of (6.40) into (6.35) leads to the output value of the PID controller

$$\bar{w}_i(k) = \frac{w_i(k)}{\sum_{i=1}^{3} w_i(k)}$$
$$u(k) = u(k-1) + K \sum_{i=1}^{3} \bar{w}_i(k) \cdot x_i(k) \tag{6.41}$$

In the design of PIDMC, we divide the PID controller into three sub-controllers: proportional (P), proportional-derivative (PD), and proportional-integral (PI) in the corresponding zones. Thus, the convergence rate of $I$ in (6.36) can be enhanced. Moreover, the objective of the division of the three functional zones can improve the controller performance, such as settling time, rising time and static error, which is maintained by an appropriate PID controller in each zone. We add the weighting factors $w_1(k)$, $w_2(k)$, $w_3(k)$. The sub-periods are redefined as a combination of the knowledge of experts, which can be generally divided into several cases. The normalized velocity is defined as

$$N(v) = \frac{v}{v_{\max}} = \begin{cases} 1 & |v| > v_{\max} \\ v/v_{\max} & |v| < v_{\max} \end{cases} \tag{6.42}$$

where $v_{\max}$ is the maximal velocity of the WMR permitted. The normalized velocity error varies between 0 and 1. According to two set-points $M_1$ and $M_2$, where $0 < M_2 < M_1 < 1$, the functional zones (error signal of step response) are divided into three parts.

(1) If $|N(e(k))| > M_1$, the PID model should be switched into P or PD model because the absolute value of velocity error is too big;
(2) If $M_1 > |N(e(k)| > M_2$, the PID model should be applied because the error is relatively big;
(3) If $M_2 > |N(e(k)|$, PI model should be used to eliminate or minimize the static error because the error is small.

The PIDMC controller, which is shown in Fig. 6.14, is designed by using a numerical P system with a hierarchical membrane structure containing ten membranes: the skin membrane $OutputControlValue$ containing three identical inner membranes $ComputeControlValue$ ($i = 1, 2, 3$). Each membrane $ComputeControlValue$ includes membrane $ComputeQuadraticof Output-Error$ and innermost membrane $ComputeQuadraticof ControlIncrement$. The design of the PIDMC membrane system are described as follows:

The skin membrane $OutputControlValue$ has five variables: the control increment $U_{inc}[0]$ with the initial value 0, the current position $U_{cur}[U_{last}]$ of the controller with the initial value being equivalent to the control value of the controller at the previous time unit, and three enzymes for controlling the algorithm flow: $E_D[6e_{max}]$ with the initial value $6e_{max}$ ($e_{max}$ is a very large value), $E_T[0]$ with the initial value

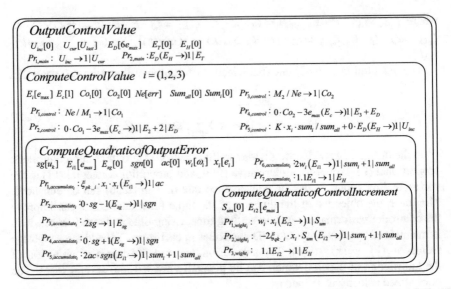

**Fig. 6.14** PIDMC controller

0 and $E_H[0]$ the initial value 0. The controller algorithm will end at $E_T \neq 0$. The control value of the controller requires to be updated and the error information is read again after the reaction of the controller $Pr_{2,control}$.

The inner membrane *ComputeControlValue* has seven variables: enzyme $E_i[e_{max}]$ for checking whether the current weights of parameters $K_P$, $K_I$ and $K_D$ should be calculated or not, two variables $Co_1$ and $Co_2$ for storing results, enzyme $E_c[1]$ for determining whether rule $Pr_{2,control}$ or $Pr_{4,control}$ should be performed or not according to the variable value of $Co_1$ or $Co_2$, the normalized velocity error $Ne[err]$ with the initial value being equivalent to the input normalized error of the PID controller at the current time, $Sum_i[0]$ denoting the current weight $w_i(k)$ and $Sum_{all}[0]$ for the sum of weights $\bar{w}_i(k)$ in (6.40) and (6.41). The normalized velocity error $Ne[err]$ is divided by $M_1$ (rule $Pr_{1,control}$) or $M_2$ (rule $Pr_{3,control}$). The computing results are stored into $Co_1$ or $Co_2$. The two summation results ($Sum_i$ and $Sum_{all}$) are decided by the relationship between the sizes of $E_D$ and $E_H$. The execution of rule $Pr_{5,control}$ follows the two summation results ($Sum_i$ and $Sum_{all}$).

The inner membrane *ComputeQuadraticofOutputError* has seven variables: the input variables $sg[u_k]$ of a sign function with the initial value being equivalent to the control value $u_k$ at the previous time unit, enzyme $E_{sg}[0]$ which is combined with $sg$ in rules ($Pr_{2,accumulate_i}$ - $Pr_{4,accumulate_i}$) to compute sign function, enzyme $E_{i1}[e_{max}]$ with the initial value $e_{max}$, a sign variable $sgn[0]$, the quadratic error $ac[0]$ in (6.36), a time-varying weighting variable $w_i[\omega_i]$ with the initial value $\omega_i$ representing parameter values $K_I$, $K_P$, $K_D$ at the previous time unit, and the values $x_i[e_i]$ of the three types errors in (6.34). The result of $ac \cdot sgn$ is assigned to $Sum_i$

and $Sum_{all}$ simultaneously in rule $Pr_{5,accumulate_i}$. The value $\omega_i$ is also assigned to $Sum_i$ and $Sum_{all}$ simultaneously in rule $Pr_{6,accumulate_i}$.

The innermost membrane *Compute Quadratic of Control Increment* has two variables: the sum $S_{um}[0]$ of three error weighting variables $S_{um}[k]$ in (6.40) and enzyme $E_{i2}[e_{max}]$ with the initial value $e_{max}$. Rule $Pr_{1,wight_i}$ is executed at $E_{i2} = 0$. Rule $Pr_{2,wight_i}$ is executed at $E_{i2} \neq 0$.

## 6.7 Experiments

In this section, the experiments are conducted on the robotic simulator, Webots. The robot uses e-puck. Both Webots and e-puck have been described in detail in Sect. 6.5. In what follows, the experiments and results on the kinematic controller and dynamic controller are first presented. Then, the robots simulation platform, Webots, is used to introduce the experiments and results on trajectory tracking control.

### 6.7.1 Experiments on Kinematic Controller

This experiment considers sine wave trajectory to test the kinematic controller. The parameter setting is as follows: $\eta_1 = \eta_2 = 0.01$, $\kappa_1 = \kappa_2 = 5$, $\mu_1 = \mu_2 = 0.6$ in (6.20); $k_1 = k_3 = 4$, $k_2 = 6$ in (6.24). The reference trajectory uses sine wave. The initial posture of the actual WMR is $p_R(0) = [x_r(0), y_r(0), \theta_r(0)]^T = [10, 0, 0]^T$. The red dotted sine wave in Fig. 6.15 is the reference trajectory.

The following three kinematic controllers are considered as benchmarks to conduct comparative experiments.

(1) classic kinematic control law widely used in various publications such as in [22];
(2) feed-forward kinematic control law designed by using a backstepping method in (6.20);
(3) the presented kinematic control law integrating feed-forward with feedback parts in (6.25).

Experimental results of the three control methods are also shown in Fig. 6.15, where the blue solid wave is the WMR actual tracking trajectory in the kinematic model. The results indicate that the introduced integration method obtains better convergence than the feed-forward method and much better than classic method. As compared with the other two methods, the introduced method has better tracking ability at the turning point of sine wave.

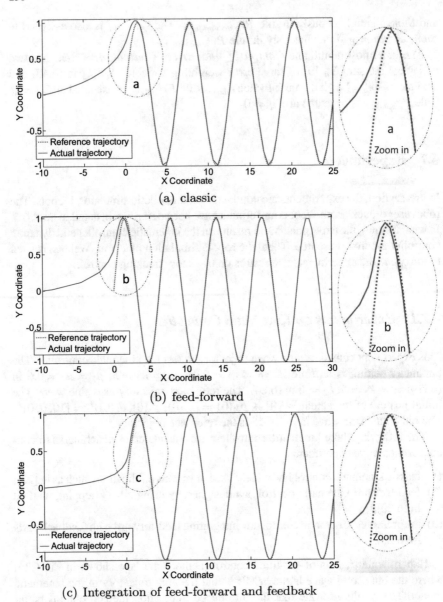

(a) classic

(b) feed-forward

(c) Integration of feed-forward and feedback

**Fig. 6.15** The sine wave trajectory results of three kinematic controllers

## 6.7.2  Experiments on PID Based Membrane Controller

In the experiments, the controller parameters in (6.40), learning rate $\zeta_1$ of proportional coefficients, integral coefficients $\zeta_2$ and derivative coefficients $\zeta_3$, are set to 12, 3 and 16, respectively; weighting factors $P = 2$ and $Q = 1$; the scale factor $K = 0.02$. The proper parameters of PIDMC are obtained by trial-and-error through experiments. The servo controller of WMR considers the actual friction model: the friction is common in the servo system and has a complex, non-linear and uncertainty characteristics. The LuGre model in [38] is chosen as the friction model. The dynamic function for a simpler servo system uses the differential equation $I\ddot{\omega} = u - F$, where $I$ is the inertia moment; $\omega$ is the rotation angle; $u$ and $F$ are the control torque and friction torque, respectively. $F$ is described by the following LuGre model [38]:

$$F = \lambda_1 \cdot s + \lambda_2 \cdot \dot{s} + \beta \cdot \omega \tag{6.43}$$

$$\dot{s} = \dot{\omega} - \frac{\lambda_1 \cdot |\omega| \cdot s}{F_c + (F_s-)e^{-\left(\omega/V_s\right)^2} + \beta \cdot \omega} \tag{6.44}$$

where $\lambda_1$ and $\lambda_2$ are the dynamic friction parameters; state variable $s$ is the average deformation of the contact surface; $F_c, F_s, \beta$ and $V_s$ are the static friction parameters; $\beta$ is the viscous friction coefficient; $V_s$ is the switching speed; $F_c$ and $F_s$ are the Coulomb's friction and static friction, respectively. The parameter values of the servo system and friction model are given in Table 6.2.

In the simulation test, a sinusoidal waveform $0.05 \sin(2\pi t)$ is used as the desired input trajectory. We compare the control performance between PIDMC and the conventional PID controller. The parameters of conventional PID controller are set to $K_P = 25$, $K_I = 4$, $K_D = 7$ through trial and error tests. The initial values of the PIDMC weight, $w_1$, $w_2$ and $w_3$, are set to be the same as those of conventional PID. The experimental results are shown in Fig. 6.16.

**Table 6.2** Parameter setting for the servo system and friction model

| Parameter | Value |
| --- | --- |
| Moment of inertia $I$ | 1.0 |
| Dynamic friction parameter $\lambda_1$ | 230 |
| Dynamic friction parameter $\lambda_2$ | 2.0 |
| Viscous friction coefficient $\beta$ | 0.025 |
| Coulomb's friction $F_c$ | 0.24 |
| Static friction $F_s$ | 0.37 |
| Switching speed $V_s$ | 0.03 |

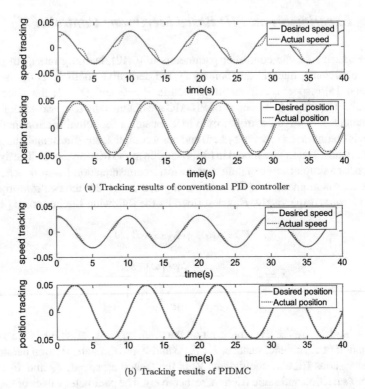

(a) Tracking results of conventional PID controller

(b) Tracking results of PIDMC

**Fig. 6.16** Position and speed tracking results of two kinds of PID controllers considering friction

It can seen from the results of the conventional PID controller shown in Fig. 6.16a that there is speed tracking deviation at the beginning and the dead zone phenomenon, and there is a flat top phenomenon in the position tracking. While a little bit deviation in speed tracking due to friction occurs in the response of PIDMC shown in Fig. 6.16b, but the deviation is smaller better than the conventional PID controller. Furthermore, the response of PIDMC in Fig. 6.16b is in accordance with that of position tracking.

We further draw a comparison of PIDMC with the sliding mode control, which is a special discontinuous control approach with strong robustness to external disturbances and system uncertainties. In [20], a hybrid approach of neural network and sliding mode control (NNSMC) was proposed to overcome the inherent deficiency of SMC with fixed larger upper boundedness.

In the experiment, a cosine $0.5\cos(2\pi t)$ is used as the S-typed trajectories. The uncertainties and the external disturbances to the control system of WMR at the third and sixth second are considered by using the parameter variations of the inertia and mass of WMR, which are described as follows:

$$\begin{cases} I = 0.5 + 0.35 \ (kg.m^2), \ t = 3 \ (s) \\ m = 2 \pm 1.25 \ (kg), \ 6 < t < 9 \ (s) \end{cases} \tag{6.45}$$

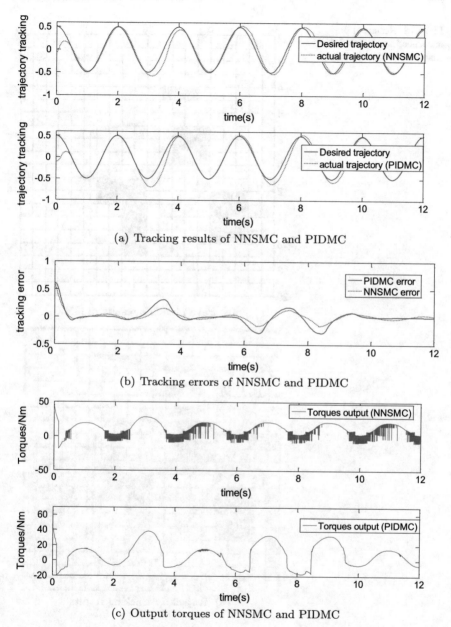

(a) Tracking results of NNSMC and PIDMC

(b) Tracking errors of NNSMC and PIDMC

(c) Output torques of NNSMC and PIDMC

**Fig. 6.17** Comparison of PIDMC and NNSMC

In Fig. 6.17a, the robustness can be ensured for both NNSMC and PIDMC large disturbances at different time instants with different types of external disturbances occur due to the self-learning ability of NN. Figure 6.17b shows that PIDMC has

**Fig. 6.18** Results of the
introduced controllers for
simulated robot trajectory

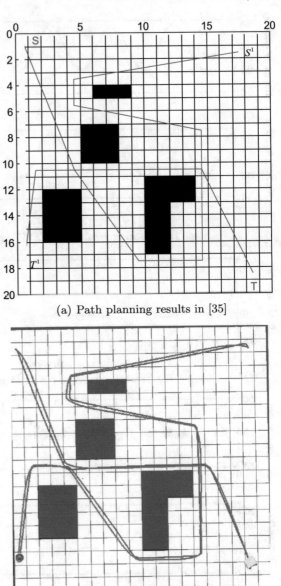

(a) Path planning results in [35]

(b) Trajectory tracking results

a slightly faster tracking speed and slightly trajectory tracking stability better than NNSMC, at the cost of a slightly larger tracking error than NNSMC. In Fig. 6.17c, PIDMC has a smoother control output than NNSMC due to the chattering phenomenon, which is the inherent deficiency of SMC.

### 6.7.3 Experiments on Trajectory Tracking Control of Wheeled Mobile Robots

We use the mobile robot simulation software Webots Sect. 6.2.1 to conduct experiments to show the feasibility and effectiveness of the introduced approach integrating outer kinematic controller with inner dynamic controller. The experiments on the simulated robot with differential wheels were carried out in an environment described in [35]. A path planning algorithm was used to find the optimal path for WMR in the environment, as shown by the red line (from S to T) in Fig. 6.18a. The same area with obstacles in the Webots robotics simulator is shown in Fig. 6.18b.

In the experiments, robots with two kinds of controllers (PIDMC controller and the controller in [42]) are used to make a comparison; the local error of the simulated robot is set to $e_l = 0.5$; the grid step $S_g = 1$; the starting position of the reference path is located at grid $(0, 0)$; both of the simulated robots are located at grid $(1, 0)$, where there is little distance error between them. At the beginning of trajectory tracking, we find that the robot with PIDMC (blue line) takes less time to find the desired path than the robot with the controller in [42] (red line). The robots controlled by the two controllers can walk along the reference path and cross the lane between two obstacles. Finally, they reach the target grid. It is worth pointing out that the robot with PIDMC can keep closer to the reference path than another one. In addition, another sub-optimal path (blue line) from $S^1$ to $T^1$ is found in Fig. 6.18a. We use E-puck as the simulated robot. Similarly, the robot with PIDMC (blue line) in Fig. 6.18b can also keep up with the desired path with much smaller tracking errors and much stronger tracking ability than the robot with the controller in [42] (red line).

## 6.8 Conclusions

This chapter discusses the design of membrane controllers for wheeled mobile robots. The membrane controllers are designed by using (enzymatic) numerical P systems. The fundamentals of (enzymatic) numerical P systems and their simulators, the simulator for robots and preliminaries of robot control, are also introduced. Some experiments and results are presented.

The future work with respect to this robot-based real world applications of P systems may focus on the following problems:

(1) the extension of the method discussed in this chapter to designing membrane controllers for more complex behaviors of a larger class of wheeled robots;

(2) combining numerical P systems with other advanced control strategies for other control engineering applications;

(3) the use of various classes of P systems, including numerical P systems, in constructing cognitive and executable architectures (planning, learning, execution) of autonomous robots [8]; more specifically, how to develop P systems-based cognitive architectures at higher levels in a control system, which can communicate with lower level membrane controllers for more complex robot applications.

# References

1. Ahn, K.K., and D.C. Thanh. 2006. Nonlinear PID control to improve the control performance of 2 axes pneumatic artificial muscle manipulator using neural network. *Mechatronics* 16 (9): 577–587.

2. Arsene, O., C. Buiu, and N. Popescu. 2011. SNUPS-a simulator for numerical membrane computing. *International Journal of Innovative Computing, Information and Control* 7 (6): 3509–3522.

3. Bennett, S. 2001. The past of PID controllers. *Annual Reviews in Control* 25: 43–53.

4. Blazic, S. 2011. A novel trajectory-tracking control law for wheeled mobile robots. *Robotics and Autonomous Systems* 59 (11): 1001–1007.

5. Buiu, C. 2009. Towards integrated biologically inspired cognitive architectures. In *Proceedings of the international conference on electronics, computers, and AI-ECAI'09*, 2–8.

6. Buiu, C., O. Arsene, C. Cipu, and M. Patrascu. 2011. A software tool for modeling and simulation of numerical P systems. *Biosystems* 103: 442–447.

7. Buiu, C., A.B. Pavel, C.I. Vasile, and I. Dumitrache. 2011. Perspectives of using membrane computing in the control of mobile robots. In *Proceedings of of the beyond AI-interdisciplinary aspect of artificial intelligence conference*, 21–26.

8. Buiu, C., C.I. Vasile, and O. Arsene. 2012. Development of membrane controllers for mobile robots. *Information Sciences* 187: 33–51.

9. Campion, G., G. Bastin, and B. Dandrea-Novel. 1996. Structural properties and classification of kinematic and dynamic models of wheeled mobile robots. *IEEE Transactions on Robotics and Automation* 12 (1): 47–62.

10. Chen, C., T. Li, and Y. Yeh. 2009. EP-based kinematic control and adaptive fuzzy sliding-mode dynamic control for wheeled mobile robots. *Information Sciences* 179 (1–2): 180–195.

11. Chen, C., T. Li, Y. Yeh, and C. Chang. 2009. Design and implementation of an adaptive sliding-mode dynamic controller for wheeled mobile robots. *Mechatronics* 19 (2): 156–166.

12. Colomer, M.A., A. Margalida, D. Sanuy, and M.J. Pérez-Jiménez. 2011. A bio-inspired computing model as a new tool for modeling ecosystems: The avian scavengers as a case study. *Ecological Modelling* 222 (1): 33–47.

13. Cyberbotics Ltd., O. Michel, F. Rohrer, and N. Heiniger. 2010. Wikibooks contributors. Cyberbotics' Robot Curriculum.

14. Cyberbotics, professional mobile robot simulation. http://www.cyberbotics.com.

15. DARPA, DARPA's call for biologically inspired cognitive architectures research program. http://www.darpa.mil/ipto/programs/bica/bica.asp.

16. Epucksite, e-puck website. http://www.e-puck.org.

17. Epucksite, e-puck website. http://www.e-puck.org/images/electronics/shematics.png.

18. Garcia-Quismondo, M., A.B. Pavel, and M.J. Pérez-Jiménez. 2012. Simulating large-scale ENPS models by means of GPU. In *Proceedings of the tenth brainstorming week on membrane computing*, 137–152.

19. Hoffmann, H., M. Maggio, M.D. Santambrogio, A. Leva, and A. Agarwal. 2011.SEEC: A general and extensible framework for self-aware computing. Technical report MIT-CSAIL-TR-2011-046.

20. Hu, H., and P.Y. Woo. 2006. Fuzzy supervisory sliding-mode and neural network control for robotic manipulators. *IEEE Transactions on Industrial Electronics* 53 (3): 929–940.

21. Kanayama, Y., Y. Kimura, F. Miyazaki, and T. Noguchi. 1990. A stable tracking control method for an autonomous mobile robot. In *Proceedings of the IEEE conference robotics and automation*, 384–389.

22. Kukao, T., H. Nakagawa, and N. Adachi. 2000. Adaptive tracking control of nonholonomic mobile robot. *IEEE Transactions on Robotics and Automation* 16 (6): 609–615.

23. Lambercy, F., and G. Caprari. Khepera III manual ver 2.2, http://ftp.k-team.com/KheperaIII/Kh3.Robot.UserManual.2.2.pdf.

24. Moubarak, P., and P. Ben-Tzvi. 2011. Adaptive manipulation of a hybrid mechanism mobile robot. In *Proceedings of IEEE international symposium on robotic and sensors environments (ROSE)*, 113–118.

25. Păun, Gh. 1999. Computing with membranes: An introduction. *Bulletin of the EATCS* 67: 139–152.

26. Păun, Gh, and R. Păun. 2006. Membrane computing and economics: Numerical P systems. *Fundamenta Informaticae* 73 (1): 213–227.

27. Păun, Gh, G. Rozenberg, and A. Salomaa (eds.). 2010. *The Oxford Handbook of Membrane Computing*. Oxford: Oxford University Press.

28. Pavel, A.B., O. Arsene, and C. Buiu. 2010. Enzymatic numerical P systems - a new class of membrane computing systems. In *Proceedings of IEEE fifth international conference on bio-inspired computing: Theories and applications (BIC-TA)*, 1331–1336.

29. Pavel, A.B., and C. Buiu. 2012. Using enzymatic numerical P systems for modeling mobile robot controllers. *Natural Computing* 11 (3): 387–393.

30. Pavel, A.B., C.I. Vasile, and I. Dumitrache. 2012. Robot localization implemented with enzymatic numerical P systems. In *Proceedings of the international conference on biomimetic and biohybrid systems*, 204–215.

31. Pavel, A.B., C.I. Vasile, and I. Dumitrache. 2013. Membrane computing in robotics. In *Beyond artificial intelligence*, Series: Topics in Intelligent Engineering and Informatics, eds. J. Kelemen, J. Romportl, E. Zackova, Vol. 4, 125–136. Berlin: Springer.

32. Vasile, C.I., A.B. Pavel, I. Dumitrache, and Gh Păun. 2012. On the power of enzymatic numerical P systems. *Acta Informatica* 49 (6): 95–412.

33. Wang, X., G. Zhang, F. Neri, J. Zhao, M. Gheorghe, F. Ipate, and R. Lefticaru. 2016. Design and implementation of membrane controllers for trajectory tracking of nonholonomic wheeled mobile robots. *Integrated Computer-Aided Engineering* 23: 15–30.

34. Wang, T., G. Zhang, J. Zhao, Z. He, J. Wang, and M.J. Pérez-Jiménez. 2015. Fault diagnosis of electric power systems based on fuzzy reasoning spiking neural P systems. *IEEE Transactions on Power Systems* 30 (3): 1182–1194.

35. Wang, X., G. Zhang, J. Zhao, H. Rong, F. Ipate, and R. Lefticaru. 2015. A modified membrane-inspired algorithm based on particle swarm optimization for mobile robot path planning. *International Journal of Computers, Communications and Control* 10 (5): 725–738.

36. Website of the simulator for numerical P Systems (SNUPS). http://snups.buiu.net/, Laboratory of Natural Computing and Robotics, Politehnica University of Bucharest.

37. Webots robot simulator. http://www.cyberbotics.com/, Cyberbotics Ltd.

38. Wit, C.C.D., H. Olsson, K.J. Astrom, and P. Lischinsk. 1995. A new model for control of systems with friction. *IEEE Transactions on Automatic Control* 40 (3): 419–425.

39. Wu, W., H. Chen, and Y. Wang. 2001. Global trajectory tracking control of mobile robots. *ACTA Automatic Sinica* 27 (3): 325–331.

40. Xiao, J., Y. Huang, Z. Cheng, J. He, and Y. Niu. 2014. A hybrid membrane evolutionary algorithm for solving constrained optimization problems. *Optik* 125 (2): 897–902.

41. Xu, D., D.B. Zhao, and J.Q. Yi. 2009. Trajectory tracking control of omnidirectional wheeled mobile manipulators: Robust neural network-based sliding mode approach. *IEEE Transactions on Systems, Man, and Cybernetics-Part B: Cybernetics* 39 (3): 788–799.

42. Ye, J. 2008. Adaptive control of nonlinear PID-based analog neural networks for a nonholonomic mobile robot. *Neurocomputing* 71 (7–9): 1561–1565.

43. Zhang, G., M. Gheorghe, and Y. Li. 2012. A membrane algorithm with quantum-inspired subalgorithms and its application to image processing. *Natural Computing* 11 (3): 701–717.

44. Zhang, G., F. Zhou, X. Huang, J. Cheng, M. Gheorghe, F. Ipate, and R. Lefticaru. 2012. A novel membrane algorithm based on particle swarm optimization for solving broadcasting problems. *Journal of Universal Computer Science* 18 (13): 1821–1841.

# Chapter 7
# Data Modeling with Membrane Systems: Applications to Real Ecosystems

**Abstract** A probabilistic approach to P systems, called *population dynamics P systems* (PDP systems, for short) is introduced for studying the dynamics of (real) ecological populations. An implementation of this approach, as part of the P-Lingua software library, called pLinguaCore, is provided in order to assist in the definition, analysis, simulation and validation of PDP-based models. Four significant case studies of (real) ecosystems - the scavenger birds, Zebra mussel, Pyrenean chamois and Giant panda - are presented.

## 7.1 Introduction

In an informal way, a *model* is a simplified description of (a part of) a real world system that allows studying, analyzing, describing, explaining and reasoning about it. The use of models is intrinsic to any scientific activity and represents a fundamental approach when tackling some real-life problems in order to capture many complex interactions between components operating simultaneously in a highly interdependent manner. Scientists regularly use abstractions of the reality such as diagrams, graphs, plots, relationships, laws, etc., with the aim of describing and understanding real-life phenomena they are examining. In one way or another, they use models trying to understand how a complex system works/evolves and partially capture its behavior. Therefore, scientists are using simplified conceptualizations of such system incorporating its mechanisms, components and their relationships by means of a precise language that allows logical reasoning and precise analysis of the system's behavior [122].

A *formal/mathematical model* is an abstraction mapping a real-world scenario onto a mathematical domain that highlights some key features while ignoring others that are assumed to be not relevant. Therefore, mathematics is used as an useful and powerful tool to describe nature. A model should not be seen or presented as a representation of the truth, but instead as a statement of all our current knowledge about the phenomenon under study [19]. In particular, it should explicitly describe which are the assumptions that were made. Formal models are often used instead of experiments when they are expensive, dangerous, large or time consuming. Thus,

G. Zhang et al., *Real-life Applications with Membrane Computing*,
Emergence, Complexity and Computation 25, DOI 10.1007/978-3-319-55989-6_7

they provide experimentalists a way to decide which experiments should actually be carried out, contributing in this way to the progress of our understanding [6]. Models are also useful when proved to be wrong, since they show that our current understanding of the studied phenomenon does not match reality.

There exists a "balance" issue when designing a model since, on the one hand, the model has to incorporate the essential or relevant features of the system. On the other hand, models have to be rich enough in order to provide insights about the unknown, interesting, helpful and significant behavior of the system.

The development of models usually requires a multidisciplinary approach. In this context, mathematics and computer science have been used by experts simply as auxiliary tools to achieve a better quantitative formulation of the phenomena and interactions involved.

Models must be experimentally validated by comparing simulation results against an independent observed data set. In this process, it is crucial for the experiments input conditions to match the input data provided to the model. Validation allows to determine the uncertainty of the model results, usually caused by limitations in our knowledge. Contrary to formal validation, experimental validation is not an objective process: experts act as oracles, validating the model when results are "close enough" to real data. Defining what is close enough is also a crucial matter, which depends on the problem domain.

The achievements during the past century in cellular and molecular biology, ecology and population dynamics in general, economics, as well as in computer science of course, both from theoretical and practical points of view, have provided the convergence of such disciplines through the use of mathematical models, with the aim of obtaining significant progress in science.

The first tools used for cellular processes and population dynamics modeling are based on differential or partial differential equations (ODEs/PDEs). Despite the satisfactory results obtained, the deterministic and continuous approach has some drawbacks. For instance, when the number of species in a model of an ecological system is greater than two, the equations system proposed is so complex that it is usually solved using numerical methods. Besides, improvements on the performance of the models are generally obtained by the addition of ingredients, which in the case of ODEs/PDEs means that the whole modeling process needs to be done again from scratch. Besides, that approach is questioned, for instance, in the case of cell systems with low number of molecules, with slow reactions or non-homogeneous structures [2, 35], and other stochastic approaches have been developed even for specific differential equations [100].

In this chapter, a bioinspired computing modeling framework within membrane computing, called *multienvironment P systems*, is presented. Two approaches are considered, stochastic and probabilistic, allowing us to model both cellular processes and population dynamics of ecological systems. This framework presents important advantages with respect to classical models based on differential equations, such as a high computational potential, modularity and ability to work in parallel.

The chapter focuses on the study of the dynamics of (real) ecological populations which are modeled following the probabilistic approach, called *population dynamics P systems* (PDP systems, for short). The flexibility of our computing framework enables to increase the number of species with no major changes in the model, by simply adding the new information into the biological parameters of the new species to be included, or updating the information already available. It also enables virtual experimentation on population dynamics under different conditions as well as the validation of its usefulness as a simulation tool.

Several ad-hoc algorithms have been proposed to capture semantics of PDP-based models, such as DNDP, DCBA and others, presenting different compromises between accuracy and resource consumption. These algorithms have been implemented as part of the P-Lingua software library, called pLinguaCore [25, 26]. In order to assist in the definition, analysis, simulation and validation of PDP-based models related to different real-world ecosystems, the general purpose application MeCoSim, which uses pLinguaCore as its simulation engine, has then been used. Also speed-up of the implemented algorithms by using parallel platforms based on GPUs are addressed. Finally, four case studies are considered, involving modeling of (real) ecosystems of scavenger birds, Zebra mussel, Pyrenean chamois and Giant panda.

## 7.2 A General Bioinspired Computing Modeling Framework

The need to solve concrete real life problems has led researchers to look for systematic procedures allowing them to obtain solutions by means of *methods* that we can denominate as *mechanical*. Basically, such procedures consist in performing a finite sequence of "elemental" tasks in such manner that any entity able to execute such tasks in an autonomous way, can implement such procedures and, consequently, solve those problems.

Computing models are mathematical theories aiming to formally define the concept of mechanical procedure, hence they provide a general framework to design mathematical models able to be handled by using the mechanisms associated with them.

The computing modeling process consists of some semi-formal stages that guide us in the design task, in the description within a formal language, as well as its implementation on a computer for evaluation and analysis. According to [96], a formal model must be *relevant*, in the sense of capturing the basic properties of the phenomena under study, in relation to its structure and its dynamics, and providing a better *understandability* of the studied system. A good model has to be easily *extensible* and *scalable* to other levels of organization and changeable with ease in order to include new knowledge or remove false hypothesis. Finally, a model has to be *computationally tractable*, in the sense that we have to be able to implement it in order to execute simulations allowing us to study the system dynamics in different scenarios.

In order to capture inherent randomness and uncertainty associated with dynamics of ecological systems, a strategy considering time-dependent probabilistic functions is used. These functions are usually assigned to certain basic elements controlling system dynamics. Their numeric values, associated to those elements in the model at any instant, represent the probability for each one of them to be used to make the system evolve in that moment. Those functions can be experimentally obtained or estimated, but unfortunately in many cases they are unknown, or their values range within an interval. Then it is necessary to calibrate the model: the different model results deduced from scenarios associated with sets of parameters must be compared with the observed values corresponding to the same scenarios. After that, an ad hoc algorithm is designed to describe the semantics of the computational model, in which the work is being developed. This semantics must incorporate the treatment of the random nature materialized by probabilistic constants or functions [53].

The simulation algorithm needs to be experimentally validated by contrasting the values produced by the algorithm with those data that were obtained by means of experiments or field work. Obviously, this contrast is not valid unless the algorithm is simulated under the same specific scenarios in which the data were obtained. To do so, it is necessary for the algorithm to be efficient enough, so that the simulations of the different scenarios of interest are executed in a reasonable time. In Sect. 7.2.3, two probabilistic simulation algorithms for P systems will be described.

## 7.2.1   Membrane Systems for Modeling Biological Data

Ecological systems consisting of multi-species communities are usually governed by many intricate processes and inter-specific interactions such as competition or predation, operating in parallel. Hence, it is essential to detect the relevant processes and also the important components of the systems in order to be able to organize these interrelations in a schematic and graphical way. Quantification of the strength of interactions between species is basic for understanding how ecological systems are organized and how they respond to human intervention. The volume of data and knowledge about processes and interactions requires more complex tools, and mathematics provide a precise framework to express these terms in a succinct way, and to manage all available information.

A formal model for dynamics of ecological populations is expected to describe how the population is going to change in time, in its composition, starting from the current status and the environmental conditions of the ecosystem under study.

Generally, the formal tools used for population dynamics modeling are based on ordinary differential equations (ODEs). For example, the Lotka–Volterra model has been one of the most frequently used ones for the modeling of two species, predator and prey [78], and Verhulst's model (logistic differential equations) has been used as a fundamental growth model in ecological studies because of its mathematical simplicity and single biological definition [101, 102]. Fuzzy set theory has been used to estimate the parameters of the ODE-based models studying the interaction between a prey and its predator [22]. Despite the satisfactory results presented by these models, ODE-based approach has some drawbacks, due to the implications of its deterministic and continuous nature. For instance, Russell et al. [100] developed a stochastic model for three species by using ordinary logistic differential equations analyzed through numerical analysis techniques. Consequently, in recent years new models based on the latest computational paradigms and technological advances have been adopted. Some examples are Petri nets [40], process algebra ($\pi$-calculus [96], bioambients [97], brane calculus [8], $\kappa$-calculus [23], etc.), state charts [41], agent based systems [45] and viability models [105, 106].

Each ecosystem has its own important peculiarities, thus trying to design a "universal" ecosystem model is not a good approach. For instance, the model should be adapted taking into account if a protected and endangered species is being studied, or if the ecosystem deals with an invasive species, or simply an endemic area. Nevertheless, there are some aspects common to most ecosystems such as:

- They contain a large number of individuals and a large number of species.
- The life cycles of species that inhabit the ecosystem display several basic processes such as: feeding, growth, reproduction and death.
- These processes are annually repeated.
- The evolution often depends on the environment: weather, soil, vegetation, etc.
- The natural dynamics suffers modifications due to human activities.

These common features yield some requisites for the model, from a computational point of view: many processes take place simultaneously; there is cooperation between individuals and elements of the ecosystem; there is partial synchronization among the dynamic evolution of sub-ecosystems (for example, there could be adverse weather conditions some year, and this does not affect a single sub-ecosystem, but has a global influence on the entire ecosystem), and situations need to be restored annually.

In the original approach, a computation in a membrane system is obtained by repeatedly applying the rules in a non-deterministic synchronous maximal parallel way: in each step, in each compartment, all the objects and strings that can evolve by means of any rule must evolve in parallel at the same time.

According to the original motivation, P systems were not intended to provide a comprehensive and accurate abstract representation of the living cell, but rather, to explore the computational nature of various features of biological membranes. Indeed, much work has been focused on *computational completeness* by studying the computational power with respect to deterministic Turing machines, and *computational efficiency* by analyzing the capability of the membrane systems to provide "efficient" solutions to hard problems by trading space for time.

Suzuki and Tanaka were the first researchers to consider membrane computing as a framework for modeling chemical systems [111–113] and ecological systems [114] by using *abstract rewriting system on multisets*. This research line has been continued by modeling different biological phenomena presenting membrane systems-based models of oscillatory systems [29], signal transduction [93], gene regulation control [98], quorum sensing [99] and metapopulations [94].

The first application in the field of (real) ecosystems modeling within the framework of membrane computing took place in 2008 [9], and was focused on the study of the population dynamics of *Bearded vulture* in the Catalan Pyrenees, Spain. In that work, a novel computational membrane system was presented to manage the cited ecosystem, studying in the target area the population dynamics of Bearded vulture, along with other five species providing the bones which vultures feed from. In order to experimentally validate the model designed in this framework, a first ad-hoc software application [3] was developed, including in the source code (in C++ language) all the data about the specific scenario under study: (a) the rules of the system; (b) the parameters and constants; (c) the rest of the elements of the model; and (c) the simulation engine itself. In [10, 15] new models were presented, adding population density regulation, limits over feeding in case of shortage, variability in terms of growth rate ("reproduction") of Bearded vulture and competition for certain resources. These works make use of P-Lingua framework [30] for specifying and simulating the model, stating a clear separation among: (a) the parser of the specification language and simulation engine, provided by P-Lingua developers; (b) the model made by the designer, specifying the structure, rules, multisets, etc. in the system; and (c) input data, provided by the end users (ecological experts).

Due to the success in the results coming from the previous modeling works, a new requirement was received from the company Endesa S.A., in order to work in the modeling of an ecosystem placed in the reservoir of Riba-roja in the area of the Ebro river basin, specifically focusing on an exotic invasive species: *Zebra mussel*. This species has produced an important damage both from an ecological and an economical point of view over the last decades. Thus, a computational model was designed, based on P systems, to provide the company with a global view of the problem, and to help them predicting, to some extent, the evolution of the density of larvae of this species [18].

After those experiences, additional models based on P systems were designed, covering a broad range: (a) ecosystems related to scavenger birds in the Navarre Pyrenees and in Swaziland (South Africa); (b) ecosystems affected by random changes in the environment related to temperature, rain, etc. (development and growth of certain amphibians in ponds or lakes); (c) reintroduction of a species of birds in the Catalan Pyrenees (*Hazel Grouse*); (d) ecosystems very sensitive to floods and heavy rainfalls (prediction of possible scenarios of extinction of species such as the Pyrenean brook newt in the Segre river, Serra del Cadí, Pyrenees) and (e) the effect of pestivirus in the population dynamics of Pyrenean chamois.

### 7.2.2  *Multienvironment P Systems*

Now, we present a general membrane computing based modeling framework to model both cellular systems and the dynamics of ecological populations.

**Definition 1** A multienvironment P system of degree $(q, m, n)$ with $q, m, n \geq 1$ and $T \geq 1$ time units, is a tuple

$$\Pi = (G, \Gamma, \Sigma, T, \mathcal{R}_E, \mu, \mathcal{R}, \{f_{r,j} \mid r \in \mathcal{R} \wedge 1 \leq j \leq m\},$$
$$\{\mathcal{M}_{i,j} \mid 1 \leq i \leq q \wedge 1 \leq j \leq m\}, \{E_j \mid 1 \leq j \leq m\})$$

where:

- $G = (V, S)$ is a directed graph with $V = \{e_1, \ldots, e_m\}$.
- $\Gamma$ and $\Sigma$ are alphabets such that $\Sigma \subsetneq \Gamma$.
- $T$ is a natural number.
- $\mathcal{R}_E$ is a finite set of rules of the form

$$(x)_{e_j} \xrightarrow{\ p\ } (y_1)_{e_{j_1}} \cdots (y_h)_{e_{j_h}} \ ; \ (\Pi_k)_{e_j} \xrightarrow{\ p'\ } (\Pi_k)_{e_{j_1}}$$

  where $x, y_1, \ldots, y_h \in \Sigma$, $(e_j, e_{j_l}) \in S$, $1 \leq l \leq h$, $1 \leq k \leq n$ and $p, p'$ are computable functions whose domain is $\{1, \ldots, T\}$.
- $\mu$ is a rooted tree with $q$ nodes (called membranes) injectively labeled by elements from the set $\{1, \ldots, q\} \times \{0, +, -\}$. If the label of a membrane is $(i, \alpha)$ then such a membrane will be denoted as $[\ ]_i^\alpha$ and we will say that the membrane has label $i$ and electrical charge $\alpha$. The root of the tree has 1 as associated label.
- $\mathcal{R}$ is a finite set of rules of the form $r \equiv u[v]_i^\alpha \longrightarrow u'[v']_i^{\alpha'}$, where $u, v, u', v'$ are finite multisets over $\Gamma$, $u + v \neq \emptyset$, $1 \leq i \leq q$ and $\alpha, \alpha' \in \{0, +, -\}$. Besides, if $(x)_{e_j} \xrightarrow{\ p\ } (y_1)_{e_{j_1}} \cdots (y_h)_{e_{j_h}}$ is a rule in $\mathcal{R}_E$, then there cannot exist any rule in $\mathcal{R}$ whose left-hand side is of the form $u[v]_1^\alpha$ with $x \in u$.
- For each $r \in \mathcal{R}$ and $1 \leq j \leq m$, $f_{r,j}$ is a computable function whose domain is $\{1, \ldots, T\}$.
- For each $i, j$ ($1 \leq i \leq q, 1 \leq j \leq m$), $\mathcal{M}_{ij}$ is a finite multiset over $\Gamma$.
- For each $j$, $1 \leq j \leq m$, $E_j$ is a finite multiset over $\Sigma$.

A multienvironment P system of degree $(q, m, n)$ and with $T$ time units can be seen as a set of $m$ environments $e_1, \ldots, e_m$ linked by arcs from a directed graph $G$ and a set of P systems, $\{\Pi_k \mid 1 \leq k \leq n\}$, having the same "skeleton", that is, the same working alphabet $\Gamma$, the same membrane structure $\mu$, and the "same" rules from $\mathcal{R}$. Each environment $e_j$ initially contains a finite multiset $E_j$ of objects from $\Sigma$ (which will be referred to as environment objects from $e_j$). Besides, each system $\Pi_k$ must be included within some environment $e_j$. If a system $\Pi_k$ is located within an environment $e_j$, then its initial multisets $\mathcal{M}_{1,k}, \ldots, \mathcal{M}_{q,k}$ depend on the environment where it resides. Finally, all membranes in any of the P systems initially have neutral charge. For the sake of simplicity, neutral polarization will be omitted.

Multienvironment P system allow rules of two types: on the one hand rules associated with the P systems contained in the environments, whose set is denoted by $\mathcal{R}$; and on the other hand rules for communicating objects among environments, as well as for P systems traveling from an environment to a neighbour one, and this set is denoted by $\mathcal{R}_E$. Every rule $r$ has computable functions associated with it ($f_{r,j}$, $p_r$, or

$p'_r$) which depend on the environment where the rule is located. These functions can be interpreted as an indication of the affinity of each rule, to be taken into account when several applicable rules compete for the available objects.

The natural number $T$ stands for the simulation time.

The *semantics* of multienvironment P systems is defined as a parallel, non deterministic, synchronized mode, in the sense that we assume that a global clock exists, marking the time steps of the system evolution. Let us describe next the semantics of multienvironment P systems.

An *instantaneous description* or *configuration* of the system at a given instant $t$ is a tuple composed by: (a) the multisets of objects over $\Sigma$ present in the $m$ environments; (b) the P systems included on each environment; (c) the multisets of objects over $\Gamma$ contained on each of the regions of such P systems; (d) the values of functions associated with the rules of the system; and (e) the polarizations of the membranes from each P system.

A rule $r \in \mathcal{R}$ of the type $u[\, v\, ]_i^\alpha \xrightarrow{f_r} u'[\, v'\, ]_i^{\alpha'}$ associated to a system $\Pi_k$ is *applicable* to a configuration $\mathcal{C}$ at a given instant $t$ if the membrane from $\Pi_k$ labeled by $i$ has electrical charge $\alpha$ in that configuration, it contains the multiset $v$, and besides its parent membrane (or the environment, in case $i$ is the skin membrane) contains multiset $u$. The execution of the rule produces the following effects: (a) multisets $v$ and $u$ are removed from membrane $i$ and its parent region, respectively; (b) at the same time, multiset $u'$ is added to the parent membrane of $i$ and multiset $v'$ is added to membrane $i$; and (c) the new electrical charge of membrane $i$ will now be $\alpha'$ (which might be the same as $\alpha$). It is interesting to note that when applying a given set of rules to a given membrane, all of the rules must have the same electrical charge on their right-hand side; that is, all rules applied simultaneously in one step on the same membrane should be *consistent*. The value of function $f_r$ at a given instant $t$ indicates the affinity which the rule has.

A rule $r \in \mathcal{R}_E$ of the type $(x)_{e_j} \xrightarrow{p_r} (y_1)_{e_{j_1}} \dots (y_h)_{e_{j_h}}$ is applicable to a configuration $\mathcal{C}$ of the system at a given instant $t$ if the environment $e_j$ contains object $x$ in this configuration. The execution of such rule produces the following effects: (a) the object $x$ is eliminated from environment $e_j$; and (b) environments $e_{j_1}, \dots, e_{j_h}$ get new objects $y_1, \dots, y_h$ respectively.

A rule $r \in \mathcal{R}_E$ of type $(\Pi_k)_{e_j} \xrightarrow{p_{r'}} (\Pi_k)_{e_{j'}}$ is applicable in environment $e_j$ if this environment contains P system $\Pi_k$. The execution of such rule makes the system $\Pi_k$ move from environment $e_j$ into environment $e_{j'}$.

From these notions, one can define in a natural way what means that the multienvironment P system goes from a configuration $\mathcal{C}$ at a given instant to another configuration $\mathcal{C}'$ in the next instant, by means of the execution of a maximal multiset of rules following the previous indications. In this way, a *computation step* is obtained; that is, a *transition* from a configuration to a *next* configuration of the system. Thus, the concept of *computation* is introduced in a natural way, as a sequence of configurations as it was defined in Sect. 1.4 of the first chapter.

### 7.2.2.1 Types of Multienvironment P Systems

We can highlight two main types of multienvironment P systems, which correspond to different frameworks of computational modeling: systems having a stochastic orientation, called *multicompartmental P systems*, and those having a probabilistic orientation, called *population dynamics P systems*. We will next describe the specific syntax of each one, as well as some simulation algorithms which allow us to implement the corresponding semantics.

The **stochastic approach** of multienvironment P systems is called *multicompartmental P systems* and constitutes a variant which provides a framework for modeling molecular and cellular processes. In these systems, each environment contains a certain number of P systems which are randomly distributed at the beginning of the computation. Besides, on each environment, the P systems residing there will get specific initial multisets associated with this environment.

In these systems, the inherent randomness of molecular and cellular processes is captured by using a semantics designed for rewriting rules that have associated a *propensity*.

**Definition 2** A *multicompartmental P system* of degree $(q, m, n)$, with $q, m, n \geq 1$, having $T \geq 1$ time units, is a multienvironment P system of degree $(q, m, n)$ and with $T$ time units

$$\Pi = (G, \Gamma, \Sigma, T, \mathcal{R}_E, \mu, \mathcal{R}, \{f_{r,j} \mid r \in \mathcal{R} \wedge 1 \leq j \leq m\},$$
$$\{\mathcal{M}_{i,j} \mid 1 \leq i \leq q \wedge 1 \leq j \leq m\}, \{E_j \mid 1 \leq j \leq m\})$$

which fulfills the following conditions:

- The computable functions associated to the rules of the environment and the rules of the P systems are the *propensities* of such rules. These functions are determined from the values of stochastic constants, by applying the *mass action law*, *exponential decay law* and *Michaelis–Menten dynamics*. The stochastic constants associated to each rule are calculated, on their turn, from the values of kinetic constants which have been experimentally calculated. The propensities are functions which depend on time, but on the other hand, they do not depend on the environment $e_j$ where the corresponding P system currently is.
- Initially, the P systems are randomly distributed among the $m$ environments of the system.
- For every rule $r$ of type $(x)_{e_j} \xrightarrow{pr} (y_1)_{e_{j_1}} \cdots (y_h)_{e_{j_h}}$, we have $h = 1$; that is, an object $x$ can just move from an environment to another one, possibly getting transformed into another object $y_1$.

The semantics of a multicompartmental P system can be simulated through an extension of the *Gillespie algorithm* [34–38], an algorithm for simulating the time-course evolution of a stochastic kinetic model and developed for a single, well mixed and fixed volume or compartment. The extension is called *multicompartmental Gillespie algorithm* (introduced in [93]) and it is an adaptation of the Gillespie algorithm

**Fig. 7.1** A
multicompartmental P
system

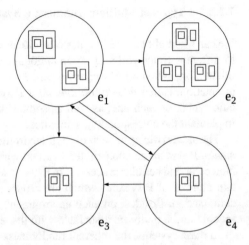

that can be applied in the different regions defined by the hierarchical and com-
partmentalised structure of a multicompartmental P system model (Figs. 7.1 and
7.2).

The **probabilistic approach** of multienvironment P systems is called *population
dynamics P systems* and constitutes a variant of multienvironment P systems which
provides a framework for modeling of processes related to population dynamics,
in general. The inherent randomness of such processes is captured by means of a
semantics designed for rewriting rules that have associated probability functions
at any instant which will also depend on the environment. In these systems, each
environment contains a single P system, and the initial multisets of their membranes
are specific for each environment.

**Definition 3** A population dynamics P system (*PDP system*, for short) of degree
$(q, m)$, where $(q, m \geq 1)$ and having $T \geq 1$ time units, is a multienvironment P
system of degree $(q, m, m)$ and with $T$ time units

$$\Pi = (G, \Gamma, \Sigma, T, \mathcal{R}_E, \mu, \mathcal{R}, \{f_{r,j} \mid r \in \mathcal{R} \wedge 1 \leq j \leq m\},$$
$$\{\mathcal{M}_{i,j} \mid 1 \leq i \leq q \wedge 1 \leq j \leq m\}, \{E_j \mid 1 \leq j \leq m\})$$

satisfying the following conditions:

- At the initial instant, each environment $e_j$ contains exactly one P system, which
  will be denoted by $\Pi_j$. Therefore, the number $n$ of P systems matches the number
  $m$ of environments.
- Function $p_r$ associated to rule $r$ from $\mathcal{R}_E$ of the type

$$(x)_{e_j} \xrightarrow{\;p_r\;} (y_1)_{e_{j_1}} \cdots (y_h)_{e_{j_h}}$$

  have their range included in [0, 1] and they verify:

  – For each $e_j \in V$ and $x \in \Sigma$, the sum of the functions associated to rules of the
    above type is the constant function 1.

**Fig. 7.2**  A population
dynamics P system

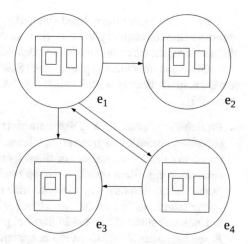

- Function $p_{r'}$ associated to rule $r'$ from $\mathcal{R}_E$ of the type $(\Pi_k)_{e_j} \xrightarrow{\ p_{r'}\ } (\Pi_k)_{e_{j'}}$ are all constant and equal to 0; that is, one may as well assume that this type of rule is forbidden, or equivalently, P systems residing in an environment cannot "travel" to any other environment.
- For each rule $r \in \mathcal{R}$ of the system $\Pi_j$ located in $e_j$, $1 \le j \le m$, the computable function $f_{r,j}$ also depends on the environment and its range is contained within $[0, 1]$. Moreover, for each $u, v \in M_f(\Gamma)$, $1 \le i \le q$ and $\alpha, \alpha' \in \{0, +, -\}$, the sum of the functions $f_{r,j}$ with $r \equiv u[v]_i^{\alpha} \rightarrow u'[v']_i^{\alpha'}$, is the constant function 1. We write $\mathcal{R}_{\Pi_j}$ instead of $\mathcal{R}$.

Note that at any instant $t$, $1 \le t \le T$, for each object $x$ in environment $e_j$, if there exist communication rules of type $(x)_{e_j} \xrightarrow{\ p_r\ } (y_1)_{e_{j_1}} \ldots (y_h)_{e_{j_h}}$, then some of them will be applied. In case several of such rules exist having $(x)_{e_j}$ as their left-hand side, then the rules will be applied according to their corresponding probabilities associated to this particular instant $t$.

## 7.2.3   Simulation Algorithms for PDP Systems: DNDP and DCBA

Simulation algorithms are necessary for reproducing the execution of computational models, since they provide the mechanisms to evolve a system from one configuration to another. The aim of simulation algorithms within the PDP systems framework is to serve as inference engines, reproducing the semantics of the model in a reliable and accurate way.

On the one hand, the syntactical elements refer to all data elements of P systems (graph associated membrane structure, multisets of objects, rules, etc.); on the other hand, the semantics of the model refers to the way the rules are applied, that is, how the system evolves. Therefore, a simulation algorithm is assumed to represent a *semantics* for a model, while dealing with all the syntactical elements.

Before introducing the defined simulation algorithms for PDP systems, we shall discuss some semantical properties that are desirable on their simulation. They are defined to adjust the behavior of the models according to how the experts and designers think about the modeled phenomena. Some of them also come from the fact that we are actually dealing with P systems and Markov Chains based probabilistic systems [65].

- *Probabilistic behavior*: A PDP system aims to reproduce the stochasticity of nature processes according to given probabilities. This random behavior is directly associated to rules by the number of times they are going to be applied. To do this, a simulation algorithm should calculate random numbers according to the defined probabilities, in such a way that each time the system is simulated the reproduced computation should be different. Thus, statistical studies over several parallel and independent simulations should fit the expected mean and variance.
- *Resource competition*: since the cooperation degree in PDP systems is greater than 1, rules are used to dictate how groups of elements and individuals evolve in the model. According to the semantics of PDP systems (and the way experts think on Population Dynamics), what is going to happen to the same group of individuals must be predefined, and the probability of each option has to sum 1. However, the same elements can participate in different groups, and so in different evolutions. This issue is also known as *competition* for resources (by the evolution of groups viewpoint). The behavior for this is not such explicitly specified, and can be carried out by several approaches. Some algorithms implement a random way to distribute the resources, while others assume that nature has a tendency towards proportionally distribute resources to such groups were less elements are required (the more required, the more energy is needed).
- *Maximality of the model*: rules are applied in a maximally parallel way, as traditionally in P systems. This property can, in fact, help to control the evolution of every element, even if they remain unchanged or disappear. We will see that, sometimes, an extra phase within the algorithms assuring maximality will be required.
- *Consistency of rules*: according to the syntactics of the model, two rules can lead to an inconsistent configuration in which a membrane can have two different charges. However, the semantics imposes that each state must be consistent, by restricting the property of maximality to those rules that produce consistent states of the system; that is, two rules producing a different charge to the same membrane cannot be simultaneously applied.

One of the first simulation algorithms for PDP systems was *Binomial Block Based algorithm* (BBB) [11]. Basically, the chosen approximation was to arrange rules into blocks, in such a way that rules having their left-hand sides exactly equal are put in a single block (see Definitions 4 and 5). The algorithm consists on a random of over those blocks, maximally selecting a number of applications for each one (according to that "common" left-hand side), and calculating a multinomial distribution of those applications to the rules, according to the probabilities. This multinomial distribution is implemented internally by using binomial random variates according to the rules.

Although this simulation algorithm is useful for the majority of models, like [10, 11], it has some disadvantages since it does not handle accurately the following semantic properties:

- Resource competition: Rules with partial (no total) overlapping on their left-hand sides are classified into different blocks, so the common objects will not be distributed since these blocks are maximally executed.
- Consistency of rules.
- Evaluation of probabilistic functions related to rules. Only constant probabilities are considered, what is not the case for forthcoming PDP systems based models.

### 7.2.3.1 DNDP

The resource competition is a well-known issue in P system simulation. It has been addressed by different approaches. For instance, in [80], a way to convey the non-determinism in reconfigurable hardware was studied. This simulation algorithm was called *Direct Non-deterministic Distribution* (DND). The procedure consists of two phases for selecting rules (a random order is assumed): a forward phase which chooses a random number of applications for each rule, and a backward phase which iterates again the rules checking and assigning the remaining applications. This mechanism was the basis for a new simulation algorithm for PDP systems, called *Direct Non-Deterministic distribution with Probabilities* (DNDP). It was first presented in [66], formally verified in [68], and compared with BBB in [16, 65].

Similarly to BBB, the transitions of the PDP system are simulated in two phases, selection and execution, in order to synchronize the consumption and production of objects. However, selection is divided in two micro-phases, following the design of the DND algorithm: first phase calculates a multiset of *consistent* applicable rules, and second phase eventually increases the multiplicity of some of the rules in the previous multiset to assure maximal application, obtaining a multiset of *maximally consistent* applicable rules. The main procedure of the DNDP algorithm is shown in Algorithm 1.

---

**Algorithm 1** DNDP main procedure

---

**Input:** A PDP system of degree $(q, m)$ with $q \geq 1$, $m \geq 1$, taking $T$ time units, $T \geq 1$.
1: $C_0 \leftarrow$ initial configuration of the system
2: **for** $t \leftarrow 0$ **to** $T - 1$ **do**
3:     $C'_t \leftarrow C_t$
4:     *INITIALIZATION*                                                              ▷ (Algorithm 2)
5:     *FIRST SELECTION PHASE*: consistency                    ▷ (Algorithm 3)
6:     *SECOND SELECTION PHASE*: maximality                  ▷ (Algorithm 4)
7:     *EXECUTION*                                                               ▷ (Algorithm 5)
8:     $C_{t+1} \leftarrow C'_t$
9: **end for**

---

Firstly, the *initialization* process (Algorithm 2) constructs two ordered set of rules, $A_j$ and $B_j$, gathering only rules from $\mathcal{R}_E$ and $\mathcal{R}_{\Pi_j}$ in environment $e_j$ having a probability greater than 0. Finally, the probabilities of rules are recalculated for each moment $t$.

---

**Algorithm 2** DNDP initialization

---

1: **for** $j \leftarrow 1$ **to** $m$ **do**
2:     $R_{E,j} \leftarrow$ ordered set of rules from $\mathcal{R}_E$ related with the environment $j$
3:     $A_j \leftarrow$ ordered set of rules from $R_{E,j}$ whose probability is $> 0$ at step $t$
4:     $LC_j \leftarrow$ ordered set of pairs $\langle label, charge \rangle$ for all the membranes from $C_t$
            contained in the environment $j$
5:     $B_j \leftarrow \emptyset$
6:     **for all** $\langle h, \alpha \rangle \in LC_j$ (following the considered order) **do**
7:         $B_j \leftarrow B_j \cup$ ordered set of rules from $R_{\Pi_j}$ whose probability is $> 0$ at
            step $t$ for the environment $j$
8:     **end for**
9: **end for**

---

As already mentioned, *first selection* phase (Algorithm 3) calculates a multiset of consistent applicable rules, denoted by $R_j$ (for each environment $j$). First of all, the ordered set $D_j$ stores a random order applied to $A_j \cup B_j$. A temporal copy of the current configuration $C_t$, called $C_t'$, is also created, that will be updating by removing the LHS of rules being selected.

A rule $r$ is considered applicable if the following holds: it is consistent with the previously selected rules in $R_j$ (according to the order in $D_j$), and the number of possible applications $M$ in $C_t'$ is non-zero. When a rule $r$ is applicable, a random number of applications $n$ is calculated according to the probability function and using a binomial distribution. On the one hand, since $C_t'$ has been updated by the previously selected rules, the number $n$ cannot exceed $M$. On the other hand, if the generated number $n$ is 0, the corresponding rule is also added to the multiset $R_j$, which gives a new chance to be selected in the next phase (maximality). In this way, the "multiset" of rules $R_j$ is managed as a set of pairs $\langle x, y \rangle$, where $x \in D_j$ and $y$ is the number of times that $x$ is going to be applied (eventually $y = 0$). Let denote $R_j^0 = \{r \in D_j : \langle x, 0 \rangle \in R_j\}$ and $R_j^1 = \{r \in D_j : \langle x, n \rangle \in R_j, n > 0\}$. Only rules from $R_j^1$ are considered for the consistency condition, since they have already consumed objects in $C_t'$.

---

**Algorithm 3** DNDP first selection phase: consistency

---

1: **for** $j \leftarrow 1$ **to** $m$ **do**
2:     $R_j \leftarrow \emptyset$
3:     $D_j \leftarrow A_j \cup B_j$ with a *random order*
4:     **for all** $r \in D_j$ (following the considered order) **do**
5:         $M \leftarrow$ maximum number of times that $r$ is applicable to $C_t'$
6:         **if** $r$ is *consistent* with the rules in $R_j^1 \wedge M > 0$ **then**
7:             $N \leftarrow$ maximum number of times that $r$ is applicable to $C_t$
8:             $n \leftarrow \min\{M, F_b(N, pr_{,j}(t))\}$
9:             $C_t' \leftarrow C_t' - n \cdot LHS(r)$
10:             $R_j \leftarrow R_j \cup \{< r, n >\}$
11:         **end if**
12:     **end for**
13: **end for**

---

In the *second selection* phase (Algorithm 4), the consistent applicable rules are checked again in order to achieve maximality. Only consistent rules are considered, and taken from $R_j$. If one rule $r \in R_j$ has still a number of applications $M$ greater than 0 in $C_t'$, then $M$ will be added to the multiplicity of the rule, and subtracted again from $C_t'$. In order to fairly distribute the objects among the rules, they are iterated in descendant order with respect to the probabilities. Moreover, if one rule from the multiset $R_j^0$ was checked, it would be possible that another rule from $R_j^1$, inconsistent to this one, had been previously selected. In this case, the consistent condition has to be tested again.

---

**Algorithm 4** DNDP second selection phase: maximality

---
1: **for** $j \leftarrow 1$ **to** $m$ **do**
2:    $R_j \leftarrow R_j$ *with an order by the rule probabilities, from highest to lowest*
3:    **for all** $< r, n > \in R_j$ (following the selected order) **do**
4:       **if** $n > 0 \vee (r$ is *consistent* with the rules in $R_j^1$) **then**
5:          $M \leftarrow$ maximum number of times that $r$ is applicable to $C_t'$
6:          **if** $M > 0$ **then**
7:             $R_j \leftarrow R_j \cup \{< r, M >\}$
8:             $C_t' \leftarrow C_t' - M \cdot LHS(r)$
9:          **end if**
10:       **end if**
11:    **end for**
12: **end for**

---

Finally, execution phase (Algorithm 5) is similar to BBB algorithm. It will iterate all the rules in every $R_j^1$ (maximal applicable consistent rules from environment $j$), and it will add the right-hand sides of them to the configuration $C_t'$. At the end of the process, $C_t'$ is actually the next configuration: the left-hand sides of rules have been removed in the first and second selection phases, and the right-hand sides are added in the execution stage.

---

**Algorithm 5** DNDP execution

---
1: **for all** $< r, n > \in R_j, n > 0$ **do**
2:    $C_t' \leftarrow C_t' + n \cdot RHS(r)$
3:    Update the electrical charges of $C_t'$ according to $RHS(r)$
4: **end for**

---

As it can be seen, DNDP included mechanisms to partially solve the drawbacks from BBB: handling probability functions, consistency of rules and resource competition. Its functionality has been verified in [68]. Although the behavior of object distribution is still not accurate enough, its efficiency makes it a good candidate to simulate many PDP system models. Section 7.2.3.3 shows a simple example and a discussion over its accuracy. However, DNDP still creates some distortion in the distribution of objects among competing (with overlapped LHS) rules. That is, instead of rules being selected according to its probabilities in a uniform manner, this selection process is biased towards those with the highest probabilities. Moreover, the probabilistic distribution of rule executions inside blocks will not eventually follow a multinomial distribution, since competing rules from other blocks might consume objects in the selection process.

### 7.2.3.2   DCBA

Both DNDP and BBB share a common drawback: a distortion in the object distribution among competing rules. This is where the latest algorithm, called *Direct distribution based on Consistent Blocks Algorithm* (DCBA) [71], come into play. The main idea behind DCBA is to implement a proportional distribution of objects among consistent block of rules (a concept similar, but not the same, to blocks in BBB), while dealing with consistency and probabilities.

In what follows, the key concepts required for DCBA are going to be described. Then, the pseudocode of the algorithm is provided, along with brief explanations. Finally, DCBA and DNDP algorithms are going to be compared by a very simple test example.

Firstly, rules in $\mathcal{R}$ and $\mathcal{R}_E$ can be classified into blocks having the same left-hand side, following the Definitions 4 and 5. This notion was also used in BBB algorithm.

**Definition 4** The left and right-hand sides of the rules are defined as follows:

(a) Given a rule $r \in \mathcal{R}_E$ of the form $(x)_{e_j} \xrightarrow{\ p\ } (y_1)_{e_{j_1}} \cdots (y_h)_{e_{j_h}}$ where $e_j \in V$ and $x, y_1, \ldots, y_h \in \Sigma$:

- The left-hand side of $r$ is $LHS(r) = (e_j, x)$.
- The right-hand side of $r$ is $RHS(r) = (e_{j_1}, y_1) \cdots (e_{j_h}, y_h)$.

(b) Given a rule $r \in \mathcal{R}$ of the form $u[v]_i^{\alpha} \to u'[v']_i^{\alpha'}$ where $1 \le i \le q$, $\alpha, \alpha' \in \{0, +, -\}$ and $u, v, u', v' \in \Gamma^*$:

- The left-hand side of $r$ is $LHS(r) = (i, \alpha, u, v)$. The charge of $LHS(r)$ is $charge(LHS(r)) = \alpha$.
- The right-hand side of $r$ is $RHS(r) = (i, \alpha', u', v')$. The charge of $RHS(r)$ is $charge(RHS(r)) = \alpha'$.

The charge of $LHS(r)$ is the second component of the tuple (idem for $RHS(r)$).

**Definition 5** Rules from $\mathcal{R}$ and $\mathcal{R}_E$ can be classified in blocks as follows: (a) the block associated to $(i, \alpha, u, v)$ is $B_{i,\alpha,u,v} = \{r \in \mathcal{R} : LHS(r) = (i, \alpha, u, v)\}$; and (b) the block associated with $(e_j, x)$ is $B_{e_j,x} = \{r \in \mathcal{R}_E : LHS(r) = (e_j, x)\}$.

Recall that, according to the semantics of our model, the sum of probabilities of all the rules belonging to the same block is always equal to 1. In particular, rules with probability equal to 1 form individual blocks. Note that rules having exactly the same left-hand side (LHS) belong to the same block, but rules with overlapping (but different) left-hand sides are classified into different blocks. The latter leads to object (resource) *competition*, which is a critical aspect to manage with simulation algorithms.

Rule consistency is easily handled in DCBA by the concept of *consistent blocks*. It is an expansion of the notion of block, together with consistent rules and set of rules, as given in Definitions 6–9.

**Definition 6** Two rules, $r_1 \equiv u_1[v_1]_{i_1}^{\alpha_1} \to u_1'[v_1']_{i_1}^{\alpha_1'}$ and $r_2 \equiv u_2[v_2]_{i_2}^{\alpha_2} \to u_2'[v_2']_{i_2}^{\alpha_2'}$, are consistent if and only if $(i_1 = i_2 \wedge \alpha_1 = \alpha_2 \to \alpha_1' = \alpha_2')$.

**Definition 7** A set of rules is consistent if every pair of rules of the set is consistent.

**Definition 8** Given $(i, \alpha, u, v)$ where $1 \leq i \leq q$, $\alpha \in EC$, $u, v \in \Gamma^*$, the block $B_{i,\alpha,u,v}$ is consistent if and only if there exists $\alpha'$ such that, for each $r \in B_{i,\alpha,u,v}$, $charge(RHS(r)) = \alpha'$.

**Definition 9** Given $i, \alpha, \alpha', u, v$ where $1 \leq i \leq q$, $\alpha, \alpha' \in EC$, $u, v \in \Gamma^*$, the block associated with $(i, \alpha, \alpha', u, v)$ is the set:

$$B_{i,\alpha,\alpha',u,v} = \{r \in \mathcal{R} : LHS(r) = (i, \alpha, u, v) \wedge charge(RHS(r)) = \alpha'\}$$

Note that the block $B_{i,\alpha,\alpha',u,v}$ determines a consistent set of rules. Then, the left-hand side of a block $B$, denoted by $LHS(B)$, is defined as the left-hand side of any rule in the block.

**Definition 10** We say that two blocks $B_{i_1,\alpha_1,\alpha_1',u_1,v_1}$ and $B_{i_2,\alpha_2,\alpha_2',u_2,v_2}$ are mutually consistent with each other, if and only if $(i_1 = i_2 \wedge \alpha_1 = \alpha_2) \Rightarrow (\alpha_1' = \alpha_2')$.

That is, two rule blocks are mutually consistent if the union of them represents a consistent set of rules.

**Definition 11** A set of blocks $\mathcal{B} = \{B^1, B^2, \ldots, B^s\}$ is self consistent (or mutually consistent) if and only if they are pairwise mutually consistent, that is $\forall i, j$ ($B^i$ and $B^j$ are mutually consistent).

In such a context, a set of blocks has an associated set of tuples $(i, \alpha, \alpha')$, that is, a relation between labels and electrical charges ($H \times EC$) in $EC$. Then, a set of blocks is mutually consistent if and only if the associated relationship $H \times EC$ in $EC$ is functional.

DCBA solves the resource competition by performing a proportional distribution of objects among competing blocks (with overlapping LHS), determining in this way the number of times that each rule in $\bigcup_{j=1}^{m} \mathcal{R}_{\Pi_j} \cup \mathcal{R}_E$ is applied. Algorithm 6 describes the main loop of the DCBA. It has the same general scheme as its predecessors, *DNDP* and *BBB*, where the simulation of a computational step is structured in two stages: selection and execution. However, in this case, selection stage consists of three phases: Phase 1 distributes objects to the blocks in a certain proportional way, Phase 2 assures the *maximality* by checking the maximal number of applications of each block, and Phase 3 translates block applications to rule applications by calculating random numbers using the multinomial distribution.

---

**Algorithm 6** DCBA main procedure

---

**Require:** A Population Dynamics P system of degree $(q, m)$, $T \geq 1$ (time units), and $A \geq 1$ (*Accuracy*). The initial configuration is called $C_0$.

1: *INITIALIZATION*                                                                      ▷ (Algorithm 7)
2: **for** $t \leftarrow 1$ **to** $T$ **do**
3:     Calculate probability functions $f_{r,j}(t)$ and $p(t)$.
4:     $C'_t \leftarrow C_{t-1}$
5:     *SELECTION* of rules:

     – *PHASE 1*: distribution                                             ▷ (Algorithm 8)
     – *PHASE 2*: maximality                                               ▷ (Algorithm 9)
     – *PHASE 3*: probabilities                                            ▷ (Algorithm 10)

6:     *EXECUTION* of rules.                                                    ▷ (Algorithm 11)
7:     $C_t \leftarrow C'_t$
8: **end for**

---

INITIALIZATION procedure (Algorithm 7) constructs a static distribution table $\mathcal{T}_j$ for each environment. Two variables, $B^j_{sel}$ and $R^j_{sel}$, are also initialized, in order to store the selected multisets of blocks and rules, respectively. It can be observed that each column label of the tables $\mathcal{T}_j$ contains the information of the corresponding block left-hand side, and each row of the tables $\mathcal{T}_j$ contains the information related to the object competitions (those columns having non-null values are competing for such a object).

---

**Algorithm 7** Initialization

---

1: Construction of the *static distribution* table $\mathcal{T}$:

   • Column labels: consistent blocks $B_{i,\alpha,\alpha',u,v}$ of rules from $\mathcal{R}$.
   • Row labels: pairs $(x, i)$, for all objects $x \in \Gamma$, and $0 \leq i \leq q$.
   • For each row, for each cell of the row: place $\frac{1}{k}$ if the object in the row label appears in its associated compartment with multiplicity $k$ in the LHS of the block of the column label.

2: **for** $j = 1$ **to** $m$ **do**                                      ▷ (Construct the *expanded static* tables $\mathcal{T}_j$)
3:     $\mathcal{T}_j \leftarrow \mathcal{T}$.                                            ▷ (Initialize the table with the original $\mathcal{T}$)
4:     For each rule block $B_{e_j,x}$ from $\mathcal{R}_E$, add a column labeled by $B_{e_j,x}$ to $\mathcal{T}_j$;
      place the value 1 at row $(x, 0)$ for that column.
5:     Initialize the multisets $B^j_{sel} \leftarrow \emptyset$ and $R^j_{sel} \leftarrow \emptyset$
6: **end for**

---

The distribution of objects among the blocks is carried out in selection Phase 1 (Algorithm 8). The expanded static tables $\mathcal{T}_j$ are used for this purpose in each environment, together with three different filter procedures. FILTER 1 discards the columns of the table corresponding to non-applicable blocks due to mismatch charges in LHS and $C'_t$. Then, FILTER 2 discards the columns with objects in the LHS not appearing in $C'_t$. Finally, in order to save space in the table, FILTER 3 discards empty rows. These three filters are applied at the beginning of Phase 1, and the result is a *dynamic table* $\mathcal{T}^t_j$ (for the environment $j$ and time step $t$).

In order to get a set of mutually consistent blocks, the consistency condition is checked after applying FILTERS 1 and 2. This checking can be implemented by a loop over the blocks. If it fails, the simulation process is halted, providing a warning message to the user. Nevertheless, the algorithm can be configured to find a way to continue the execution by non-deterministically constructing a subset of mutually consistent blocks. Since this method can be exponentially expensive in time, this is optional.

---

**Algorithm 8** Selection phase 1: distribution

---

1: **for** $j = 1$ **to** $m$ **do** ▷ (For each environment $e_j$)
2:     Apply filters to table $T_j$, using $C'_t$ and obtaining $T^t_j$, as follows:

        a. $T^t_j \leftarrow T_j$
        b. FILTER 1 $(T^t_j, C'_t)$.
        c. FILTER 2 $(T^t_j, C'_t)$.
        d. Check *mutual consistency* for the blocks remaining in $T^t_j$. **If** there is at least one inconsistency **then** report the information about the error, and optionally halt the execution (in case of not activating step 3).
        e. FILTER 3 $(T^t_j, C'_t)$.

3:     *(OPTIONAL)* Generate a set $S^t_j$ of sub-tables from $T^t_j$, formed by sets of
        *mutually consistent* blocks, in a maximal way in $T^t_j$ (by the inclusion
        relationship). Replace $T^t_j$ with a randomly selected table from $S^t_j$.
4:     $a \leftarrow 1$
5:     **repeat**
6:         **for all** rows $X$ in $T^t_j$ **do**
7:             $RowSum_{X,t,j} \leftarrow$ total sum of the non-null values in the row $X$.
8:         **end for**
9:         $TV^t_j \leftarrow T^t_j$ ▷ (A temporal copy of the dynamic table)
10:        **for all** non-null positions $(X, Y)$ in $T^t_j$ **do**
11:           $mult_{X,t,j} \leftarrow$ multiplicity in $C'_t$ at $e_j$ of the object at row $X$.
12:           $TV^t_j(X, Y) \leftarrow \lfloor mult_{X,t,j} \cdot \frac{(T^t_j(X,Y))^2}{RowSum_{X,t,j}} \rfloor$
13:        **end for**
14:        **for all** not filtered column, labeled by block $B$, in $T^t_j$ **do**
15:          $N^a_B \leftarrow \min_{X \in rows(T^t_j)}(TV^t_j(X, B))$ ▷ (The minimum of the column)
16:          $B^j_{sel} \leftarrow B^j_{sel} + \{B^{N^a_B}\}$ ▷ (Accumulate the value to the total)
17:          $C'_t \leftarrow C'_t - LHS(B) \cdot N^a_B$ ▷ (Delete the LHS of the block.)
18:        **end for**
19:        FILTER 2 $(T^t_j, C'_t)$
20:        FILTER 3 $(T^t_j, C'_t)$
21:        $a \leftarrow a + 1$
22:     **until** $(a > A) \vee$ *(all the selected minimums at step 15 are* 0)
23: **end for**

---

Once the columns of the *dynamic table* $T^t_j$ represent a set of mutually consistent blocks, the distribution process starts. This is carried out by creating a temporal copy of $T^t_j$, called $TV^t_j$, which stores the following products:

- The normalized value with respect to the row: this is the way to *proportionally* distribute the corresponding object along the blocks. It relays on the multiplicities in the LHS of the blocks; in fact, blocks requiring more copies of the same object are penalized in the distribution. This is inspired in the amount of energy required to gather individuals from the same species.
- The value in the dynamic table (i.e. $\frac{1}{k}$): this indicates the number of possible applications of the block with the corresponding object.
- The multiplicity of the object in the configuration $C'_t$: this performs the distribution of the number of copies of the object along the blocks.

The number of applications for each block is calculated by selecting the minimum value in each column of $\mathcal{TV}^t_j$. This number is then used to consume the LHS from the configuration. However, this application could not be maximal. The distribution process can eventually deliver objects to blocks that are restricted by other objects. As this situation may occur frequently, the distribution and the configuration update process is performed $A$ times, where $A$ is an input parameter referring to *accuracy*. The more the process is repeated, the more accurate the distribution becomes, while the performance of the simulation decreases. $A = 2$ gives experimentally the best accuracy/performance ratio. In order to efficiently repeat the loop for $A$, and also before going to the next phase (maximality), it is interesting to apply FILTERS 2 and 3 again.

After phase 1, it may be the case that some blocks are still applicable to the remaining objects. This may be caused by a low $A$ value or by rounding artifacts in the distribution process. Due to the requirements of P systems semantics, a maximality phase is now applied (Algorithm 9). Following a random order, the maximal number of applications is calculated for each block still applicable (remaining columns in table $T^t_j$).

---

**Algorithm 9** Selection phase 2: maximality

---

1: **for** $j = 1$ **to** $m$ **do**                                            ▷ (For each environment $e_j$)
2:     Set a random order to the blocks remaining in the last updated table $T^t_j$.
3:     **for all** block $B$, following the previous random order **do**
4:         $N_B \leftarrow$ number of possible applications of $B$ in $C'_t$.
5:         $B^j_{sel} \leftarrow B^j_{sel} + \{B^{N_B}\}$                        ▷ (Accumulate the value to the total)
6:         $C'_t \leftarrow C'_t - LHS(B) \cdot N_B$                             ▷ (Delete the LHS of block $B$, $N_B$ times.)
7:     **end for**
8: **end for**

---

After the application of phases 1 and 2, a maximal multiset of selected (mutually consistent) blocks has been computed. The output of the selection stage has to be, however, a maximal multiset of selected rules. Hence, Phase 3 (Algorithm 10) passes from blocks to rules, by applying the corresponding probabilities (at the local level of blocks). The rules belonging to a block are selected according to a multinomial distribution $M(N, g_1, \ldots, g_l)$, where $N$ is the number of applications of the block, and $g_1, \ldots, g_l$ are the probabilities associated with the rules $r_1, \ldots, r_l$ within the block, respectively.

---

**Algorithm 10** Selection phase 3: probability

1: **for** $j = 1$ **to** $m$ **do**  $\qquad\qquad\qquad\qquad\qquad\qquad$ ▷ (For each environment $e_j$)
2: $\quad$ **for all** block $B^{N_B} \in B_{sel}^j$ **do**
3: $\qquad$ Calculate $\{n_1, \ldots, n_l\}$, a random multinomial $M(N_B, g_1, \ldots, g_l)$ with
$\qquad\quad$ respect to the probabilities of the rules $r_1, \ldots, r_l$ within the block.
4: $\qquad$ **for** $k = 1$ **to** $l$ **do**
5: $\qquad\quad$ $R_{sel}^j \leftarrow R_{sel}^j + \{r_k^{n_k}\}$.
6: $\qquad$ **end for**
7: $\quad$ **end for**
8: $\quad$ Delete the multiset of selected blocks $B_{sel}^j \leftarrow \emptyset$. $\qquad\qquad$ ▷ (Useful in next step)
9: **end for**

---

Finally, the execution stage (Algorithm 11) is applied. This stage consists in adding the RHS of the previously selected multiset of rules, as the objects present on the LHS of these rules have already been consumed. Moreover, the indicated membrane charge is safely set (given that the consistency condition is assured).

---

**Algorithm 11** Execution

1: **for** $j = 1$ **to** $m$ **do** $\qquad\qquad\qquad\qquad\qquad\qquad\qquad$ ▷ (For each environment $e_j$)
2: $\quad$ **for all** rule $r^n \in R_{sel}^j$ **do** $\qquad\qquad\qquad$ ▷ (Apply the RHS of selected rules)
3: $\qquad$ $C_t' \leftarrow C_t' + n \cdot RHS(r)$
4: $\qquad$ Update the electrical charges of $C_t'$ from $RHS(r)$.
5: $\quad$ **end for**
6: $\quad$ Delete the multiset of selected rules $R_{sel}^j \leftarrow \emptyset$. $\qquad$ ▷ (Useful for the next step)
7: **end for**

---

### 7.2.3.3 A Test Example: DCBA Versus DNDP

Let us consider a test example, with no biological meaning, in order to show the different behaviors of DNDP and DCBA algorithms. This test PDP system is of degree $(2, 1)$, and of the following form:

$$\Pi_{test} = (G, \Gamma, \Sigma, T, \mathcal{R}_E, \mu, \mathcal{R}, \{f_{r,1} \mid r \in \mathcal{R}\}, \mathcal{M}_{1,1}, \mathcal{M}_{2,1}, E_1)$$

where:

- $G$ is an empty graph.
- $\Gamma = \{a, b, c, d, e, f, g, h\}$ and $\Sigma = \{b\}$.

- $T = 1$, only one time step.
- $\mathcal{R}_E = \emptyset$.
- $\mu = [\, [\, ]_2\, ]_1$ is the membrane structure, and the corresponding initial multisets are:

  - $E_1 = \{b\}$ (in the environment)
  - $\mathcal{M}_1 = \{a^{60}\}$ (in membrane 1)
  - $\mathcal{M}_2 = \{a^{90}, b^{72}, c^{66}, d^{30}\}$ (in membrane 2)

- The rules $\mathcal{R}$ to apply are:

  $r_{1.1} \equiv [\, a^4\, b^4\, c^2\, ]_2 \xrightarrow{\ 0.7\ } e^2\, [\, ]_2$

  $r_{1.2} \equiv [\, a^4\, b^4\, c^2\, ]_2 \xrightarrow{\ 0.2\ } [\, e^2\, ]_2$

  $r_{1.3} \equiv [\, a^4\, b^4\, c^2\, ]_2 \xrightarrow{\ 0.1\ } [\, e\, f\, ]_2$

  $r_2\ \ \equiv [\, a^4\, d\, ]_2 \xrightarrow{\ 1\ } f^2[\, ]_2$

  $r_3\ \ \equiv [\, b^5\, d^2\, ]_2 \xrightarrow{\ 1\ } g^2[\, ]_2$

  $r_4\ \ \equiv b\, [\, a^7\, ]_1^- \xrightarrow{\ 1\ } [\, h^{100}\, ]_1^-$

  $r_5\ \ \equiv a^3\, [\, ]_2 \xrightarrow{\ 1\ } [\, e^3\, ]_2$

  $r_6\ \ \equiv a\, b\, [\, ]_2 \xrightarrow{\ 1\ } [\, g^3\, ]_2^-$

We can construct a set of six consistent rule blocks $B_{\Pi_{test}}$ (of the form $b_{h,\alpha,\alpha',u,v}$) from the set $\mathcal{R}$ of $\Pi_{test}$ as follows:

- $b_1 \equiv b_{2,0,0,\emptyset,a^4b^4c^2} = \{r_{1.1}, r_{1.2}, r_{1.3}\}$
- $b_2 \equiv b_{2,0,0,\emptyset,a^4d} = \{r_2\}$
- $b_3 \equiv b_{2,0,0,\emptyset,b^5d^2} = \{r_3\}$
- $b_4 \equiv b_{1,-,-,b,a^7} = \{r_4\}$
- $b_5 \equiv b_{2,0,0,a^3,\emptyset} = \{r_5\}$
- $b_6 \equiv b_{2,0,-,ab,\emptyset} = \{r_6\}$

It is noteworthy that the set $B_{\Pi_{test}}$ is not mutually consistent. However, only the blocks $b_1$, $b_2$, $b_3$ and $b_5$ are applicable in the initial configuration, and they, in fact, conform a mutually consistent set of blocks. Block $b_4$ is not applicable since the charge of membrane 1 is neutral, and block $b_6$ cannot be applied because there are no $b$'s in membrane 1.

Table 7.1 shows five different runs for one time step of $\Pi_{test}$ using DNDP algorithm. The values refers to the number of applications for each rule, which is actually the output of the selection stage (and the input of the execution stage). Note that for simulation 1, the applications for $r_{1.1}$, $r_{1.2}$ and $r_{1.3}$ follows the multinomial distribution. The applications of these rules are reduced because they are competing with rules $r_2$ and $r_3$. However, this competition leads to situations where the applications of the block $b_1$ does not follow a multinomial distribution. It comes from the fact of using a random order over the rules, but not over the blocks. Rules having a probability equals to 1 are more restrictive on the competitions because they are applied in a maximal way in their turn. This is the reason because on simulations 4 and 5, none of the rules $r_{1.i}$, $1 \le i \le 3$ are applied.

This behavior could create a distortion of the reality described in the simulated model. But it is usually appeased running several simulations and making a statistical study, at expenses of a larger variance. Finally, rules not competing for objects are applied as is, in a maximal way. For example, rule $r_5$ is always applied 20 times because its probability is equal to 1.

Table 7.2 shows the corresponding results using DCBA. It is noteworthy that the selection of rules belonging to block 1 $\{r_{1.i}, 1 \leq i \leq 3\}$ always follows a multinomial distribution with respect to the 3 probabilities. This solves the drawback we showed on Table 7.1. Moreover, it can be seen that the maximality sometimes can give one more application to blocks 2 and 3, in spite of keeping the original 10 applications for block 1 from phase 1. In any case, the number of applications is proportionally distributed, avoiding the distortion of using a random order over the blocks (or rules), as made in the DNDP algorithm.

Besides, DNDP is usually more efficient than DCBA, since it is based on a direct method to calculate the selection of rules, whereas DCBA performs 1 more step and has to handle a big table.

**Table 7.1** Simulating $\Pi_{test}$ using the DNDP algorithm

| Rules | Simulation 1 | Simulation 2 | Simulation 3 | Simulation 4 | Simulation 5 |
|---|---|---|---|---|---|
| $r_{1.1}$ | 11 | 0 | 0 | 0 | 0 |
| $r_{1.2}$ | 4 | 4 | 3 | 0 | 0 |
| $r_{1.3}$ | 1 | 0 | 0 | 0 | 0 |
| $r_2$ | 6 | 18 | 6 | 22 | 2 |
| $r_3$ | 1 | 6 | 12 | 4 | 14 |
| $r_4$ | – | – | – | – | – |
| $r_5$ | 20 | 20 | 20 | 20 | 20 |
| $r_6$ | – | – | – | – | – |

**Table 7.2** Simulating $\Pi_{test}$ using the DCBA algorithm

| Rules | Simulation 1 | Simulation 2 | Simulation 3 | Simulation 4 | Simulation 5 |
|---|---|---|---|---|---|
| $r_{1.1}$ | 7 | 10 | 7 | 6 | 7 |
| $r_{1.2}$ | 3 | 0 | 4 | 1 | 2 |
| $r_{1.3}$ | 1 | 1 | 5 | 3 | 1 |
| $r_2$ | 11 | 11 | 11 | 12 | 12 |
| $r_3$ | 5 | 5 | 5 | 6 | 6 |
| $r_4$ | – | – | – | – | – |
| $r_5$ | 20 | 20 | 20 | 20 | 20 |
| $r_6$ | – | – | – | – | – |

## 7.3   Simulation Platform

### 7.3.1   P-Lingua: A General Framework to Simulate P Systems

Since the introduction of Membrane Computing in [88], many variants of its associated computational devices, called P systems, have been designed, and their theoretical properties and possible practical applications studied. Unfortunately, P systems cannot be implemented yet on their "natural" substrate, biological cells. Consequently, in order to further research on P systems capabilities, simulation tools, working on conventional electronic devices, come into scene. In this sense, P-Lingua framework, was introduced in [26]. This framework provides a general programming language for P systems, called `P-Lingua` itself, and a Java [116] based open source library called `pLinguaCore`. On the one hand, P-Lingua language provides a common syntax for specifying different kinds of P systems, belonging to the main variants that can be found within the Membrane Computing paradigm: cell-like systems, tissue-like systems and neural-like systems. The number of supported variants is increased with each revision of the language. On the other hand, pLinguaCore library provides both parsers and simulators for the variants supported in the P-Lingua language. P-Lingua framework has been a software platform of many conference and journal papers, as well as discussed in four Ph.D. theses (see [119]).

In terms of real-life applications, PDP systems constitute a specially relevant variant which have been successfully applied to study the dynamics of interesting real ecological systems. P-Lingua framework has provided the required specification and simulation tools to assist in these tasks, in conjunction with `MeCoSim` [117], a general purpose application to model, design, simulate, analyze and verify different types of models based on P systems, which uses pLinguaCore as its inference engine.

This section aims to the following points: (a) provide a general overview of P-Lingua framework and (b) cover the details of specifying and simulating PDP systems with P-Lingua.

#### 7.3.1.1   P-Lingua Framework Official Versions and Variants

The latest official version of the framework is 4.0. Since this release, different variants have been developed and presented to the scientific community. One of the most widely used variant is that included into `MeCoSim` featuring, among other things, some syntax extensions to define PDP systems in a easier way. As specification and simulation of PDP systems by means of P-Lingua is a core part of this section, we will consider our "working version" of P-Lingua the `MeCoSim` variant released at the time of writing of this book.

### 7.3.1.2 Supported Models

P-Lingua framework provides support for the many kinds of models belonging to three main variants of P systems: cell-like systems, tissue-like systems and neural-like systems. In `MeCoSim` version, support for simple kernel P systems is also provided.

*Cell-like systems*

Cell-like systems are inspired by the hierarchical structure of eukaryotic cells [88]. The following cell-like P system variants are supported in P-Lingua:

- Transition P systems. The basic P systems were introduced in [88] and allow definition of priority-based rules.
- Symport/antiport P systems. These systems were introduced in [87] and allow only communication rules of the symport or antiport kind (a change of the places of objects with respect to the membranes of a system takes place along computations but not a change/evolution of the objects themselves). P-Lingua does not support arbitrary multiplicity of objects in the environment for this variant.
- Active membranes with division rules. These systems were introduced in [89] and allow membrane duplication by means of division of membranes. P-Lingua supports object-fired membrane division rules for both elementary and non-elementary membranes.
- Active membranes with creation rules. These systems were first considered in [51, 79] and allow membrane multiplication by means of creation of membranes.
- Probabilistic P systems. Multienvironment probabilistic functional extended P systems with active membranes, also called Population Dynamics P systems, were introduced with this denomination in [15]. This variant takes its name from its original application to model real-life ecosystems, although this model has proved successful to deal with other phenomena such like gene networks.

*Tissue-like systems*

Tissue-like systems take inspiration from the way in which cells organize and communicate within a net-like structure in tissues [64]. The following tissue-like P system variants are supported in P-Lingua:

- Tissue P systems with communication and division rules. These systems were introduced in [90] and allow only communication of objects among cells (or the environment) and multiplication of cells by means of the process of cell division.
- Tissue P systems with communication and separation rules. These systems were introduced in [84] and allow only communication of objects among cells (or the environment) and multiplication of cells by means of the process of cell separation.

*Neural-like systems*

Neural-like systems are inspired from the way in which neurons in the brain exchange information by means of the propagation of spikes [50]. The following features of Spiking Neural P systems (SN P systems for short) are supported:

- SN P systems with neuron division and budding rules, introduced in [85].
- SN P systems with functional astrocytes, introduced in [58].
- SN P systems with excitatory/inhibitory astrocytes, introduced in [86].
- SN P systems with anti-spikes, introduced in [83].

*Kernel P systems*

Kernel P systems (kP systems for short), introduced in [32] constitute a formalism combining features of different P systems introduced and studied so far. The following kP systems variants are supported in P-Lingua:

- Simple kP systems, also introduced in [32]. In this variant, a reduced subset of features of kP systems are considered.

**Supported Formats**

P-Lingua supports specification of P systems in the following formats:

- P-Lingua format. This format is the "native" format in P-Lingua framework. All supported P system variants can be defined in this format. Besides, this format allows parametrization, thus enabling definition of P system families.
- XML format. This format is intended for interoperability of the framework with other general-purpose systems. Contrary to P-Lingua format, only cell-like variants are supported in XML format. Parametrization is not supported either.
- Binary format. This format is intended for interoperability of the framework with CUDA simulators. As these simulators are expected to work with really huge systems, a compact way to specify them becomes necessary. At present, P systems with active membranes are supported, with a recent development to support PDP systems to be incorporated in the next version of the framework.

**Input and Output Formats**

The previously defined formats can be categorized into input or output formats. **Input formats** are intended for P-Lingua framework to accept P systems specifications and either parse/simulate the related system definitions or transform them to an **output format**. At present, P-Lingua and XML are the supported input formats, while XML and binary are the supported output formats.

**pLinguaCore Library**

pLinguaCore is JAVA based open source library enabling definition, parsing, simulation and compilation for P systems within P-Lingua framework. It is released under GNU GPL [115] license, thus users are encouraged to download, use and extend it and become part of the P-Lingua community.

In pLinguaCore supported P system variants, input/output formats are declared in XML files. For each input format, a parser is defined while for each output format a compiler is declared. For each variant, specific syntax checkers can also be specified, while a list of simulators is declared.

In this way, pLinguaCore can be easily extended to consider other models and formats. Additionally, the following command-line tools are provided:

- Compilation command-line tool. This tool allows translating P systems specified in an input format, such as P-Lingua format or XML format. This specification is then converted to an output format, such as XML format or binary format.
- Simulation command-line tool. This tool allows defining a P system in an input format, such as P-Lingua format or XML format, parsing (with error detection) such definition and simulating the corresponding system execution. An output is produced containing the resulting computation.

Specific calling method for these tools can be consulted in [31].

The compilation and simulation procedures can also be called by invoking the corresponding Java methods directly.

For additional information about pLinguaCore, please refer to [119].

### 7.3.1.3 An Introduction to P-Lingua Language

P-Lingua framework aims to provide a unified programming language for P systems, which is called P-Lingua itself. This language has evolved to allow specification of an increasing number of P system variants. The main features of P-Lingua language are the as follows:

- Specification is provided in plain text files, which favors interoperability.
- P-Lingua specification language is purposely close to P systems mathematical specification language, reducing the learning curve for users familiar with the Membrane Computing paradigm.
- P-Lingua specification language is parametric, allowing specification of P system families.
- P-Lingua specification language is modular, favoring reusability of code fragments that can be called more than once during a program execution.

**P-Lingua Syntax**

P-Lingua language syntax has evolved through different versions of the framework to include more supported variants and features. At present, there is not any updated unique documentation detailing the full syntax of P-Lingua. In its place, details of the syntax are spread over a series of conference and journal papers and Ph.D. theses. A detailed list of publications can be found at [119]. The following is an essential list of such publications:

- *P-Lingua: A Programming Language for Membrane Computing* [25]. This paper deals with the first version of the language. Only active membranes P systems with cell division are supported.
- *An Overview of P-Lingua 2.0* [31]. This paper deals with version 2.0 of the language, which incorporates support for transition P systems, symport/antiport P systems, active membranes P systems with cell creation, a first implementation for probabilistic P systems and stochastic P systems.

- *A P-Lingua based simulator for tissue P systems* [67]. This paper deals with version 2.1 of the language, which incorporates support for tissue-like P systems with cell division.
- *DCBA: Simulating Population Dynamics P Systems with Proportional Object Distribution* [71]. This paper deals with a new simulation algorithm, called DCBA, developed for PDP systems and included in version 3.0 of the framework. Also in this version, stochastic variant is discontinued in favor of Infobiotics Workbench [5].
- *A P-Lingua based Simulator for Tissue P Systems with Cell Separation* [92]. This paper introduces support for tissue-like P systems with symport/antiport rules and cell separation starting from version 4.0 of the framework.
- *A P-Lingua based simulator for Spiking Neural P systems* [56] and *On recent developments in P-lingua based simulators for Spiking Neural P Systems* [57]. These two papers introduce support for SN P systems starting from version 4.0 of the framework.
- *Kernel P Systems - Version 1* [32]. This paper introduces support for simple kernel P systems that can be found in the `MeCoSim` version of P-Lingua.

**Structure of a P-Lingua Program**

A P-Lingua program file is mainly composed of a sequence of modules that are used to specify a P system or family of P systems. The general structure is as follows:

```
model specification
global variables declaration
main module declaration
other modules declaration
```

The first line of a P-Lingua file declares the P system variant. Next global variables are declared, accessible from any program module. With respect to modules, at least a module, called `main` has to be declared. This module will be executed in the first place and may contain calls to other modules that have also to be declared. The order in which modules are declared is not important.

### 7.3.1.4   Defining PDP Systems in P-Lingua

In this section we provide an exemplified guideline to define PDP systems in P-Lingua text files.

*Model specification*

PDP systems are probabilistic models. Consequently, any PDP system definition in P-Lingua must start with the following sentence:

```
@model<probabilistic>
```

*Membrane structure specification*

In P-Lingua, to define the initial membrane structure of a P system, the reserved word @mu is used, along with a sequence of matching square brackets representing the membrane structure, including some identifiers that specify the label and the electrical charge of each membrane. Neutral charges are omitted.

Examples:

- $[[\ ]_2^0]_1^0 \equiv$ @mu = [ [ ] '2] '1
- $[[\ ]_b^0[\ ]_c^-]_a^+ \equiv$ @mu = +[ [ ] 'b,  -[ ] 'c] 'a

In PDP systems, membranes are arranged in a set of environments, each one containing the same inner membrane structure (a rooted tree), which is called the skeleton.

In P-Lingua, environments are contained within a virtual membrane representing the whole system. This virtual membrane has neutral charge and its label is not important, but a good convention is to use the global identifier. Each environment is also represented by a virtual membrane, containing the corresponding skeleton membranes.

Virtual membranes for environments have neutral charge and a double label, made of two labels separated by a comma. The first one corresponds to the label of the virtual membrane, and the second one to the identifier of the environment being represented by that virtual membrane. Any skeleton membrane automatically inherits the environment identifier of the virtual environment membrane that contains it.

The following constraints apply when specifying a membrane structure for a PDP system in P-Lingua: (a) each environment must have a unique identifier, and (b) all the membranes related to a specific environment, the virtual membrane representing the environment itself and the inner skeleton membranes, must have unique membrane labels. To easily satisfy these constraints, the following design guidelines are recommended when defining PDP systems in P-Lingua:

1. The label for the virtual membrane representing the whole system takes the global identifier.
2. Membrane labels and environment identifiers are represented using natural numbers.
3. The skeleton skin membrane is labeled by 0. Other skeleton membranes take labels with correlative numbers starting from that initial number.
4. The label for the first environment membrane corresponds to a number much higher than the number of inner skeleton membranes. This label usually is 101, 1001, etc. Other environment labels take correlative numbers from that initial value.
5. The environment identifier for an environment membrane is equal to its membrane label.

Example:

```
@mu = [ [ []'1 ]'0 ]'101,101 [ [ []'1 ]'0]'102,102]'global;
```

In this example a membrane structure for a PDP system is defined. The system contains two environments, identified as `101` and `102`. Each environment contains a skeleton with the structure `[ [ []'1 ]'0`. The aforementioned syntax becomes too tedious and repetitive when having to define PDP systems with several environments each one containing a complex skeleton, which happens to be always the same. Alternatively, membrane structures can be defined in a recursive way.

As an example, the sentence:

- `@mu = [ [ []'1 []'2 ]'0 ]'101,101 [ [ []'1 []'2]'0]'`
  `102,102 ... [ [ []'1 []'2]'0]'140,140]'global;`

can be written with the following sentences (using iterators):

- `@mu = []'global;`
- `@mu(global) += [[]'0]'{k},{k}: 101 <= k <= 140;`
- `@mu(0,{k}) += []'{m}: 1 <= m <= 2, 101 <= k <= 140;`

The first sentence specifies a PDP system with an external virtual membrane labeled with `global`. The second sentence specifies that within the `global` membrane, 140 environments will be created, each one containing a skeleton with a skin membrane labeled with `0`. The third sentence specifies that inside the skin membrane of each environment, two membranes `[]'1 []'2` will be created.

### Initial multisets specification

In P-Lingua, to define the initial multisets of a P system, the reserved word `@ms` is used, along with the list of objects contained in the multiset. In the case of PDP systems, the two following ways to specify the initial multisets are possible:

- `@ms(label) = list_of_objects;`
- `@ms(label,environment) = list_of_objects;`

where `label` is a membrane label, `environment` is an environment identifier and `list_of_objects` is a comma-separated list of objects. The character # is used to represent an empty multiset. In the first case, membranes with label `label` belonging to any environment are initialized with the specified object list. In the second case, only the membrane with label `label` belonging to the environment `environment` is initialized. The operator += (union) can be used to enable adding of objects to the multisets through several sentences.

Examples:

- `@ms(2) = c,d*16;`
- `@ms(2,101) = a,b*12;`
- `@ms(1,101) = a;`
- `@ms(1,101) += b,c;`

*Rules specification*

In PDP systems, there are two kind of rules, skeleton rules, that define the rewriting of objects within the skeleton membranes, and environment rules, that define the communication of objects among environments.

1. Skeleton rules have the signature $u[\, v\, ]_h^\alpha \xrightarrow{p} u_1[\, v_1\, ]_h^\beta$. They can be defined in P-Lingua as follows:

   - u$\alpha$[v]'h --> u1$\beta$[v1]'h::p;
   - u$\alpha$[v]'h --> u1$\beta$[v1]::p;
   - u$\alpha$[v]'h,e --> u1$\beta$[v1]'h,e::p;
   - u$\alpha$[v]'h,e --> u1$\beta$[v1]::p;

2. Environment communication rules have the signature $(x)_j \xrightarrow{p} (y)_k$. They can be defined in P-Lingua as follows:

   - [ [x]'j --> [y]'k ]'global::p;

where $x$, $y$ are objects; $u$, $v$, $u_1$, $v_1$ are multisets of objects; $h$ is a membrane label; $j$, $k$ are membrane labels corresponding to environments; $e$ is an environment identifier; *global* is the label corresponding to the virtual membrane representing the whole PDP system; $\alpha$, $\beta$ are charges; and $p \in [0, 1]$ is a real number representing the probability of the rule.

Examples:

- $a, b^2[c, d]_2^+ \xrightarrow{0,8} x, y[z]_2^- \equiv$ a,b*2 +[c,d]'2 --> x,y -[z]'2 :: 0.8;

- $(x)_i \xrightarrow{p\{i,j\}} (y)_j : 1 \leq i \leq E, 1 \leq j \leq E \equiv$
  [ [x]'i []'j --> []'i [y]'j ]'global :: p{i,j}
  : 1<=i<=E,1<=j<= E

## 7.3.2 MeCoSim: *A Graphical User Interface to Convey Virtual Experimentation*

Membrane Computing has been proved to be a suitable paradigm for modeling real-life phenomena that are considered complex dynamical systems, involving a significant number of elements interacting, subject to many different processes. These models are interesting on their own from a theoretical point of view, but it is more enriching for science if these models can also be somehow simulated, so that their dynamics can be carefully analyzed. In the previous Section, P-Lingua framework has been presented, providing a specification language for several types of P systems, as long as the corresponding parsers for these P systems and some simulators capturing their dynamics. This framework provides a powerful platform to work with P systems as a designer, and it comes with some command-line tools to parse the specified files and perform simulations.

A step further is to bring closer these conceptual and software tools to the practical solution of real problems, not only within the area of Membrane Computing, but as a more affordable platform for end users, without a deep knowledge about this paradigm, interested in solving their problems. MeCoSim (Membrane Computing Simulator) [91, 117, 121], is built on top of P-Lingua as a higher level visual environment, more oriented to the end user, with the aim of providing them with final, end-user visual applications acting as black boxes, such a way that they can enter the data about their particular problems and get the desired results. It also provides P systems designers with visual tools to model, edit, debug, analyze, simulate and verify their models, as well as to deliver end-user applications based on P systems.

In Sect. 7.3.2.1, MeCoSim simulation environment is described, highlighting the main contributions of the software developed. Then, in Sect. 7.3.2.2 a proposed methodology is depicted to take us from the study of a problem to the delivery of a practical solution for end users. Last, the main contributions of this approach within the area of real ecosystems and population dynamics are outlined in Sect. 7.3.2.3.

### 7.3.2.1 MeCoSim **Simulation Environment**

P-Lingua framework provides a complete programming environment for *Membrane Computing*. Given an abstract problem, the P system describing a solution by a specific variant will be written in a file with *.pli* extension. For instance, for a PDP system, the following elements must be given:

- Environments and membrane structure.
- Sets of rules.
- Initial multiset present inside each membrane and environment.

The simulation of different instances can be performed by providing specific values for the parameters in the P-Lingua file, and running through command line. This specific values determine the set of rules, alphabet, initial multisets and membrane structure.

This approach is quite useful to define P systems when the designer is the person who performs the simulations. However, it could be an *end user* that needed to solve a problem (e.g. analyzing the population dynamics of an ecosystem, by using a model designed by an expert in P systems). He might not be familiar with this framework, but he usually needs to provide data about the specific instance whose answer is trying to solve (e.g. the initial situation and population for a species and environment). From this data, the parameters for instantiating a P system of the family to solve the specific instance of the problem could be internally calculated.

In addition, it would be worth for P systems designers having facilities to enter data about specific instances visually, easing the debugging process. Thus, it would be interesting to have a high level simulation environment, instead of coding approach for introducing data in *.pli* file for each different instance and running from command line. Besides, it would be useful to provide some mechanisms for the visual analysis of P systems complementing the text output produced by pLinguaCore.

It would not be feasible to develop and maintain a software application to manage each designed model. Instead, MeCoSim is a software environment able to provide mechanisms for defining simulation applications (apps) customizable for any ecosystem, with user interfaces adapted for each problem. Thus, this environment acts as a meta-simulation app, allowing the definition of a simulation app for each problem, by setting the following elements:

1. *Input* definition to specify the data the user needs to introduce for each instance of the problem.
2. Generation of *parameters* of the P system from the data of the instance just introduced.
3. *Output* definition, to show the end user the desired information about each target application (output tables, charts, graphs, etc.), including filtering, grouping and post-processing of the information from the simulation of the P systems.
4. Some arrangements of the elements in the visual application.

Detailed information about this approach can be found at [121]. A clear separation of the roles involved in modeling and simulation process is stated in that document: (a) software developer; (b) P systems designer; and (c) end user of a simulation app. What does MeCoSim provide within this scope? Figure 7.3 illustrates the main actions performed by each role.

Thus, taking into account these roles, some main functionalities of MeCoSim are listed below:

- *General environment for the simulation of solutions based on P systems.*
  It is supported by P-Lingua framework, enabling the parsing, debugging and simulation from MeCoSim environment.

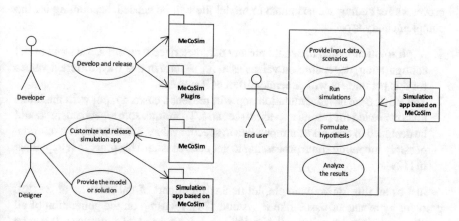

**Fig. 7.3** Roles and uses of MeCoSim

- A visual hierarchical structure determined by the user.
- A definition of input tables for the user to enter information and outputs to show the desired results of the simulation.
- Output charts of different types (lines, bars, stacked columns, pie, and lists of charts).
  Along with the visual elements described so far, two essential components for the app have to be defined, as illustrated in the following items.
- Parameters: variable elements of the model based on P systems depending on the instance of the problem, according to the data entered by the user from the input tables. It defines the way each parameter is generated from the input tables, possibly affected by some processing from these data. To achieve this goal, a new parameters generation language has been defined, as described in [121].
- Results: before showing output tables and charts, the corresponding data to show should be obtained, essentially coming from computation data, getting the desired objects from the stated configurations. The set of data available as part of the simulation, along with the process to get these results, is described in [121].

MeCoSim environment enables the load and store of simulation apps, and includes a list of all the previously loaded apps. They are ready to run, thus generating the corresponding visual application (i.e., its tabs hierarchy, input tables, outputs, etc.) according to its defined configuration.

In order for the simulation to be performed from the loaded P-Lingua model or solution, the data about a specific scenario must be given, by entering the data from the visual input tables or loading a previously saved scenario data, so that the corresponding P system can be generated. Then, the simulation can be performed from the initial situation given by the entered data. When the simulation starts, the first step is the instantiation of the P system corresponding to the given solution and scenario, including membrane structure, initial configuration and set of rules for each compartment. This information could depend on the specific instance, so some processes for adding scenario data to model file will be needed, accounting for the complementary aspects:

1. A *.pli* solution/model file could present parameterized structures, rules and initial configuration, depending on values as $n$, $m$, $p_i$, etc. In this way, different values for this parameters will generate different P systems.
2. The MeCoSim based simulation app will present a custom input with the information needed for setting a specific scenario. Therefore, some mechanism should be designed to convert user input data in specific values for the model or solution, possibly involving some processing to get derived data. This process is explained in [121].

Just as an illustrative example, let us analyze a part of the solution of 3-COL problem by means of tissue-like P systems [24], focusing on the generation of all possible colorings by using cell division rules, for a graph of $n$ vertices. It can be described as follows:

Let $\Pi(n) = (\Gamma(n), w_1, w_2(n), \varepsilon, R(n), i_0)$ family of tissue-like P systems with cell division of degree 2, where:

1. $\Gamma(n) = \{A_i, R_i, T_i, B_i, G_i : 1 \leq i \leq n\}$
2. $w_1 = \emptyset$, $w_2(n) = \{A_1, \ldots, A_n\}$
3. $R(n)$ is a set of division rules:

- $r_{1,i} \equiv [A_i]_2 \rightarrow [R_i]_2[T_i]_2$ para $i = 1, \ldots, n$
- $r_{2,i} \equiv [T_i]_2 \rightarrow [B_i]_2[G_i]_2$ para $i = 1, \ldots, n$

Thus, the solution *.pli* file representing the corresponding family of P systems presents some instructions of the type:

```
/* Multiset in region 2 */
@ms(2)  += A{i} : 1<=i<=n;

/* Division rules: */
/* r1 */ [A{i}]'2 --> [R{i}]'2 [T{i}]'2 : 1<=i<=n;
/* r2 */ [T{i}]'2 --> [B{i}]'2 [G{i}]'2 : 1<=i<=n;
```

The solution *.pli* file represents the whole family of P systems solving this abstract problem, but the user will be interested in simulating a specific instance of the problem; that is, analyzing the P system associated to each different value of $n$. To provide an app accepting values for $n$ and generating and simulating the corresponding P system for that solution for that specific instance, an input table is defined in the configuration file, with the information shown in Fig. 7.4 for tabs hierarchy, table and columns, and parameters generation.

As shown in Fig. 7.4, parameter $n$ is defined, as referenced in P-Lingua file. In this case, $n$ takes its value from the data entered in the first row, first column of table with Id 5. When running the solution, if value 3 has been entered in the visual table, MeCoSim based simulation app generates the parameters ($n$, as defined in *SimulationParams* tab of the configuration file). Then, the specific P system can be instantiated from the solution *.pli* file and start the simulation of a computation.

| Tab Id | Tab Name | Tab Parent Id |
|---|---|---|
| 1 | Tissue Example | 0 |
| 2 | Input | 1 |
| 3 | Size (n) | 2 |

| Table Id | Table Name | Tab Id | Columns | Init Rows | Save To File | Input / Output |
|---|---|---|---|---|---|---|
| 1 | Size (n) | 3 | 1 | 1 | TRUE | Input |

| Column Id | Column Name | Default Value | Editable | Tooltip | GraphicRole |
|---|---|---|---|---|---|
| 1 | Size | 3 | TRUE | n | |

| Param Name | Param Value |
|---|---|
| n | <5,1,1> |

**Fig. 7.4** App definition - parameters generation

Along with the general mechanism for simulating, simulation apps based on `MeCoSim` provide debugging functionalities to:

- Parse the given model file (along with the parameters generated from the input scenario data), leading to:

1. A model or solution with a valid scenario, ready to simulate step by step or a number of steps.
2. A non valid solution or scenario, thus showing the detected problems.

Both cases will show, along with the previous information:

– *ParsingInfo*: shows the parameters being generated, along with the unfolded rules (single rules, generated from – folded – rule schemes that can be expressed in P-Lingua to iterate over some set of possible indexes) for the P system.
– *Warnings*: shows warnings related to the parsing of the solution file for the current scenario, not avoiding the simulation or debugging of the solution.

- Simulate step by step. Two different options are available:

– *Step*: moves the computation a step forward from the current configuration.
– *Run steps*: takes the specified number of steps from the current configuration.

In both cases the information about each configuration is shown in *SimulationInfo* tab, along with the information about the selected rules for executing every step. To complement this text output, some plugins show the alphabet, the membrane structure and the multisets inside each compartment.

**Plugins**

Functionalities described so far constitute the CORE of `MeCoSim`, but many others provide additional features and abilities to interact with other software. `MeCoSim` **plugins architecture**, in conjunction with the customization elements, permits adding Java-based functionality or external programs being called from `MeCoSim`. Following some existing plugins taking advantage of these mechanisms are listed:

- *MeCoSimBasics plugin*: includes a P-Lingua (*.pli*) files editor, along with viewers for alphabets, structures and multisets.
- *Processes plugin*: permits connecting `MeCoSim` to external programs, directly invoked by operating system or command line calls transparently, from `MeCoSim` menu options. Besides, additional custom menu options can be included for sending emails or opening web sites.
- *Graphs plugin*: enables the definition of graphs from parameters information, coming from input/output tables through parameters generation mechanism. Options for showing graphs, trees or graphs in a tree hierarchical structure are available.
- *Daikon plugin*: plugin for invariants extraction from the trace of the models or solutions computation. A plugin was developed to connect `MeCoSim` with `Daikon` by means of a *daikonapplication* program, developed by collaborators of the *University of Pitesti*, responsible for calling Daikon with the required input parameters.

This way, `MeCoSim` generates a file with the trace of the computation in a format expected for *daikonapplication*, which is then called from `MeCoSim`. This interface loads the trace file and generates an output with the detected invariants.

- *Promela plugin*: integrates `MeCoSim` with tools for formal verification of properties of the models by model checking techniques. Developed in collaboration with the Department of Mathematics and Computer Science, University of Pitesti, and the Department of Computer Science, University of Sheffield, this plugins enables the generation of a *.pml* file in Promela format from the model and scenario loaded in `MeCoSim`. This file could then be edited to include properties to verify, and run from `MeCoSim` by calling Spin Model checker [4, 46, 107].

The details for setting `plugins-properties` are given in `MeCoSim` web site [117], in `Getting started> User guide> Plugins` section. Further details are given in [121].

**Repositories**

`MeCoSim` based simulation apps, models, scenarios and plugins developed could be interesting for other P systems designers or end users of the apps. To make them available publicly, a repositories system has been designed, developed and released for:

- *Plugins (.jar)*.
- *Apps (.xls)*.
- *Models (.pli)*.
- *Scenarios (.ec2)*.

Any `MeCoSim` user may provide a repository of plugins, apps, etc. generating depending on the type of repository. For instance, the configuration of a models repository would be something like this:

```
<models>
<model name="Tricolor Simple Kernel 1" pli="tricolor.pli"
path="http://www.gcn.us.es/redmine/dmsf/files/2952/download"/>
...
</models>
```

Anybody can have his own repository by placing his configuration and associated files in some public URL, and any user can add that repository in `MeCoSim` and access all the available files by simply adding the repository to `MeCoSim` indicating the URL. The goal of this mechanism is to provide the community an easy and fast way of accessing the available models. In addition, it permits other members to share their own models, enforcing the collaborative work and knowledge sharing.

### 7.3.2.2   A Methodology for the Practical Solution of Problems

The infrastructure provided by P-Lingua and `MeCoSim` supports the application of an integral methodology for the practical resolution of real life or theoretical problems by means of P systems. A first draft involving some steps of this methodology was first presented in [49] for simulation, invariants extraction and verification of properties over models based on P systems. This approach was later extended in [33]. Following the main aspects of the different stages of this methodology are described:

### Modeling

From the initial analysis of the problem under study, an abstraction process is performed to capture the essential data relevant to the studied phenomena, until a P systems based model or solution is designed to answer the questions posed in that analysis, including the necessary processes and data. Some of the main steps involved in this sense, specifically under the scope of the so called Population Dynamics P systems (PDP systems), are described in the protocol detailed in [17].

Once a model capturing the essence of a phenomena is designed, this model can be translated to P-Lingua [30] language and be stored as a *.pli* file. If the data corresponding to a specific instance of the problem are introduced in the same file (that is, the specific values for the parameter of the corresponding scenario of interest), the model would be ready to work with it in pLinguaCore or MeCoSim [91].

### Simulation

The previous model could be simulated by P-Lingua framework, or in a graphical way by MeCoSim interface, delegating in pLinguaCore the parsing and simulation process (also other external simulators can be used). Once a model file is loaded in MeCoSim, a number of available simulators for its corresponding variant of P systems will appear in a submenu in MeCoSim, ready for the user to choose the desired one, as shown in Fig. 7.5.

### Customization

MeCoSim mechanism for defining custom simulation apps allows the definition of input tables for the user to enter the custom data for the studied problem, ready to introduce data about each particular scenario of that problem. From these data, the corresponding parameters values will complement the model to be simulated, instead of being those values hardcoded in the model file. This way, a single model file can define a family of P systems including a number of parameters, such a way that the different instances of the problem and their corresponding scenarios lead to different P systems and inputs for them in simulation time.

### Debugging

MeCoSim includes a mechanism to perform the debugging of models and solutions based on P systems. The main options include (*Set Model*) to load a *.pli* file, (*Init Model*) to initialize the model, *Errors* tab to show possible errors detected when initializing and validating, and *Warnings* tab to inform over non critical possible issues detected. If the model along with the parameters generated for

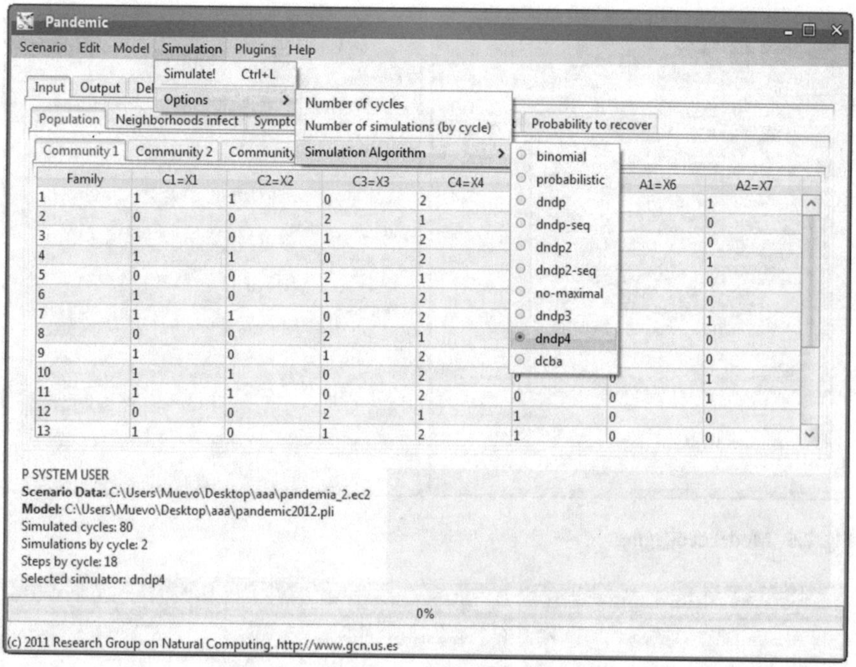

**Fig. 7.5** Available simulators for a loaded P systems variant

the specific instance introduced are valid, the values for the parameters and the unfolded rules will be shown in *Parsing Info* tab. Then functionalities for step by step simulation (*Step*) or the simulation of a fixed number of steps (*Run steps*) will be enabled, thus showing the data about the sequence of configurations and rules appliances in *Simulation Info* tab. An example of debugging is shown in Fig. 7.6.

The debugging of the models permits detecting errors, not meeting their syntactic or semantic constraints, as well as mismatches in the values of the parameters used by the model, depending on the loaded scenario.

**Visualization and data analysis**

- Definition of custom post-processed outputs: the definition of MeCoSim outputs permits showing specific information in the form of tables or charts. This extends the functionalities to visualize information from the computation, focusing on the specific elements that we are interested in. Besides, it permits showing more complex outputs, post-processing the data from the computation, including filtering and grouping techniques broadening the possible analysis over data resulting from simulations. These tools are devoted to provide more information to the P systems designer and the end user of the simulation app, interested in his application domain and not in the underlying P systems.

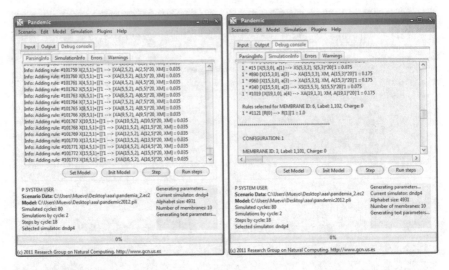

**Fig. 7.6**  Model debugging

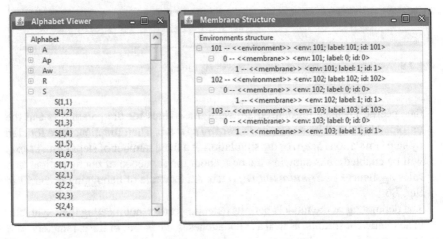

**Fig. 7.7**  Alphabet (*left*) and structure (*right*) viewers

– Advanced visualization of structures and elements of P systems configurations: MeCoSim basics plugin provide viewers oriented to the designer user, for alphabet (Fig. 7.7, left), structure (Fig. 7.7, right) and multisets inside each compartment.
– Graphs visualization: graphs plugin enables showing graphs with their nodes and edges coming from the information generated as parameters from the input or output tables. An example is illustrated in Fig. 7.8.
  Further details are given in [121].

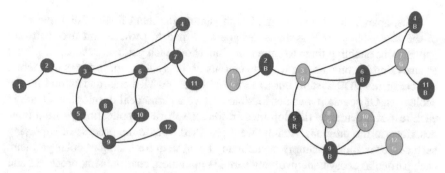

**Fig. 7.8**  Graphs visualization plugin

**Invariants detection**

The joint power of customizable outputs in MeCoSim and Daikon [28, 75, 76] extraction plugin permits the detection of invariants from the models and solutions designed based on P systems. The outputs definition is used to select the desired elements from the computation, and Daikon plugin is used to extract traces, select the desired one and detect properties by delegating in Daikon invariant detector. This process is described in a deeper way in [49, 121].

**Properties verification** *PromelaPlugin* enables the generation of a file in Promela format (*.pml*) from the model and scenario loaded in MeCoSim to simulate. This file can be edited to include properties to verify, and run from MeCoSim by calling Spin Model checker [4, 46, 107]. Some details are commented in [33], taking into consideration the basis for automated verification of P systems through Spin stated in [47, 48]. The visual interface for the generation of Promela code and Spin execution from MeCoSim, along with further details, can be found in [121].

### 7.3.2.3  Working with Real Ecosystems

The previous sections have widely described a software platform and conceptual model to bring closer the paradigm of membrane computing to the systematic and practical solution of the problems a user aims to study. The need of delivering end user applications to make this abstract devices really solve practical problems, as black boxes internally working with P systems, was first detected after the study and modeling of different real ecosystems. Ecologists were not familiar with P systems, and this was even unnecessary, they only required some practical tools to manage and reason about the inherent complexity of the phenomena under study. So it emerged the need of developing visual tools abstracting them from the details of the underlying computational paradigm, allowing them to focus on the study of the processes involved in the biology of the species, the environment, and all the related parameters involved in the interactions of the elements inside the ecosystems.

Thus, apart from designing models to capture the main features and dynamics of the ecosystems, as P systems designers we had to provide end user software applications enabling them to enter their data about each particular scenario, as well as analyzing in some usable way the outputs of the system, not in terms of internal objects in the PDP system but in terms of facts and elements understood by the ecologists. Of course it was not feasible to have a permanent development team in each research center to develop these different software applications, so the tools described in the previous sections were provided to ease the process of supplying such end-user apps by simply customizing the system for each particular problem, each particular ecosystem involving certain population, environment, processes and requiring different outputs.

In this context, many real ecosystems have been studied, presenting certain common features:

- A series of species where the study is focused on. Some facts related to the biology of these species have to be taken into account.
- A specific geographical environment under study, with certain relevant features to consider.
- Several processes affecting the evolution of the system.

Thus, the design of a model for this kind of complex systems should capture the essential elements regarding the species, environment, processes and possibly several other ingredients, depending on the goals of each particular model. As an abstraction of the too complex underlying reality, a model of this kind should not be designed to be restricted to a particular scenario, but being abstract enough to be applied to many different possible situations that may result interesting for the ecologists. Therefore, when these models are designed they do not come in the form of specific P systems, but they are given by a **family of PDP systems**.

For each particular scenario that the ecologists aim to study, each situation subject to analysis of its evolution, a specific PDP system of the family is instantiated, finally generating a structure, final set of rules and initial multisets that may vary depending on each specific instance. Besides, this PDP system will accept certain data as an input, taking different values depending on the particular scenario under study, possibly involving different initial populations and parameters affecting certain processes. In addition, for each situation to analyze the timeframe where the evolution of the species is focused should be determined.

The practical handling of these models by the ecologists requires, as stated in the previous sections, the translation of the model to a language that the machine can understand. In this case, the model is specified in P-Lingua language, described in Sect. 7.3.1. As just mentioned, it is advisable for this model to be useful for many particular situations (past, present or future, allowing the formulation of hypothesis of the possible causes of past events or the prediction of possible effects of future events or actions in advance). Therefore, P-Lingua file should capture the description of model for a family of PDP systems, not a particular one, thus presenting some input points that will vary depending on each particular scenario.

In order for the ecologists to work with the model, they will be required to provide the data for each particular scenario, and it will be desirable for them to have some infrastructure available to perform this task in a suitable and usable way. It is here that MeCoSim plays a very significant role, easing the task of supplying these end-user applications, that in the case of ecosystems will usually present at least the following elements:

- Input tables to enter data about:

  - Biology of the species.
  - Properties of the different areas included in the model.
  - Environmental conditions.
  - Initial populations of each species.

- Output information from the virtual experiments:

  - Tables to quantify the population of the species along time.
  - Charts providing a fast view of evolution of certain elements and its trend.

- Controls to set:

  - Number of cycles to simulate (this number will represent years, months, days, etc. depending on the system under study)
  - Number of simulations to perform (given the probabilistic nature of PDP systems, a number of simulations should run, such a way that the output for the end user will present the average, deviations, etc. of the data across the different simulations).

These elements will slightly differ from one application to another, but MeCoSim provides, as described in the previous sections, a mechanism to define the custom inputs, outputs and internal conversion from inputs to the parameters used to instantiate each specific PDP system to be parsed by P-Lingua and simulated by means of the simulators available (P-Lingua based ones or external simulators).

The corresponding visual application will appear in MeCoSim when providing the proper customization file, as shown in Fig. 7.9.

Several output results can be defined according to the needs identified by the end users. The more general ones are of two types: output tables, as the one shown in Fig. 7.10, and output charts, as the one shown in Fig. 7.11.

## 7.3.3  Parallel Simulators on Multi-core and Many-Core Platforms

As discussed in the previous section, P-Lingua framework enables to conduct PDP systems simulations through both DNDP and DCBA algorithms. However, the run times offered by the framework are high for some scenarios involving large and complex models. This lack of efficiency is mainly given from the facts of using Java Virtual Machine and implementing sequential algorithms. Indeed, simulating

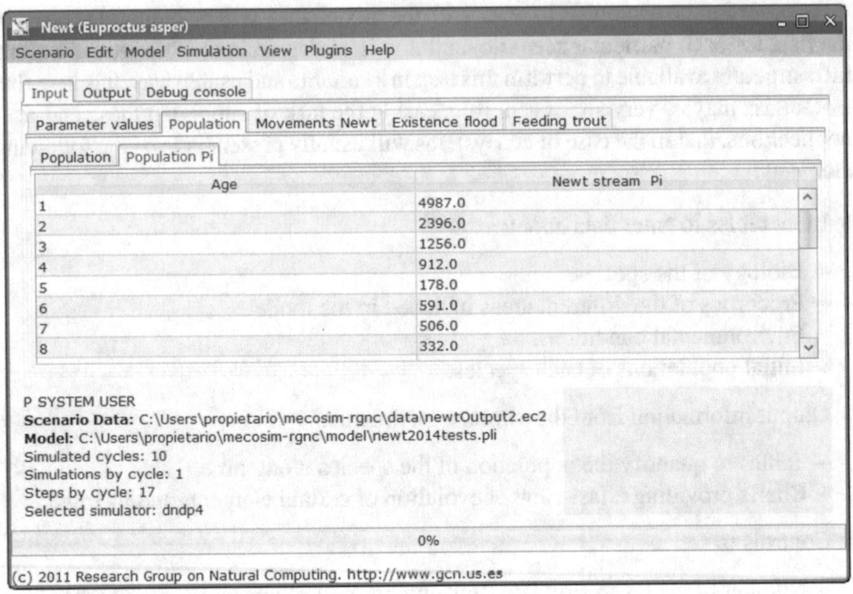

**Fig. 7.9** `MeCoSim` based application

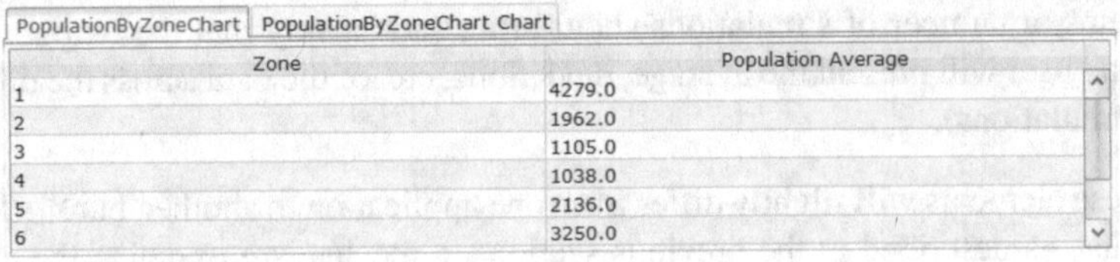

**Fig. 7.10** `MeCoSim` app. Output table

massively parallel devices like P systems in a sequential fashion is twice inefficient. A solution to outcome this issue is by harnessing the highly parallel architecture within modern processors to map the massively parallelism of P systems.

Current commodity processors offer from 2 to 16 computing cores, or even more if they are high end. These cores are complex enough to run threads simultaneously, each one with its own context, exploiting a coarse grain level of parallelism. For example, OpenMP is a threading library for multicore processors, which can be used in C/C++.

Alternatively, current graphic processors (GPUs) provide thousands of computing cores that can be programmed using general purpose frameworks such as CUDA, OpenCL and OpenAcc [81]. GPUs exploit data parallelism by using a very fast memory and simplifying the cores. In fact, a GPU consists in SIMD multiprocessors interconnected to a fast bus with the main memory system. Each multiprocessor has a set of computing cores that execute instructions synchronously (they always perform the same instruction over different data) and a small portion of sketchpad memory (similar to caches in CPUs, but manually managed by programmers), among other

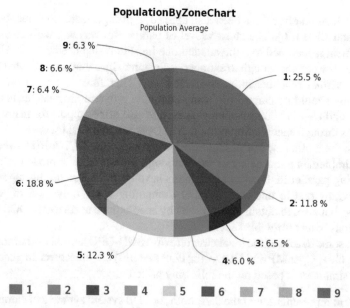

**PopulationByZoneChart**

Population Average

9: 6.3 %
8: 6.6 %
7: 6.4 %
1: 25.5 %
6: 18.8 %
2: 11.8 %
5: 12.3 %
3: 6.5 %
4: 6.0 %

1  2  3  4  5  6  7  8  9

**Fig. 7.11** MeCoSim app. Output chart

elements. In this case, CUDA provides an abstraction of the GPU with a SPMD model (Single Program Multiple Data), in which threads execute the same code (called kernel) over different pieces of data. Moreover, threads are arranged in blocks, in such a way that threads belonging to the same block can cooperate and easily be synchronized.

Given the high level of parallelism within modern GPUs, they have provided an interesting platform to implement real parallelism of P systems in a natural way. Many P system models have been considered to be simulated with CUDA [72]: P systems with active membranes, SAT solutions with families of P systems with active membranes and of tissue P systems with cell division, Enzymatic Numerical P systems, Spiking Neural P systems without delays, and Population Dynamics P systems, among others. Most of these simulators are within the scope of PMCGPU (*Parallel simulators for Membrane Computing on the GPU*) software project [118], which aims to gather efforts on parallelizing P system simulators with GPU computing. As mentioned, there is already a subproject for PDP systems, called ABCD-GPU, that will be discussed below.

When comparing the efficiency of simulation algorithms, DNDP has been shown to be faster than BBB in some scenarios, when a substantial amount of rules is involved [66]. However, DCBA is much slower than DNDP, since it involves the usage of a large sparse matrix in phase 1, whereas DNDP implements a more direct method. For example, using the implementation in pLinguaCore for DCBA in order to simulate the model of the scavenger birds presented in [15] was estimated to take 6 months for 100 simulations. In any case, the simulation of PDP systems should be as efficient as possible in order to ease their usage for virtual experimentation.

Parallelization efforts has been also applied to pLinguaCore implementations of DNDP and BBB [66]. The chosen level of parallelism was at environments: each environment is assigned to a thread, and one transition step is simulated. However, Java threads were not enough to extract conclusions, since they are executed through the Java Virtual Machine in an abstracted way. Therefore, further efforts of parallelization should be carried out using compiled programming languages. In this sense, C and C++ are compiled languages, that also offer support for many parallel platforms through specific libraries, such as OpenMP and CUDA.

Given the inefficiency of pLinguaCore when executing DCBA [65], even avoiding the implementation of a real table through a hash table approach, efforts for developing parallel PDP-system simulators have been focused on this algorithm. It is noteworthy that in this case, Parallel Computing is not only used to get faster solutions, but also, to obtain better results by enabling the users to run DCBA-based simulations in an affordable time.

Next, some details of the inner structure of ABCD-GPU simulator (parallel simulators in C++, OpenMP and CUDA for PDP systems) are presented. In general, this parallel simulator is based on the following principles:

- Efficient representation of the data, both for PDP system syntactical elements and auxiliary structures of DCBA (virtual table [70]).
- Efficient implementation of the algorithm.
- Exploiting levels of parallelism presented in the simulation of PDP systems: processing of rule blocks and rules, evolution of environments, and conducting several simulations to extract statistical data from the probabilistic model.

### 7.3.3.1   A Virtual Table for DCBA

The first challenge of the DCBA implementation in C/C++ was to save memory by avoiding the creation of the expanded static table $T_j$, used in phase 1. The implementation of this table can be inefficient in systems with a large number of rule blocks and/or objects. Specifically, the size of this table is $O(|B| \cdot |\Gamma| \cdot (q + 1))$, where $|B|$ is the number of rule blocks, $|\Gamma|$ is the size of the alphabet (amount of different objects), and $q + 1$ corresponds to the number of membranes plus the space for the environment.

Moreover, a real implementation of $T_j$ can lead to a sparse matrix, having null values in the majority of the positions: competitions for one object appears for a relatively small number of blocks. This problem was overcome in pLinguaCore by using a hash table storing only non-null values. For the sequential version in C++, the idea was to avoid the construction of $T_j$ by translating the operations over the table to others directly applied to the rule blocks information, as follows [70]:

- Operations over columns: they can be transformed to operations for each rule block and the objects appearing in the multisets of the LHS.
- Operations over rows: they can be translated similarly to operations over columns, but the partial results are stored into a global array (one position per row).

Phase 1 can be implemented as described in Algorithm 12. Note that FILTER 3 is not needed any more. Although the full table is not created, some auxiliary data structures are used to virtually simulate it (called a *virtual table*):

- *activationVector*: the information of filtered blocks is stored here as boolean values. The full global size is $O(|B| * m)$, where $m$ is the number of environments. This vector is actually implemented passing from boolean to bits.
- *addition*: the total calculated sums for rows are stored here, one number per each pair object and region. Its size is of order $O(|\Gamma| * (q + 1) * m)$.
- *MinN*: the minimum numbers calculated per column are stored here. This is needed in order to subtract the corresponding number of applications to $C'_t$ in each loop for the $A$ value. The total global size is $O(|B| * m)$.
- *BlockSel*: the total number of applications for each rule block is stored here. The total global size is $O(|B| * m)$.
- *RuleSel*: the total number of applications for each rule is stored here. The total global size is $O((|\mathcal{R}| * m) + |\mathcal{R}_E|)$, where $|\mathcal{R}|$ is the number of skeleton rules and $|\mathcal{R}_E|$ the number of communication rules.

---

**Algorithm 12** Implementation of selection phase 1 with virtual table

---

1: **for** $j = 1, \ldots, m$ **do**                                                      ▷ For each environment
2:  **for all** block $B$ **do**
3:    $activationVector[B] \leftarrow true$
4:    **if** $charge(LHS(B))$ is different to the one presented $C'_t$ **then**
5:      $activationVector[B] \leftarrow false$                                           ▷ (Apply FILTER 1)
6:    **else if** one of the objects in $LHS(B)$ does not exist in $C'_t$ **then**
7:      $activationVector[B] \leftarrow false$                                           ▷ (Apply FILTER 2)
8:    **end if**
9:  **end for**
10:  Check the mutual consistency of blocks.
11:  **repeat**
12:    **for all** block $B$ having $activationVector[B] = true$ **do**                   ▷ (Row sums)
13:      **for** each object $o^k$ appearing in $LHS(B)$, associated to region $i$ **do**
14:        $addition[o, i] \leftarrow addition[o, i] + \frac{1}{k}$
15:      **end for**
16:    **end for**
17:    **for all** block $B$ having $activationVector[B] = true$ **do**                   ▷ (Col. min.)
18:      $MinN[B] \leftarrow Min_{[o^k]_i \in LHS(B)}(\frac{1}{k^2} * \frac{1}{addition[o,i]} * C'_t[o, i]).$
19:      $BlockSel[B] \leftarrow BlockSel[B] + MinN[B].$
20:    **end for**
21:    **for all** block $B$ having $activationVector[B] = true$ **do**                   ▷ (Updating)
22:      $C'_t \leftarrow C'_t - LHS(B) * MinN[B]$
23:    **end for**
24:    Apply FILTER 2 again (as described in step 6).
25:    $a \leftarrow a + 1$
26:  **until** $a = A$ **or** *for each active block $B$, $MinN[B] = 0$*
27: **end for**

---

### 7.3.3.2    Parallel Simulation over Multicore Platforms

Current multicore CPUs can be easily programmed using OpenMP. The parallel design of the simulator is based on three approaches [70]: (1) simulations, (2) environments and (3) a hybrid approach, having the following pros and cons:

- The advantage of running *simulations* in parallel is that there are no data dependencies between simulations, and, therefore, the problem is embarrassingly parallel. Also, running 50–100 simulations is enough to fully use all cores in a CPU. However, some disadvantages of this approach are the increase of amount of memory, load balancing issue when the number of simulations is not divisible by the number of processors, and resource conflicts as cores compete for shared resources.
- The advantage of parallelizing *environments* over simulations is that memory usage does not increase. However, dependencies occur twice in each transition step requiring synchronization, and that modern machines have usually more cores than environments and just parallelizing environments cannot take advantage of all computing resources. In addition, as with simulations, load balancing can be an issue if the number of environments is not divisible by the number of cores, or if the runtime of environments varies.
- By combining both forms of parallelism, the amount of each resource can be balanced. This will become more important as the number of cores within a node increases. For example, the number of simulations can be increased until available memory is used and then environments within each system can be parallelized.

Conducted experiments [65, 70] indicate the simulations are memory bound. The OpenMP simulator achieved speedups of up to 2.5x on a 4-core Intel i7. Parallelizing by simulations or hybrid techniques yields the largest speedups. For this reason, ABCD-GPU simulator has the parallel simulation approach by default, as shown in Fig. 7.12.

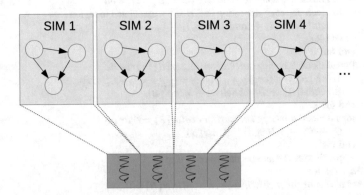

**Fig. 7.12**  Multicore selected design, distribution of simulations along the cores/threads

### 7.3.3.3 Parallel Simulation over Manycore Platforms

Manycore processors, such as GPUs, can be easily programmed using CUDA. As mentioned above, the threading model is based on a hierarchical arrange in blocks. Generally speaking, threads belonging to a block can easily cooperate through synchronization points and fast shared memory. Warps, which are groups of 32 threads, are the parallelism unit, and they execute in a SIMD mode. In reality, blocks are scheduled to each multiprocessor, and each warp to the cores in it. A good design is to define as many blocks (at least twice or three times the number of multiprocessors) and threads (from 64 to 1024) as possible [65, 81].

At first glance, simulations and environments are two levels of parallelism that could fit the double parallelism of the CUDA architecture (thread blocks and threads). For example, we could assign each simulation to a block of threads, and each environment to a thread (since they require synchronization at each time step). However, the number of environments depends inherently on the model. Typically, 2 or more environments are considered, which is not enough for fulfilling the GPU resources. Number of simulations typically ranges from 50 to 100, which is sufficient for thread blocks, but still a poor number compared to the several hundred cores available on modern GPUs.

Therefore, the selection and execution of rule blocks is also considered [69]. Hence, the CUDA simulator can utilize a huge number of thread blocks by distributing simulations and environments in each one, and process each rule block by each thread. Since there are normally more rule blocks (thousand of them) than threads per thread block (up to 1024), a constant number of threads is launched which iterate over the rule blocks in tiles. This design is graphically shown on Fig. 7.13. Each phase of the algorithm has been designed following the general CUDA design explained

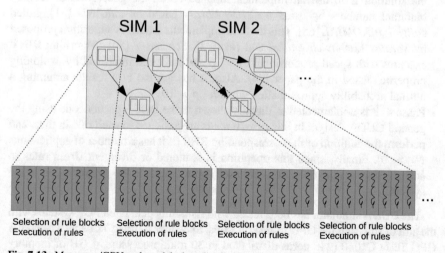

**Fig. 7.13** Manycore/GPUs selected design, distribution of simulations and environments along the multicores/blocks, and ruleblocks/rules along cores/threads

above, and implemented separately as individual kernels. Thus, simulations and environments are synchronized by the successive calls to the kernels [69].

The parallel design of each DCBA phase is as follows [65, 69]:

- Phase 1: the virtual table algorithm for this phase (as described in Algorithm 12) is used here as well. As environments are distributed along thread blocks, the outermost loop of the algorithm is already inherently performed in parallel. Within each thread block, firstly, each thread is performing the filters to the rule blocks in parallel. Secondly, each thread calculates the additions of rows for the corresponding rule block, if it is active after applying the filters. This is made in a scatter approach, since threads are adding the rows using atomic operations. Thirdly, threads calculate the corresponding column minimums, and eventually, update the temporal copy of the configuration.
- Phase 2: it is the most challenging part when parallelizing by blocks. The selection of blocks at this phase is performed in an inherently sequential way: we need to know how many objects a block can consume before selecting the next one. The random order to the blocks is *simulated* by the CUDA thread scheduler: each thread calculates the position in the order of its rule block by using the *atomicInc* operation. Since it does not perform a real random order, random numbers are going to be used soon in future versions. The chosen solution dynamically checks the blocks that are really competing for objects, and calculates which blocks can be selected in parallel, and which depend on the selection of the others. To do this, some previous computations are needed. Two arrays are used, one storing the information of the LHS, and another the selection order. Rule blocks having the same selection order number will be selected in parallel. Both arrays are implemented using the GPUs shared memory to speedup this computation.
- Phase 3: for the random multinomial number generation performed in this phase, the solution is to use an implementation based on the generation of random binomial numbers by using a *CUDA library* based on *cuRAND* [81], called *cuRNG_BINOMIAL* [65]. This module implements the BINV algorithm proposed by *Voratas Kachitvichyanukul* and *Bruce W. Schmeiser* [54]. Algorithm BINV executes with speed proportional to $n \cdot p$ and has been improved by exploiting properties listed in the paper [54]. Also, it has got the best results assuming a normal probability approximation when $n \cdot p > 10$.
- Phase 4: it is implemented as directly shown in the DCBA pseudocode using the general CUDA design. In this case, threads are assigned to each rule in tiles, and perform the addition of the corresponding RHS (if it has a number of applications $N_r > 0$). Finally, since this operation is scattered or divergent (from rules to add objects), atomic operations are used again to update the configuration of the system.

The CUDA simulator has been tested with randomly generated PDP systems with the aim of stressing the GPU in several ways. Speedups of up to 7x using a NVIDIA GPU Tesla C1060 (240 cores distributed in 30 multiprocessors, 4 GB of memory without cache) were reported [65, 69]. Moreover, it has been shown that Phase 2 is normally the bottleneck, as it can be seen given the lack of parallelism.

Furthermore, when using a real PDP system based model related to the Bearded vulture in the Pyrenees [10], phase 2 doesn't make sense at all, since blocks barely compete for objects. However, a speedup of up to 4.3x is reported using the same GPU, given the low number of rule blocks and environments, and the low rate of rule blocks applications in each transition step.

### 7.3.3.4  ABCD-GPU Simulator

ABCD-GPU simulator is a stand-alone application for conducting the DCBA over a desired model. The aim is to serve as an engine for more general simulation frameworks, such as P-Lingua and MeCoSim. To date, the latest version of ABCD-GPU is 0.9, and is freely and openly available at the PMCGPU website [118].

Figure 7.14 shows the general structure of the simulation framework. The core of the simulator is the engine, which can be the multicore (CPU or OpenMP) or the manycore (GPU or CUDA) version. The simulator is able to receive as input a set of randomly generated PDP systems, and so to output corresponding debugging and profiling information. But in the latest version, new input and output modules have been attached. Similar to the Active Membranes simulator, the input is given by a binary file which defines a PDP system model. This binary file can be generated using the new version of pLinguaCore, from a P-Lingua file, the standard format to define PDP systems. The reason of receiving a binary file, instead of a P-Lingua or XML file, is for compression and efficiency. The usage of the parallel simulator is only deserved when using a large PDP system model (many rule blocks and environments). Therefore, the communication should be efficient enough in order

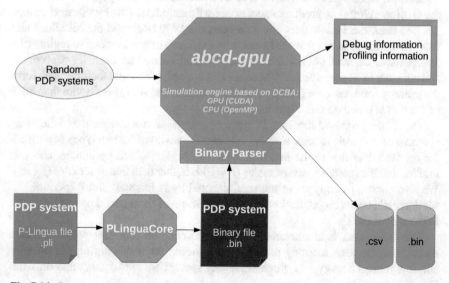

**Fig. 7.14**  Structure of the ABCD-GPU simulation framework

not to hide the fast execution of the simulator. The output of the simulator is in turn, given by a .csv file or plain text. An output binary file version is also available, but it is still experimental. Future developments will allow to load the output directly into a database.

## 7.4  Case Studies

In this section, we analyze the suitability of PDP systems, a membrane computing based modeling framework, as a singular tool for simulating complex ecosystems dynamics. This enables assisting managers, conservationists and policy-makers to make appropriate decisions related to the improvement of management and conservation programs. Specifically, four relevant/interesting case studies involving real ecosystems populations are illustrated: (a) scavenger birds at the Catalan Pyrenees; (b) Pyrenean chamois inhabiting the Catalan Pyrenees; (c) the zebra mussel at Riba-roja reservoir (Spain), managed by the company Endesa S.A.; and (d) giant pandas in captivity at the Giant Panda Breeding Base, Chengu (China).

### 7.4.1  Scavenger Birds

Next, following [10], we present a study carried out in the Pyrenean and Prepyrenean mountains of Catalonia (NE Spain). The ecosystem to be modeled is composed of 13 species: three avian scavengers (the Bearded vulture, the Egyptian vulture and the Griffon vulture) as predator species, six wild ungulates (the Pyrenean chamois, the Red deer, the Fallow deer, the Roe deer, the Wild boar, and the Mouflon) and four domestic ungulates that are found in an extensive or semi-extensive regime (the Sheep, the Goat, the Cow, and the Horse) providing carrion for the avian scavengers and considered as prey species. These species are herbivores and their remains form the primary food resource for the avian scavengers in the study area (more than 80% of the diet is based on these species [27, 63]).

The model proposed involves 17 types of animals as a consequence of the management of domestic species mainly. Specifically, we consider two types of animals for the Red deer due to the fact that males are highly valued by hunters and this implies that the mortality rate of males ($i = 6$) is higher than that of females ($i = 5$). We also consider two types of animals, denoted by A (annual) and P (periodical), for domestic ones (except for horses) because some of them spend only six months in the mountain.

The study area is characterized by the presence of abundant wild ungulates throughout the year, a strong presence of domestic ungulates during the summer and low human density. The three scavenger species are cliff-nesting and only the Egyptian vulture is migratory.

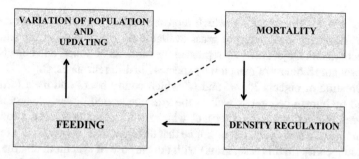

**Fig. 7.15** Modules of the P system

The algorithmic scheme of the proposed model is shown in Fig. 7.15. The algorithm has been sequenced, but all animals evolve in parallel.

- It is assumed that the population growth rate of the Bearded vulture varies depending on the surface and orography of the system as well as on the existing population.
- The mortality module considers that after an animal dies, in addition to the bones it leaves at the ecosystem, its meat serves as food for other animals.
- The feeding module considers that the feeding resources for the species at the ecosystem have been modeled.
- The population density of the ecosystem is regulated by a module that has been incorporated between the mortality and the feeding modules.

When a cycle is performed, all objects which are not associated with species are eliminated, except the biomass generated by the animals that have died due to the regulation process.

### 7.4.1.1  Model

We consider a PDP system of degree $(2, 1)$

$$\Pi = (G, \Gamma, \Sigma, T, \mathcal{R}_E, \mu, \mathcal{R}, \{f_{r,1} \mid r \in \mathcal{R}\}, \mathcal{M}_{1,1}, \mathcal{M}_{2,1}, E_1)$$

with one environment and $\Pi_1 = (\Gamma, \mu, \mathcal{M}_{1,1}, \mathcal{M}_{2,1}, \mathcal{R})$ is a P system of degree 2 with (only) two electrical charges (neutral and positive) taking T time units, defined as follows:

- $G$ is an empty graph.
- The working alphabet is

$$\Gamma = \{X_{ij}, \ Y_{ij}, \ V_{ij}, \ Z_{ij} : 1 \le i \le 17, \ 0 \le j \le g_{i,6}\} \cup$$
$$\{B, \ G, \ M, \ B', \ G', \ M', \ C, \ C'\} \cup \{h_s : 1 \le s\} \cup$$
$$\{H_i, \ H'_i, \ F_i, \ F'_i, \ T_i, \ a_i, \ b_{0i}, \ b_i, \ d_i, \ e_i : 1 \le i \le 17\}$$

Symbols $X, Y, V$ and $Z$ represent the same animal but in different states. Index $i$ is associated with the type of animal and index $j$ is associated with their age.

$B$, $B'$ are auxiliary symbols which represent bones, $M$, $M'$ represent meat and $G$, $G'$ represent the amount of grass available for the feeding of the animals in the ecosystem. Objects $H_i$, $H_i'$ represent the biomass of bones, and objects $F_i$, $F_i'$ represent the biomass of meat left by species $i$ in different states. Object $C$ enables the creation of objects $B'$, $M'$ and $G'$ which codify bones and meat (artificially added by human beings) as well as the grass generated by the ecosystem itself. Besides, object $C$ produces objects $C'$ which in turn generate object $C$ allowing the beginning of a new cycle. Let us notice that different objects (i.e. $G$, $G'$) represent the same entity (in this case, grass) with the purpose of synchronizing the model. $T_i$ is an object used for counting the existing animals of species $i$. If a species overcomes the maximum density, values will be regulated. Objects $b_{0i}$, $b_i$ and $e_i$ allow us to control the maximum number of animals per species in the ecosystem. At the moment when a regulation takes place, object $a_i$ allows us to eliminate the number of animals of species $i$ that exceeds the maximum density. Object $d_i$ is used to put under control domestic animals that are withdrawn from the ecosystem for their marketing.

- The alphabet of the environment is $\Sigma = \emptyset$.
- $T$ represents the number of years to be simulated in the evolution of the ecosystem.
- $\mu = [\ [\ ]_2\ ]_1$ is the membrane structure. The skin membrane is important to control that the densities that every species does not overcome the threshold of the ecosystem. Animals reproduce, feed and die in the inner membrane.
- $\mathcal{M}_{1,1}$ and $\mathcal{M}_{2,1}$ are finite multisets over $\Gamma$ describing the objects initially placed in regions of $\mu$ (encoding the initial population and the initial food);

   - $\mathcal{M}_{1,1} = \{b_{0i},\ X_{ij}^{q_{ij}},\ h_t^{q_{1j}} : 1 \leq i \leq 17,\ 0 \leq j \leq g_{i,6}\}$, where $q_{ij}$ indicates the number of animals of species $i$ initially present in the ecosystem whose age is $j$, and $t = \max\{1,\ \lceil \frac{\sum_{j=8}^{21} q_{1j} - 6}{1.352} \rceil\}$. The mathematical expression given for $t$ was obtained using the lineal regression (see Fig. 7.16);
   - $\mathcal{M}_{2,1} = \{C\}$

- The set $\mathcal{R}$ of evolution rules of the only P system of $\Pi$ consists of:

   - The first rule represents the contribution of energetic resources to the ecosystem at the beginning of each cycle and it is essential for the system to evolve.
     $$r_0 \equiv [C \rightarrow B'^\alpha M'^\beta G'^\gamma C']_2^0,$$
     where $\alpha$ and $\beta$ are the double of kilos of bones and meat that are externally introduced to the ecosystem, and $\gamma$ is the amount of grass produced by the ecosystem.
     The second rule is useful to synchronize the process.
     $$r_1 \equiv [b_{0,i} \rightarrow b_i]_1^0.$$

   - *Variation rules of the population.*
     We consider two cases due to the fact that in nomadic species the said variation is influenced by animals from other ecosystems.

**Fig. 7.16** Lineal regression between numbers of pairs and years for the bearded vulture

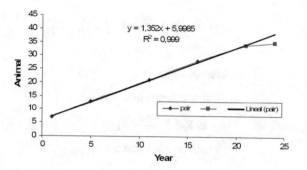

- Case 1. Non-nomadic species (No Bearded vultures).

  ★ Adult males:

  $$r_2 \equiv [X_{ij} \xrightarrow{(1-k_{i,1})} Y_{ij}]_1^0, \ 2 \leq i \leq 17, \ i \neq 5, \ g_{i,4} \leq j < g_{i,5}.$$

  ★ Adult females that reproduce:

  $$r_3 \equiv [X_{ij} \xrightarrow{k_{i,2} \cdot k_{i,1}} Y_{ij} Y_{i0}^{k_{i,3}}]_1^0, \ 2 \leq i \leq 17, \ i \neq 5, 6, \ g_{i,4} \leq j < g_{i,5}.$$

  $$r_4 \equiv [X_{5j} \xrightarrow{0.5 \cdot k_{5,2}} Y_{5j} Y_{50}^{k_{5,3}}]_1^0, \ g_{5,4} \leq j < g_{5,5}.$$

  $$r_5 \equiv [X_{5j} \xrightarrow{0.5 \cdot k_{5,2}} Y_{5j} Y_{60}^{k_{5,3}}]_1^0, \ g_{5,4} \leq j < g_{5,5}.$$

  ★ Adult females that do not reproduce:

  $$r_6 \equiv [X_{ij} \xrightarrow{(1-k_{i,2}) \cdot k_{i,1}} Y_{ij}]_1^0, \ 2 \leq i \leq 17, \ i \neq 6, \ g_{i,4} \leq j < g_{i,5}.$$

  ★ Old females and males that do not reproduce:
  $$r_7 \equiv [X_{ij} \longrightarrow Y_{ij}]_1^0, \ 2 \leq i \leq 17, \ g_{i,5} \leq j \leq g_{i,6}.$$

  ★ Young animals that do not reproduce:

  $$r_8 \equiv [X_{ij} \xrightarrow{1-k_{i,4}} Y_{ij}]_1^0, \ 2 \leq i \leq 17, \ 1 \leq j < g_{i,4}.$$

- Case 2. Nomadic species (Bearded vultures).

  $$r_9 \equiv [X_{1j}h_s \xrightarrow{v_s} Y_{1(g_{1,4}-1)}Y_{1j}h_{s+1}^2]_1^0, \ \begin{cases} g_{1,4} \leq j \leq g_{1,6}, \\ t \leq s \leq D_1 \end{cases}$$
  being $v_s = 1.352/(1.352s + 6)$ and $D_1 = \min\{21, \ T+t-1\}$.

  $$r_{10} \equiv [X_{1j}h_s \xrightarrow{0.01} Y_{1(g_{1,4}-1)}Y_{1j}h_{s+1}^2]_1^0, \ \begin{cases} g_{1,4} \leq j \leq g_{1,6} \\ D_3 \leq s \leq D_2 \end{cases}$$
  where $D_2 = \max\{21, \ T+t-1\}$ and $D_3 = \max\{21, \ t\}$.

  $$r_{11} \equiv [X_{1j}h_s \xrightarrow{1-v_s} Y_{1j}h_{s+1}]_1^0, \ \begin{cases} g_{1,4} \leq j \leq g_{1,6} \\ t \leq s \leq D_1 \end{cases}$$

$$r_{12} \equiv [X_{1j}h_s \xrightarrow{\ 0.99\ } Y_{1j}h_{s+1}]_1^0, \quad \begin{cases} g_{1,4} \leq j \leq g_{1,6} \\ D_3 \leq s \leq D_2 \end{cases}$$

– *Mortality rules.*

  ★ Young animals that survive:

$$r_{13} \equiv Y_{ij}[\ ]_2^0 \xrightarrow{\ 1-m_{i,1}-m_{i,3}\ } [V_{ij}T_i]_2^+, \ 1 \leq i \leq 17, \ 0 \leq j < g_{i,3}.$$

  ★ Young animals that die:

$$r_{14} \equiv Y_{ij}[\ ]_2^0 \xrightarrow{\ m_{i,1}\ } [H_i'^{f_{i,1} \cdot g_{i,2}} F_i'^{f_{i,2} \cdot g_{i,2}} B'^{f_{i,1} \cdot g_{i,2}} M'^{f_{i,2} \cdot g_{i,2}}]_2^+, \quad \begin{cases} 1 \leq i \leq 17 \\ 0 \leq j < g_{i,3} \end{cases}$$

  ★ Young animals that are retired from the ecosystem:

$$r_{15} \equiv [Y_{ij} \xrightarrow{\ m_{i,3}\ } \lambda]_1^0, \ 1 \leq i \leq 17, \ 0 \leq j < g_{i,3}.$$

  ★ Adult animals that do not reach an average life expectancy and survive:

$$r_{16} \equiv Y_{1j}h_s[\ ]_2^0 \xrightarrow{\ 1-m_{1,2}\ } [V_{1j}T_1h_s]_2^+, \quad \begin{cases} g_{1,3} \leq j < g_{1,6} \\ t+1 \leq s \leq D+t \end{cases}$$

$$r_{17} \equiv Y_{ij}[\ ]_2^0 \xrightarrow{\ 1-m_{i,2}\ } [V_{ij}T_i]_2^+, \quad \begin{cases} 2 \leq i \leq 17 \\ g_{i,3} \leq j < g_{i,6} \\ t+1 \leq s \leq D+t \end{cases}$$

  ★ Adult animals that do not reach an average life expectancy and die:

$$r_{18} \equiv Y_{1j}h_s[\ ]_2^0 \xrightarrow{\ m_{1,2}\ } [H_1'^{f_{1,3} \cdot g_{1,2}} F_i'^{f_{1,4} \cdot g_{1,2}} B'^{f_{1,3} \cdot g_{1,2}} M'^{f_{1,4} \cdot g_{1,2}} V_{1,g_{1,4}-1}h_s T_1]_2^+,$$
where $g_{i,3} \leq j < g_{i,6}, \ t+1 \leq s \leq T+t.$

$$r_{19} \equiv Y_{ij}[\ ]_2^0 \xrightarrow{\ m_{i,2}\ } [H_i'^{f_{i,3} \cdot g_{i,2}} F_i'^{f_{i,4} \cdot g_{i,2}} B'^{f_{i,3} \cdot g_{i,2}} M'^{f_{i,4} \cdot g_{i,2}}]_2^+,$$
where $1 \leq 2 \leq 17, \ g_{i,3} \leq j < g_{i,6}, \ t+1 \leq s \leq T+t.$

  ★ Animals that reach an average life expectancy and die in the ecosystem:

$$r_{20} \equiv Y_{1g_{1,6}}h_s[\ ]_2^0 \longrightarrow [H_1'^{f_{1,3} \cdot g_{1,2}} F_i'^{f_{1,4} \cdot g_{1,2}} B'^{f_{1,3} \cdot g_{1,2}} M'^{f_{1,4} \cdot g_{1,2}} V_{1,g_{1,4}-1}h_s T_1]_2^+,$$
where $t+1 \leq s \leq T+t.$

$$r_{21} \equiv Y_{ig_{i,6}}[\ ]_2^0 \xrightarrow{\ c_{21}\ } [H_i'^{f_{i,3} \cdot g_{i,2}} F_i'^{f_{i,4} \cdot g_{i,2}} B'^{f_{i,3} \cdot g_{i,2}} M'^{f_{i,4} \cdot g_{i,2}}]_2^+,$$
where $2 \leq i \leq 17, c_{21} = m_{i,4} + (1 - m_{i,4}) \cdot m_{i,2}$ and $t+1 \leq s \leq T+t.$

  ★ Animals that reach an average life expectancy and are retired from the ecosystem:

$$r_{22} \equiv [Y_{ig_{i,6}}h_s^{k_{i,4}} \xrightarrow{\ (1-k_{i,4}) \cdot (1-m_{i,4}) \cdot (1-m_{i,2})\ } \lambda]_1, \quad \begin{cases} 2 \leq i \leq 17 \\ t+1 \leq s \leq T+t \end{cases}$$

– *Density regulation rules.*

  ★ Creation of objects that are going to enable the control of the maximum number of animals in the ecosystem:

$$r_{23} \equiv b_i[\ ]_2^0 \rightarrow [b_i a_i^{\lceil 0,9 * g_{i,7}\rceil} e_i^{\lceil 0,2 * g_{i,7}\rceil}]_2^+, \ 1 \le i \le 17.$$

⋆ Evaluation of the density of the different species in the ecosystem:
$$r_{24} \equiv [T_i^{g_{i,7}} a_i^{(g_{i,7} - g_{i,8})} \rightarrow \lambda]_2^+, \ 1 \le i \le 17.$$
⋆ Generation of randomness in the number of animals:
$$r_{25} \equiv [e_i \xrightarrow{\ 0,5\ } a_i]_2^+, \ 1 \le i \le 17.$$

$$r_{26} \equiv [e_i \xrightarrow{\ 0,5\ } \lambda]_2^+, \ 1 \le i \le 17.$$
⋆ Change of the names of the objects which represent animals:
$$r_{27} \equiv [V_{ij} \rightarrow Z_{ij}]_2^+, \ 1 \le i \le n, \ 0 \le j < g_{i,6}.$$
⋆ Change of the names of the objects which represent food resources:
$$r_{28} \equiv [G' \rightarrow G]_2^+.$$
$$r_{29} \equiv [B' \rightarrow B]_2^+.$$
$$r_{30} \equiv [M' \rightarrow M]_2^+.$$
$$r_{31} \equiv [C' \rightarrow C]_2^+.$$
$$r_{32} \equiv [H_i' \rightarrow H_i]_2^+, \ 1 \le i \le 17.$$
$$r_{33} \equiv [F_i' \rightarrow F_i]_2^+, \ 1 \le i \le 17.$$

– *Feeding rules.*

$$r_{34} \equiv [Z_{ij} h_s^{k_{i,4}} a_i B^{f_{i,5} \cdot g_{i,2}} G^{f_{i,6} \cdot g_{i,2}} M^{f_{i,7} \cdot g_{i,2}}]_2^+ \rightarrow X_{i(j+1)} h_s^{k_{i,4}}[\ ]_2^0,$$
where $1 \le i \le 17, 0 \le j \le g_{i,6}, t + 1 \le s \le D + t.$

– *Updating rules.*
Next rules implement a balance at the end of the year, that is, the leftover food is not useful for the next year, so it is necessary to eliminate it. But if the amount of food is not enough, some animals die.

⋆ Elimination of the remaining bones, meat and grass:
$$r_{35} \equiv [G \rightarrow \lambda]_2^0.$$

$$r_{36} \equiv [M \rightarrow \lambda]_2^0.$$

$$r_{37} \equiv [B \rightarrow \lambda]_2^0.$$

$$r_{38} \equiv [T_i \rightarrow \lambda]_2^0, \ 1 \le i \le 17.$$

$$r_{39} \equiv [a_i \rightarrow \lambda]_2^0, \ 1 \le i \le 17.$$

$$r_{40} \equiv [e_i \rightarrow \lambda]_2^0, \ 1 \le i \le n.$$

$$r_{41} \equiv [b_i]_2^0 \rightarrow b_i[\ ]_2^0, \ 1 \le i \le 17.$$

$$r_{42} \equiv [H_i]_2^0 \rightarrow H_i[\ ]_2^0, \ 1 \le i \le 17.$$

$$r_{43} \equiv [F_i]_2^0 \rightarrow F_i[\ ]_2^0, \ 1 \le i \le 17.$$

⋆ Young animals that die because of a lack of food:

$$r_{44} \equiv [Z_{ij} \xrightarrow{g_{i,1}} H_i'^{f_{i,1}} F_i'^{f_{i,2}} B'^{f_{i,1}} M'^{f_{i,2}}]_2^0, \quad \begin{cases} 1 \leq i \leq 17 \\ 0 \leq j < g_{i,3} \end{cases}$$

$$r_{45} \equiv [Z_{ij}]_2^0 \xrightarrow{1-g_{i,1}} d_i[\ ]_2^0, \ 1 \leq i \leq n, \ 0 \leq j < g_{i,3}.$$

⋆ Adult animals that die because of a lack of food:

$$r_{46} \equiv [Z_{ij} h_s^{k_{1,4}} \xrightarrow{g_{i,1}} H_i'^{f_{i,3}} F_i'^{f_{i,4}} B'^{f_{i,3}} M'^{f_{i,4}}]_2^0, \quad \begin{cases} 1 \leq i \leq 17 \\ g_{i,3} \leq j \leq g_{i,6} \\ t+1 \leq s \leq D+t \end{cases}$$

$$r_{47} \equiv [Z_{ij} h_s^{k_{1,4}} \xrightarrow{1-g_{i,1}} \lambda]_2^0, \quad \begin{cases} 1 \leq i \leq 17, \\ g_{i,3} \leq j \leq g_{i,6}, \\ t+1 \leq s \leq D+t \end{cases}$$

The purpose of these rules is to eliminate objects $H$ and $F$ associated with the quantity of biomass left by every species.

$$r_{48} \equiv [H_i \rightarrow \lambda]_1^0, \ 1 \leq i \leq 17.$$

$$r_{49} \equiv [F_i \rightarrow \lambda]_1^0, \ 1 \leq i \leq 17.$$

The constants associated with the rules have the following meanings (index $i$, $1 \leq i \leq 17$, represents the type of animal):

- $g_{i,1}$: 1 for wild animals and 0 for domestic animals.
- $g_{i,2}$: proportion of time they remain in the mountain during the year.
- $g_{i,3}$: age at which adult size is reached (age at which the animal eats like an adult, and at which if the animal dies, the amount of biomass it leaves is similar to the total one left by an adult). At this age it will have surpassed the critical early phase during which the mortality rate is high.
- $g_{i,4}$: age at which it starts to be fertile.
- $g_{i,5}$: age at which it stops being fertile.
- $g_{i,6}$: average life expectancy in the ecosystem.
- $g_{i,7}$: maximum density of the ecosystem.
- $g_{i,8}$: number of animals that survive after reaching maximum density of the ecosystem.
- $k_{i,1}$: proportion of females in the population (per one).
- $k_{i,2}$: fertility rate (proportion of fertile females that reproduce).
- $k_{i,3}$: number of descendants per each fertile female that reproduces.
- $k_{i,4}$: it is equal to 0 when the species go through a natural growth and it is equal to 1 when animals are nomadic (the Bearded vulture moves from one place to another until it is 6–7 years old, when it settles down).
- $m_{i,1}$: natural mortality rate in the first years, $age < g_{i,3}$ (per one).
- $m_{i,2}$: mortality rate in adult animals, $age \geq g_{i,3}$ (per one).
- $m_{i,3}$: percentage of domestic animals belonging to non-stabilized populations which are withdrawn in the first years.

- $m_{i,4}$: is equal to 1 if the animal dies at the age of $g_{i,6}$ and it is not retired, and it is equal to 0 if the animal does not die at the age of $g_{i,6}$ but it is retired from the ecosystem.
- $f_{i,1}$: amount of bones from young animals when they die, $age < g_{i,3}$.
- $f_{i,2}$: amount of meat from young animals when they die, $age < g_{i,3}$.
- $f_{i,3}$: amount of bones from adult animals when they die, $age \geq g_{i,3}$.
- $f_{i,4}$: amount of meat from adult animals when they die, $age \geq g_{i,3}$.
- $f_{i,5}$: amount of bones necessary per year and animal (1 unit is equal 0.5 kg of bones).
- $f_{i,6}$: amount of grass necessary per year and animal.
- $f_{i,7}$: amount of meat necessary per year and animal.

The values of these constants have been obtained experimentally, except for $k_{i,4}$ and $m_{i,4}$ (see [7, 27, 62, 63] for details). Constants $k$, $m$ and $f$ are associated with reproduction, mortality and feeding rules, respectively. Constants $g$ are associated with the remaining rules.

### 7.4.1.2 Results

In order to experimentally validate the model, an ad-hoc simulator was developed based on P-Lingua 2.0 [30], starting after data from Table 7.3.

We have focused on the evolution of wild species populations for a period of 14 years, since 1994. The scavenger birds populations at the initial year has been considered according to the data in Table 7.3 by means of a logarithmic (respectively, exponential) regression (see Fig. 7.17).

At the validation process, values obtained from the simulator runs have been compared to those obtained experimentally. It is also worth noting that we have focused on the population dynamics at wild species from which there are only data about the initial (1994) and final (2008) years, except for scavengers birds which we have more information about (see Table 7.3).

The ecosystem evolution throughout the period under study has been obtained by running the simulator for 100 times having the same input data. The simulator execu-

Table 7.3 Number of animals in the Catalan Pyrenees (1979–2009)

| Specie | 79 | 84 | 87 | 89 | 93 | 94 | 95 | 99 | 00 | 05 | 08 | 09 |
|---|---|---|---|---|---|---|---|---|---|---|---|---|
| Bearded V. | – | 7 | – | 13 | – | – | 21 | – | 28 | 34 | 35 | – |
| Egyptian V. | – | – | 29 | – | 34 | – | – | – | 40 | – | 66 | – |
| Griffon V. | 29 | – | – | 106 | – | – | – | 377 | – | – | – | 842 |
| Pyrenean C. | – | – | – | – | – | 9000 | – | – | – | – | 12000 | – |
| Red deer | – | – | – | – | – | 1000 | – | – | – | – | 5500 | – |
| Fallow deer | – | – | – | – | – | 600 | – | – | – | – | 1500 | – |
| Roe deer | – | – | – | – | – | 1000 | – | – | – | – | 10000 | – |

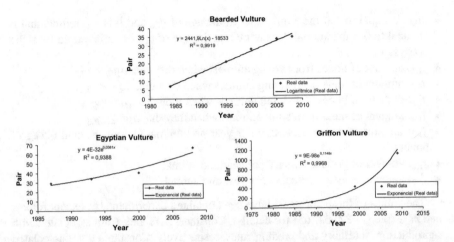

**Fig. 7.17** Regression relationships between numbers of pairs and years

tions have allowed us to estimate the standard deviation and compute the population confidence intervals of the different species. The results presented in Fig. 7.18 are the average of the 100 simulator executions. These results suggest that the PDP system based model presented correctly simulates the population dynamics in the period of time analyzed.

Next picture shows the energetic requirements of the Bearded vultures and the corresponding contributions of three ungulates (see Fig. 7.19).

### 7.4.2  Pyrenean Chamois

Ungulates have life histories that are conducive to sustainable harvests [12] and, for a variety of reasons (population control, meat production, and recreational and trophy hunting), are intensively exploited. Legislation has ensured that appropriate hunting quotas are set and that thereby the standards for the sustainable use of the species are met. Pyrenean chamois is a small ungulate living in the three Pyrenean countries (Andorra, France and Spain) that represents an important economic and social resource in rural communities. At present the existing population in the Pyrenees is estimated at about 53,000 individuals. The status of the species has not always been so favorable, in the late 60s the population decreased down to the edge of extinction due to indiscriminate hunting. Fortunately National Hunting Reserves managed by the regional administration were created in order to save the species.

This species has no major predators in the Pyrenees, except the brown bear (*Ursus arctos*), and the golden eagle (*Aquila chrysaetos*). However, it is a species with a small growth rate, compared with other species of ungulates and it has a life expectancy of 20 years, though the mortality rate is high for animals older than

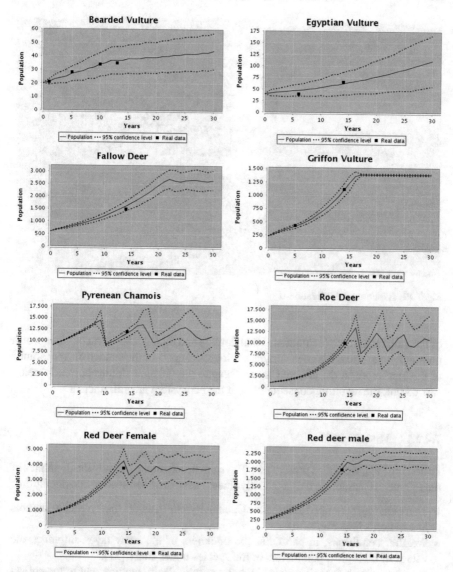

**Fig. 7.18** Experimental validation-I

11 years. In recent years, two outbreaks of pestivirus and keratoconjunctivitis have affected some Pyrenean chamois populations of the Catalan Pyrenees. In fact, the outbreak of pestivirus has caused a decline of 80% of the local population [95].

The pestivirus disease is having a very important impact at the social and economic scale in the Pyrenees. The suspension of Pyrenean chamois hunting in the affected areas has led to major loss of economic income. This loss is due not only to the lack of direct income through payment of hunting licenses, but also by the disappearance

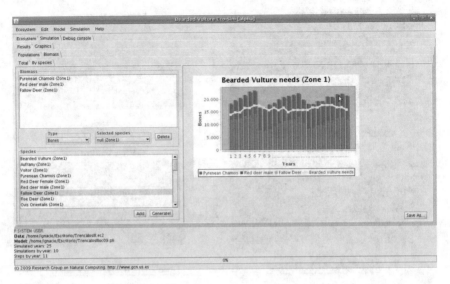

**Fig. 7.19** Experimental validation-II

of the indirect income (ecotourism) that hunters and their guests bring out. It is worth pointing out to stress the considerable ecological impact of the sudden disappearance of this herbivore in the affected areas. Despite the detailed studies currently being carried out, the resulting consequences in the ecosystem are still unclear.

### 7.4.2.1  Model

Following [14], we aim to present a PDP system based model to study the dynamics of the Pyrenean Chamois in the Catalan Pyrenees based on the following considerations:

- There are four separated areas in the Catalan Pyrenees where the Pyrenean chamois live in herds (see Fig. 7.20).
- Weather conditions, in particular the thickness of the snow layer, influence the values of biological parameters of the species under study [21].
- Causes of death for this species include: natural death, hunting and to the spread of disease. We assume that it is unlikely that this species moves between areas, and hence wherever there is a lack of resources the animals die. Only pestivirus disease will be taken into account, while other diseases will not be considered yet.

The algorithmic scheme of the proposed model is shown in Fig. 7.21. The processes to be modeled will be the weather conditions (snow), reproduction, regulation of density, food, natural mortality, hunting mortality and mortality from the disease. In order to model these processes for each species, some biological, geographical and human factors, shown in Table 7.4, are needed.

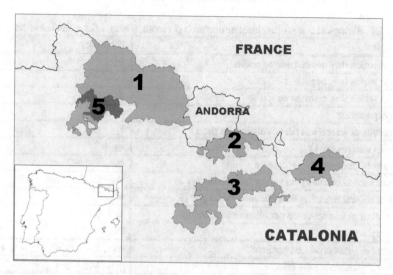

**Fig. 7.20** Study area in the Catalonia Pyrenees. Area 1: Reserva Nacional de Caça de l'Alt Pallars-Aran. Area 2: RNC Cerdanya-Alt Urgell. Area 3: RNC Cadí. Area 4: RNC Freser-Setcases. Area 5: Parc Nacional, not included in the study

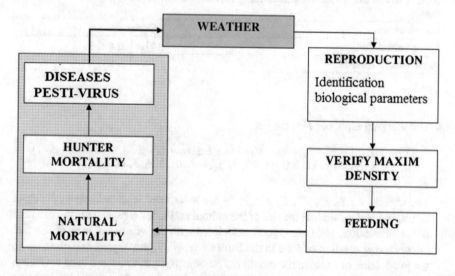

**Fig. 7.21** Scheme model of the Pyrenean chamois model

The proposed model consists of a PDP system of degree $(11, 4)$, taking $T$ times units $(G, \Gamma, \Sigma, T, \mathcal{R}_E, \mu, \mathcal{R}, \{f_{r,j} \mid r \in \mathcal{R} \wedge 1 \le j \le 4\}, \{\mathcal{M}_{i,j} \mid 0 \le i \le 10 \wedge 1 \le j \le 4\}, \{E_j \mid 1 \le j \le 4\})$, where:

- The graph of the system is $G = (V, S)$, where $V = \{e_1, e_2, e_3, e_4\}$ and $S = \{(e_1, e_i) : 1 \le i \le 4\}$.

**Table 7.4** Biological and geographical information ($i$ month, $\nu$ area, $l$ Snow thickness category)

| Biological | Parameter |
|---|---|
| Age at which they are considered adults | $g_0$ |
| Age at which they begin to be fertile | $g_1$ |
| Age at which they cease to be fertile | $g_2$ |
| Life expectancy | $g_3$ |
| Proportion of females in the population (as per 1) | $k_1$ |
| Fertility rate (as per 1) | $k_2$ |
| Number of descendants per female | $k_3$ |
| Rate of natural mortality on young animals (as per 1) | $m_1$ |
| Rate of natural mortality on adult animals (as per 1) | $m_2$ |
| Amount of grass consumed per month and animal | $\beta_i \; 1 \le i \le 10$ |
| **Geographical** | **Parameter** |
| Amount of grass produced per month | $\alpha_{i,\nu}, 1 \le i \le 10, 1 \le \nu \le 4$ |
| Probability of having the disease | $ms_\nu, 1 \le \nu \le 4$ |
| Probability of dying from a disease | $md_\nu, 1 \le \nu \le 4$ |
| Maximum density of the ecosystem | $d1_\nu, 1 \le \nu \le 4$ |
| Number of animals that survive after reaching the maximum density | $d2_\nu, 1 \le \nu \le 4$ |
| **Human factors** | **Parameter** |
| Young animals hunted | $h1_\nu, 1 \le \nu \le 4$ |
| Adult animals hunted | $h2_\nu, 1 \le \nu \le 4$ |

- The working alphabet $\Gamma$ is the set

$$\{X_{jy}, Y_{jy}, Y'_{jy}, Y''_{jy}, Z_{jy}, V_{jy}, W_{jy} : 0 \le j \le g_3, \; 1 \le y \le T\} \cup \{t_i : 1 \le i \le 3\} \cup$$
$$\{G_i : 4 \le i \le 10\} \cup \{R_i : 0 \le i \le 7\} \cup \{a, c, d, e, t, h, d_1 \; F, D, S, N\}$$

The objects $X$, $Y$, $Y'$, $Y''$, $Z$, $V$ and $W$ are associated with animals in different states, index $j$ represents the age of the animal and index $y$ represents the moment of the simulation. The $t$ are objects associated with the weather. $F$ is an object that allows the generation of food in the form of grass. $G_i$ are objects associated with the production of grass in the month $i$. The objects $D$, $c$, and $e$ are used to control the density of animals of each species. The objects $h_1$ and $h$ are used in order to know the state of pestivirus. The objects $S$ and $N$ indicate presence or absence of the disease, respectively, and finally there is the counter $R$ that will allow us to synchronize the system.

- The environment alphabet is $\Sigma = \{t, t_i : 1 \le i \le 10\}$.
- $T$ is a natural number.
- The set $\mathcal{R}_E$ of environment rules tries to select the weather conditions for the year, and to distribute this information to all environments. This is done because there

are some biological parameters that vary depending on weather conditions (we are able to simulate the snow thickness).

$$r_{e_1} \equiv (t)_{e_1} \xrightarrow{1/10} (t_i)_{e_1} (t_i)_{e_2} (t_i)_{e_3} (t_i)_{e_4}, \quad 1 \leq i \leq 10$$
$$r_{e_2} \equiv (t)_{e_\nu} \rightarrow (\#)_{e_\nu}, \quad 1 < \nu \leq 4$$

- For each $\nu$, $1 \leq \nu \leq 4$, $\Pi_\nu = (\Gamma, \mu, \mathcal{M}_{0,\nu}, \mathcal{M}_{1,\nu}, \ldots, \mathcal{M}_{10,\nu}, \mathcal{R})$ is a P system of degree 11 whose membrane structure is $\mu = [ \, [ \, ]_1 [ \, ]_2 \cdots [ \, ]_{10} \, ]_0$.

The set $R$ of rules of $\Pi_\nu$ is the following (where the probabilistic constants associated with the rules have been incorporated):

*\* Preparation of the system to start a cycle.*
$r_1 \equiv t_i [ \, ]_0^0 \rightarrow [t_i]_0^0, \ 1 \leq i \leq 10.$
$r_2 \equiv t_i [ \, ]_i^0 \rightarrow [t]_i^-, \ 1 \leq i \leq 10.$

After applying the environment rules, $r_{e_1}$ the object $t_i$ enters from the environment carrying the information about the climatic condition of the next year to be simulated.

Each of the inner membranes labeled with $1, 2, \ldots, 10$, stores information on biological parameters for each one of the ten different climatic scenarios that the model envisages. The objects associated with animals should then enter the same membrane as the object $t_i$.

$$r_3 \equiv X_{j,y} [ \, ]_k^- \rightarrow [X_{j,y}]_k^0, \quad \begin{cases} 1 \leq j \leq g_3, \\ 1 \leq y \leq T, \\ 1 \leq k \leq 10. \end{cases}$$

Each geographic area in which the species lives has a monthly production of food (grass).

$$r_4 \equiv (F [ \, ]_k^- \rightarrow [G_4^{\alpha_4(\nu)}, \ldots, G_{10}^{\alpha_{10}(\nu)}]_k^0)_{e_\nu}, \quad \begin{cases} 1 \leq k \leq 10, \\ 1 \leq \nu \leq 4. \end{cases}$$

where $\alpha_j(\nu)$ corresponds to the amount of grass produced in the area $\nu$ for the month $j$.

When the pestivirus appears in an area, object $h$ is produced. It will be present in the following configurations.
$r_5 \equiv h [ \, ]_k^- \rightarrow [h]_k^0, \ 1 \leq k \leq 10.$

The amount of animals in the ecosystem is controlled by the objects $a$ because it can not exceed a maximum load
$$r_6 \equiv (c [ \, ]_k^- \rightarrow [a^{0.9d1_\nu} e^{0.2d1_\nu}]_k^0)_{e_\nu}, \quad \begin{cases} 1 \leq k \leq 10, \\ 1 \leq \nu \leq 4. \end{cases}$$

The following rules simulate the presence or absence of disease.

$r_7 \equiv d [ \, ]_k^- \rightarrow [d]_k^0, \ 1 \leq k \leq 10.$

$r_8 \equiv [d \, h \rightarrow d_1]_k^0, \ 1 \leq k \leq 10.$

$$r_9 \equiv ([d_1 \xrightarrow{ms_\nu} S]_k^0)_{e_\nu}, \quad \begin{cases} 1 \leq k \leq 10, \\ 1 \leq \nu \leq 4. \end{cases}$$

$$r_{10} \equiv \left([d_1 \xrightarrow{1-ms_\nu} N]_k^0\right)_{e_\nu}, \quad \begin{cases} 1 \le k \le 10, \\ 1 \le \nu \le 4. \end{cases}$$

Then we have counter $R_i$ that will allow us to synchronize the P system

$$r_{11} \equiv R_0[\ ]_k^- \to [R_0]_k^0, \ 1 \le k \le 10.$$

$$r_{12} \equiv [R_i \to R_{i+1}]_k^0, \quad \begin{cases} 0 \le i \le 4, \\ 1 \le k \le 10. \end{cases}$$

Finally, we introduce some randomness in the density control

$$r_{13} \equiv [e \xrightarrow{0.5} a]_k^0, \ 1 \le k \le 10.$$

$$r_{14} \equiv [e \xrightarrow{0.5} \#]_k^0, \ 1 \le k \le 10.$$

* *Reproduction rules*

Males of childbearing age

$$r_{15} \equiv \lfloor X_{j,y} \xrightarrow{1-k_1} Y_{j,y} D]_k^0, \quad \begin{cases} g_1 \le j < g_2, \\ 1 \le y \le T, \\ 1 \le k \le 10. \end{cases}$$

Females of childbearing age that reproduce

$$r_{16} \equiv [X_{j,y} \xrightarrow{k_1 \cdot k_2_l} Y_{j,y} Y_{0,y}^{k_3} D^{k_3+1}]_k^0, \quad \begin{cases} g_1 \le j < g_2, \\ 1 \le y \le T, \\ 1 \le k \le 10. \end{cases}$$

Females of childbearing age that do not reproduce

$$r_{17} \equiv [X_{j,y} \xrightarrow{k_1 \cdot (1-k_2_l)} Y_{j,y} D]_k^0, \quad \begin{cases} g_1 \le j < g_2, \\ 1 \le y \le T, \\ 1 \le k \le 10. \end{cases}$$

Animals that are not fertile

$$r_{18} \equiv [X_{j,y} \to Y_{j,y} D]_k^0, \quad \begin{cases} g_2 \le j \le g_3, \\ 1 \le y \le T, \\ 1 \le k \le 10. \end{cases}$$

Young animals that do not reproduce

$$r_{19} \equiv [X_{j,y} \to Y_{j,y} D]_k^0, \quad \begin{cases} 1 \le j < g_2, \\ 1 \le y \le T, \\ 1 \le k \le 10. \end{cases}$$

* *Density rules*

Checking if the maximum density has been reached

$$r_{20} \equiv \left([D^{d1_\nu} a^{d1_\nu - d2_\nu}]_k^0 \to [h_0]]_k^0\right)_{e_\nu}, \quad \begin{cases} 1 \le k \le 10, \\ 1 \le \nu \le 4. \end{cases}$$

$$r_{21} \equiv [d\, h_0]_k^0 \rightarrow [d_0]_k^0, \ 1 \leq k \leq 10.$$

Transformation of objects that represent animals

$$r_{22} \equiv [Y_{j,y} \rightarrow Y'_{j,y}]_k^0, \ \begin{cases} 0 \leq j \leq g_3, \\ 1 \leq y \leq T, \\ 1 \leq k \leq 10 \end{cases}$$

\* *Feeding rules*

$$r_{23} \equiv [Y'_{j,y}\, a\, G_4^{\beta_4}\, G_5^{\beta_5}\, G_6^{\beta_6}\, G_7^{\beta_7}\, G_8^{\beta_8}\, G_9^{\beta_9}\, G_{10}^{\beta_{10}} \rightarrow Z_{j,y}]_k^0, \ \begin{cases} 0 \leq j \leq g_3, \\ 1 \leq y \leq T, \\ 1 \leq k \leq 10. \end{cases}$$

where $\beta_i$ represents the need of food in month $i$.

\* *Natural mortality rules*

Young animals that survive

$$r_{24} \equiv \left([Z_{j,y} \xrightarrow{1-m1_{k,\nu}} V_{j,y}]_k^0\right)_{e_\nu}, \ \begin{cases} 0 \leq j < g_0, \\ 1 \leq y \leq T, \\ 1 \leq k \leq 10, \\ 1 \leq \nu \leq 4. \end{cases}$$

Young animals that leave the ecosystem or die

$$r_{25} \equiv \left([Z_{j,y} \xrightarrow{m1_{k,\nu}} \#]_k^0\right)_{e_\nu}, \ \begin{cases} 0 \leq j < g_0, \\ 1 \leq y \leq T, \\ 1 \leq k \leq 10 \\ 1 \leq \nu \leq 4. \end{cases}$$

Adult animals that survive

$$r_{26} \equiv [Z_{j,y} \xrightarrow{1-m2} V_{j,y}]_k^0, \ \begin{cases} g_0 \leq j < g_3, \\ 1 \leq y \leq T, \\ 1 \leq k \leq 10 \end{cases}$$

Adult animals that die

$$r_{27} \equiv [Z_{j,y} \xrightarrow{m2} \#]_k^0, \ \begin{cases} g_0 \leq j < g_3, \\ 1 \leq y \leq T, \\ 1 \leq k \leq 10. \end{cases}$$

Animals that reach the maximum age of the species

$$r_{28} \equiv [Y_{g_3,y} \rightarrow \#]_k^0, \ \begin{cases} 1 \leq y \leq T, \\ 1 \leq k \leq 10. \end{cases}$$

\* *Hunting mortality*

Young animals that survive hunting

$$r_{29} \equiv \left([V_{j,y} \xrightarrow{1-h1_\nu} W_{j,y}]_k^0\right)_{e_\nu}, \ \begin{cases} 0 \leq j < g_0, \\ 1 \leq y \leq T, \\ 1 \leq k \leq 10, \\ 1 \leq \nu \leq 4. \end{cases}$$

Young animals that are hunted

$$r_{30} \equiv \left([V_{j,y} \xrightarrow{h1_\nu} \#]_k^0\right)_{e_\nu}, \quad \begin{cases} 0 \le j < g_0, \\ 1 \le y \le T, \\ 1 \le k \le 10, \\ 1 \le \nu \le 4. \end{cases}$$

Adult animals that survive hunting

$$r_{31} \equiv \left([V_{j,y} \xrightarrow{1-h2_\nu} W_{j,y}]_k^0\right)_{e_\nu}, \quad \begin{cases} g_0 \le j < g_3, \\ 1 \le y \le T, \\ 1 \le k \le 10, \\ 1 \le \nu \le 4. \end{cases}$$

Adult animals that are hunted

$$r_{32} \equiv \left([V_{j,y} \xrightarrow{h2_\nu} \#]_k^0\right)_{e_\nu}, \quad \begin{cases} g_0 \le j < g_3, \\ 1 \le y \le T, \\ 1 \le k \le 10, \\ 1 \le \nu \le 4. \end{cases}$$

\* *Disease mortality*

$r_{33} \equiv [R_5 S]_k^0 \to [R_6 \, h]_k^-, \; 1 \le k \le 10.$

$r_{34} \equiv [R_5 N \to R_6 \, h]_k^0, \; 1 \le k \le 10.$

$r_{35} \equiv [R_5 d_0 \to R_6 \, h]_k^0, \; 1 \le k \le 10.$

$r_{36} \equiv [R_5 d \to R_6]_k^0, \; 1 \le k \le 10.$

$r_{37} \equiv [R_6]_k^- \to [\#]_k^+, \; 1 \le k \le 10.$

$r_{38} \equiv [R_6]_k^0 \to [\#]_k^+, \; 1 \le k \le 10.$

$$r_{39} \equiv \left([W_{j,y}]_k^- \xrightarrow{md_\nu} [\#]_k^+\right)_{e_\nu}, \quad \begin{cases} 0 \le j < g_3, \\ 1 \le y \le T, \\ 1 \le k \le 10, \\ 1 \le \nu \le 4. \end{cases}$$

$$r_{40} \equiv \left([W_{j,y}]_k^- \xrightarrow{1-md_\nu} [W_{j,y}]_k^+\right)_{e_\nu}, \quad \begin{cases} 0 \le j < g_3, \\ 1 \le y \le T, \\ 1 \le k \le 10, \\ 1 \le \nu \le 4. \end{cases}$$

\* *Updating rules*

$$r_{41} \equiv [W_{j,y}]_k^+ \to X_{j+1,y+1}[\;]_k^0, \quad \begin{cases} 0 \le j < g_3, \\ 1 \le y \le T, \\ 1 \le k \le 10. \end{cases}$$

$$r_{42} \equiv [Y'_{j,y}]_k^+ \to [\#]_k^0, \quad \begin{cases} 0 \le j < g_3, \\ 1 \le y \le T, \\ 1 \le k \le 10. \end{cases}$$

$$r_{43} \equiv [t]_k^+ \rightarrow R_0, F, t, c, d[\ ]_k^0, \ 1 \leq k \leq 10.$$

$$r_{44} \equiv [h]_k^+ \rightarrow h\ [\ ]_k^0, \ 1 \leq k \leq 10.$$

$$r_{45} \equiv [a]_k^+ \rightarrow [\#]_k^0, \ 1 \leq k \leq 10.$$

$$r_{46} \equiv [G_i]_l^+ \rightarrow [\#]_k^0, \ \begin{cases} 4 \leq i \leq 10, \\ 1 \leq k \leq 10. \end{cases}$$

$$r_{47} \equiv [t]_0^0 \rightarrow t[\ ]_0^0$$

- The initial multisets $\mathcal{M}_{0,\nu}, \ \mathcal{M}_{1,\nu}, \ldots \mathcal{M}_{10,\nu}$ of the P system associated with the environment $\nu$ $(1 \leq \nu \leq 4)$ are the following:

$$\mathcal{M}_{0,\nu} = \{F, R_0, c, d\} \cup \{X_{j,1}^{q_{\nu,j}} : 1 \leq \nu \leq 4, \ 1 \leq j \leq g_3\}, \ \text{for } 1 \leq \nu \leq 4.$$
$$\mathcal{M}_{k,\nu} = \emptyset, \ \text{for } 1 \leq k \leq 10 \text{ and } 1 \leq \nu \leq 4.$$

### 7.4.2.2 Results

Now, in order to experimentally validate the model, a MeCoSim based custom simulator for the designed PDP model is generated. For that, the designer uses a spreadsheet to edit a MeCoSim configuration by introducing the following information:

- *General Data*: The designer user introduces an application identifier and an application name to define the ecosystem under study, the number of years to simulate, the total number of simulations to generate and the computational steps per year. He also introduces the mode to use in the program (designer or end-user).
- *Tabs Hierarchy*: The designer establishes the desired structure of tabs to contain the input and output views of the system to simulate (Fig. 7.22).
- *Input tables*: The designer lists the tables to include in the application to contain the data of the ecosystem, specifying the identifiers of the tabs in which the tables will be put, the number of columns, the initial rows and an indicative meaning if the content of the table must be saved into the data file (Fig. 7.23).

**Fig. 7.22** Tabs hierarchy

| Tab Id | Tab Name | Parent Tab Id |
|---|---|---|
| 1 | Pyrinean Chamois | 0 |
| 2 | Input | 1 |
| 3 | Snow | 2 |
| 4 | Population | 2 |
| 5 | Max density population | 2 |
| 6 | Biological parameters | 2 |
| 7 | Constants | 6 |
| 8 | Variables | 6 |
| 9 | Antropical parameters | 2 |
| 10 | Disease parameters | 2 |
| 11 | Values | 3 |
| 12 | Ranges | 3 |
| 13 | Grass | 2 |
| 14 | Output | 1 |

| Table Id | Table Name | Tab Id | Columns | Init Rows | Save To File |
|---|---|---|---|---|---|
| 1 | Values | 11 | 3 | 8 | TRUE |
| 2 | Population | 4 | 22 | 8 | TRUE |
| 3 | Max density population | 5 | 4 | 8 | TRUE |
| 4 | Constants | 7 | 9 | 3 | TRUE |
| 5 | Variables | 8 | 4 | 3 | TRUE |
| 6 | Antropical parameters | 9 | 5 | 8 | TRUE |
| 7 | Disease parameters | 10 | 4 | 8 | TRUE |
| 8 | Ranges | 12 | 2 | 1 | TRUE |
| 9 | Grass | 13 | 4 | 56 | TRUE |

**Fig. 7.23**  Input tables

| Column Id | Column Name | Default Value | Editable | Tooltip |
|---|---|---|---|---|
| 1 | Snow thickness | | TRUE | Snow thickness |
| 2 | Specie | | TRUE | Specie |
| 3 | Age adults | | TRUE | Age adults |
| 4 | Age start fertile | | TRUE | Age start fertile |
| 5 | Age leave fertile | | TRUE | Age leave fertile |
| 6 | Life spectancy | | TRUE | Life spectancy |
| 7 | Proportion females | | TRUE | Proportion females |
| 8 | Fertility ratio | | TRUE | Fertility ratio |
| 9 | Number of descendents | | TRUE | Number of descendents |

**Fig. 7.24**  Table columns configuration

- *Table columns configuration*: For each table, there must be listed the set of columns, including an identifier, a name, a tool-tip to show when the user passes the mouse over each column of the table and a boolean value indicating if the data of the column is editable or not (Fig. 7.24)
- *P-Lingua parameters*: In the parameters tab, the designer lists the sets of parameters to use in the simulation of the model, with their name, value and up to 4 indexes for each parameter to produce final parameters to serve as part of the input of the simulation (Fig. 7.25)

When the designer user of the P system has introduced the required information in the configuration file, it can be loaded into the general purpose application, generating the custom simulator that matches the specific necessities of the designer and end-user. The custom simulator will show the input tables which permit the end-user to introduce the data for the development of virtual experiments as shown in Fig. 7.26.

From the input data and the parameters introduced in the configuration file, the custom application will simulate the virtual experiments and will show the outputs of the simulation.

There are experimental data available from 1988, although censuses where not carried out annually so that experimental series is not a continuous one. Using the censuses in 1988 as input for the model, 22 years have been simulated repeating the process 50 times for each of the years simulated. Figure 7.27 shows the results

| Param Name | Param Value | Index 1 | Index 2 | Index 3 | Index 4 |
|---|---|---|---|---|---|
| medw | <1,$1$,2> | [1..8] | | | |
| desw | <1,$1$,3> | [1..8] | | | |
| N1 | <8,1,1> | | | | |
| N2 | <8,1,2> | | | | |
| p1 | <@ncdf,medw{1},desw{1},N1> | | | | |
| p3 | 1-<@ncdf,medw{1},desw{1},N2> | | | | |
| p2 | 1-p1-p3 | | | | |
| q | <2,$1$,$2$+2> | [1..<@r,2>] | [1..20] | | |
| d1 | <3,$1$,3> | [1..8] | | | |
| d2 | <3,$1$,4> | [1..8] | | | |
| g3 | <4,1,3> | | | | |
| g4 | <4,1,4> | | | | |
| g5 | <4,1,5> | | | | |
| g6 | <4,1,6> | | | | |
| k1 | <4,1,7> | | | | |
| k2 | <4,$1$,8> | [1..3] | | | |
| k3 | <4,1,9> | | | | |
| m1 | <5,$1$,3> | [1..3] | | | |
| m2 | <5,$1$,4> | [1..3] | | | |
| m3 | <6,$1$,3> | [1..8] | | | |
| h1 | <6,$1$,4> | [1..8] | | | |
| h2 | <6,$1$,5> | [1..8] | | | |
| ms | <7,$1$,3> | [1..8] | | | |
| md | <7,$1$,4> | [1..8] | | | |
| f | <9,($1$-1)*7+$2$-3,3> | [1..8] | [4..10] | | |
| b | <9,$1$,4> | [4..10] | | | |

**Fig. 7.25**  P-Lingua parameters

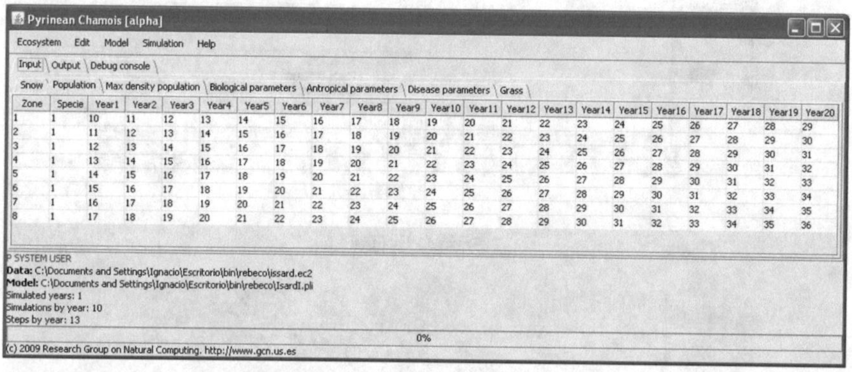

**Fig. 7.26**  Edition process for the initial parameters of the model

and more specifically, the continuous line represents the average value of the 50 simulations whereas the broken lines correspond to 95% interval and the dots are the values obtained experimentally. In both Alt Pallars-Aran and Cerdanya-Alt Urgell areas, animals have suffered from pestivirus infections whereas in Freser-Setcases, the population dynamics have not suffered the disease caused by this infection.

In general, the model behaves well in all cases. The model considers the main processes and dynamics of the species although some of them have been omitted because they are considered to be less important. This may explain the differences between the values obtained with the model and experimental ones. Among these factors, we should highlight the influence of domestical animals living in the area on the spread of pestivirus infection. In addition, there are few data regarding the thickness of the snow layer and those used in the model have been obtained from

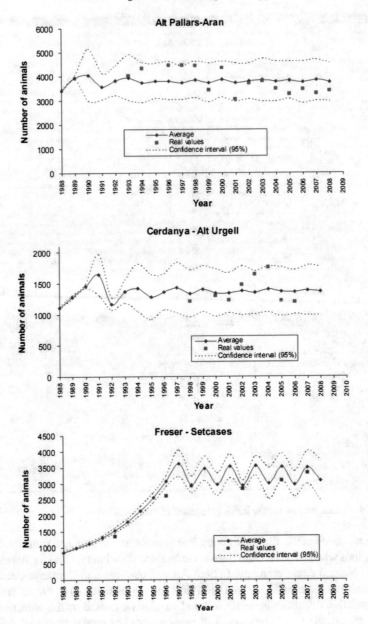

**Fig. 7.27** Results

ski resorts so that they may be overdimensioned and then, may affect the results significantly. It would be interesting to study the relationship between the thickness of the snow layer and other available climatic data in the area such as temperature and the length of the winter interval.

### 7.4.3 Zebra Mussel

Zebra mussel is an exotic invasive species causing an important damage from an environmental and economical point of view, wherever it settles. Its presence in Spain was first detected in 2001, in the reservoir of Riba-roja, managed by the company Endesa S.A.

Zebra mussel are considered as true engineers of the ecosystem, due to the significant ecological change they produce when invading an aquatic ecosystem, altering its structure and functioning [55], including changes in the water composition and increase in operating and maintenance costs, obstructing drains and pipes [52, 60].

Numerous studies have been focused on the biology of the species, its geographical dispersion and its influence on the invaded environment, including essays of methods to control or eradicate the species [42, 44, 74, 103, 104, 109, 110, 120].

In this section, the study on modeling and simulation of the ecosystem of Zebra mussel in the reservoir of Riba-roja is presented. In Sect. 7.4.3.1, the main facts related with the species, the environment and the processes included in our study are described. A summary of the more relevant aspects of the model is provided in Sect. 7.4.3.2. Finally, a software application using the model to manage the ecosystem is shown in Sect. 7.4.3.3.

Further explanation about this model is given in [18], and more technical details and files for simulating its evolution by means of the proposed software tools is available at [123].

#### 7.4.3.1 Biology of the Species

Zebra mussel (*Dreissena polymorpha*) is an aquatic bivalve mollusc presenting a fast life cycle with a huge reproductive potential, a high mobility of larvae and a great dispersal capacity [1, 13, 61]. It presents two main stages: a larval phase, suspended in the water column (planktonic state), and a latter phase, settled to the substrate (benthonic state).

Zebra mussel is a dioecious species with a proportion of genders around 1:1. The number of eggs emitted by female individual per reproductive cycle depends on its size, influenced by its age. In [18, 121] the discussion and expressions to calculate the number of eggs are detailed, referencing the corresponding sources.

Larvae remain in the water column around 3–5 weeks [13]. Some of them are depredated by the adults themselves. The mortality rate from the spawning period to the beginning of juvenile phase can reach up to 98% [59]. After this larval stage, they fall over the substrate by the effect of gravity. According to [108], the recruitment of young mussels depends on the density of adult mussels [18].

Two reproductive cycles are observed in the reservoir along the year: from March to June, and from October to November. The individuals produced in the first cycle could reach sexual maturity before the start of the second one. The production of larvae is bigger when the temperature raises 15–17 °C [52]. Given the size of the reservoir, the different temperatures arising and the movements among different zones, mussels in different stages of development coincide in the same area of the reservoir.

Zebra mussel can form aggregates of individuals, forming different layers. Maximum densities above 100.000 individuals per $m^2$ has been observed [73], reaching up to 250.000 in the reservoir of Riba-roja [82].

The reservoir is placed in the Ebro river basin, in the Northeast of Spain. With a length of 35 km and a variable depth (up to 28 m), it receives water from Ebro river, the reservoir of Mequinenza, and rivers Segre and Matarraña. Different thermal conditions and features of the substrates are considered.

The river is longitudinally structured in 9 different zones, each of them divided into 2 parts depending on its depth (U, upper y L, lower), summarizing 17 areas (see Fig. 7.28). This division is relevant, due to the different development rates of mussels and different intensity depending on such environmental factors.

In order to perform the study of the population dynamics of Zebra mussel in the reservoir of Riba-roja, the following factors are considered:

1. The basic biological processes involved in the species life cycle.
2. The special features related to the habitat, an artificial reservoir with water currents and changes in the renewal process over them, managed for the generation of hydroelectric power, being affected by the features of the incoming water.
3. The possibility for external larvae to enter from reservoirs and close tributaries, and the transference of individuals from external rivers to the reservoir through the traffic of vessels.

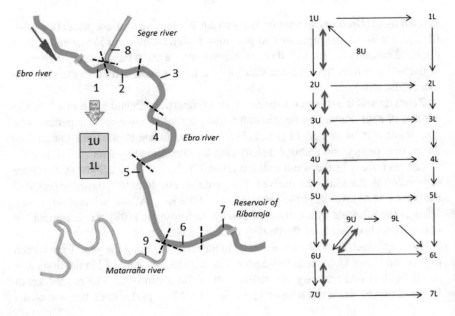

**Fig. 7.28** Division in zones in the reservoir

**Table 7.5** Percentage of mussels starting reproduction per week, given the thermal conditions

| Week | 1 | 2 | 3 | 4 | 5 | 6 | 7 | 8 | 9 | 10 | 11 |
|---|---|---|---|---|---|---|---|---|---|---|---|
| Cycle 1 (%) | 4 | 4 | 7 | 15 | 20 | 20 | 15 | 10 | 3 | 1 | 1 |
| Cycle 2 (%) | 80 | 15 | 5 | | | | | | | | |

It is worth noting the influence of the water thermal conditions. They affect the probability for an individual to reproduce, as shown in Table 7.5. In addition, the nature and composition of the substrate, specially its toughness and porosity, will influence drastically the density of mussels and mortality rate of larvae.

Another decisive factor is the effect of human actions over the water flow, affecting the movement of larvae among different zones or outside the environment.

In [17] a standardized protocol was presented for computational modeling based on P systems, briefly presented below.

### Stage 1: Purpose

The modeling of the population of Zebra mussel in the reservoir of Riba-roja and the development of tools to perform virtual experiments assisting managers in the decision making process.

### Stage 2: Processes

The main processes to model are as follows:

- *Biology of the species*: eggs release, larval phase, settlement to the substrate, mortality depending on state and depredation.
- *Environmental processes*: generation of thermal conditions, features of the substrate per zone and control of carrying capacity depending on its soundness.
- *Human intervention*: displacement of larvae among zones due to the water management, and of adults due to the movement of vessels.

### Stage 3: Input of the model and parameters to consider

The input of the model is the size of the initial population, along with some parameters regarding the biology of the species, the environment and the possible human interventions.

### Stage 4: Sequencing and parallelization of processes

In order to have a bigger control over the model, the processes to be represented are sequenced. In this case, a partial sequencing of the process is proposed, as shown in Fig. 7.29.

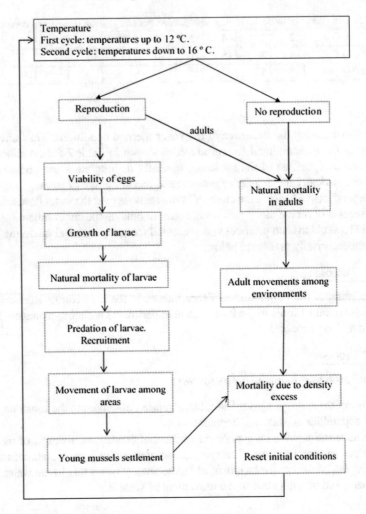

**Fig. 7.29** Model blocks sequencing

The presence of young individuals reaching sexual maturity later than adults increases the difficulty of a sequencing of processes. This way, processes are said to be unsynchronized.

A natural year involves the execution of two iterations of the basic cycle, corresponding to the two phases that take place in the two reproductive cycles of Zebra mussel along the year.

## Stage 5: The model

The design of the model should reflect all the processes and input elements commented in the previous stages descriptions, by using a Population Dynamics P system.

## Stage 6: Graphical analysis of a cycle in the execution of a model

This stage implies the depiction of an execution trace of a cycle of the model, analyzing the evolution of the objects in the system along the cycle to deeply analyze the correction of the model and its accuracy with respect to the initial purpose addressed. See [121] for further details.

## Stage 7: Design of the simulator

The application of this stage to our model by using P-Lingua and by MeCoSim is described in Sect. 7.4.3.3.

### 7.4.3.2  Model

The details about the notations used in the model are described in [121]. A number of parameters for the model are defined, concerning:

- Environmental conditions and influence over the species, as $PI_{s,c}$ (proportion of larvae released during week $s$ of reproductive cycle $c$) or $p_{s,j}$ (probability for the reproduction to get enabled for individuals in compartment $j$ during week $s$), influenced by $IT_c$ (temperature for starting reproductive cycle $c$).
- Biological parameters, as $g_1$ (percentage of eggs released in the first cycle).
- Properties of the compartment, as $\varphi_j$ (capacity of compartment $j$, calculated from the proportion of each type of soil present in the compartment and its capacity $C_j$ in cubic meters).
- Human intervention, as $PR_{s,j,j'}$ (probability for larvae to be displaced during week $s$, from compartment $j$ to compartment $j'$ due to the hydraulic regime for water renewal).

Many objects appear in this model, including objects representing individuals, as $X_{sem}$ (adult mussels aged $sem$ in semesters), $Q_s$ (young mussels aged $s$ weeks), $O_s$ (eggs produced of reproductive cycle), or $L_{is}$ for larvae in different stages, and $T_s$ (auxiliary object to trigger the process of pseudo-random generation of temperatures during the weeks of each reproductive cycle), among others.

We consider a PDP system of degree (40, 18) taking $T \geq 1$ time units defined as follows:

$$\Pi = (G, \Gamma, \Sigma, T, \mathcal{R}_E, \mu, \mathcal{R}, \{f_{r,j} \mid r \in \mathcal{R} \land 1 \leq j \leq 18\} \cup$$
$$\{\mathcal{M}_{i,j} \mid 0 \leq i \leq 39 \land 1 \leq j \leq 18\} \cup \{E_j \mid 1 \leq j \leq 40\})$$

where:

- $G = (V, S)$ is the directed graph containing nodes $V = \{e_1, \ldots, e_{18}\}$. The movements among environments are shown in [121].
- The working alphabet, $\Gamma$, and the alphabets of the environment, $\Sigma$, are also detailed in [18, 121].
- $\mu = [\ [\ ]_1 \ldots [\ ]_{39}\ ]_0$ is the membrane structure, and the initial multisets for the environment $j$ include the $X_sem$ with multiplicity $q_{j,sem}$ inside environment $j$, and $Q_d$ with multiplicity $q_{j,d}$.
- $T = 51 \cdot 2 \cdot Years$. Each year is simulated by 2 cycles of 51 steps.
- $E_1 = \cdots = E_{18} = \{T_0, I_1\}$
- The set of rules in the system is detailed in [18, 121]. Let us show some significant rules illustrating a very relevant process: the movement of larvae, due to the hydraulic regime.

$$r_{e_{13}} \equiv (L_{m,i,j,j_1} \xrightarrow{PR(m,j_1,j_2)} L_{m+1,i+1,j,j_2})e_j \begin{cases} 1 \le j \le 18 \\ 1 \le m \le 42 \\ 0 \le i \le 3 \\ 1 \le j_1 \le 7 \\ 1 \le j_2 \le 7 \end{cases}$$

$$r_{e_{15}} \equiv (L_{m,i,j,j_1} \xrightarrow{PR(m,j_1,j_2)} L_{m+1,i+1,j,j_2})e_j \begin{cases} 1 \le j \le 18 \\ 1 \le m \le 42 \\ 0 \le i \le 3 \\ 8 \le j_1 \le 14 \\ 8 \le j_2 \le 14 \end{cases}$$

These rules capture the effect of the water renewal cycle, enabled by the managers of the reservoir, causing the movement of larvae among different compartments. Specifically, these rules represent the possible movement of a larvae (released during week $m$ and suspended in the water column water for $i + 1$ weeks) from a compartment $j_1$, to another compartment $j_2$, in central areas of the reservoir. Rule $r_{e_{13}}$ represents this movement for the upper compartments, being $r_{e_{15}}$ devoted to the lower ones. These movements depend on probabilities $PR(m, j_1, j_2)$; from the input data introduced by the managers about the amount of water entering the reservoir in week $m$, an algorithm calculates these probabilities taking into account the proportion of water being displaced to each compartment due to the incoming water. The overall process involves more rules representing vertical fall and movements to side areas, as described in [18].

The model presents the modules shown in Fig. 7.29. The simulation of a natural year implies the running of two cycles of the model.

### 7.4.3.3   The Model as a Tool for the Management of the Ecosystem

The model presented in the previous section captures many of the processes affecting the population dynamics of Zebra mussel in the reservoir of Riba-roja, including natural and human elements in the environment influencing its evolution.

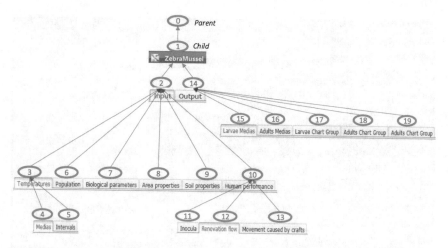

**Fig. 7.30** MeCoSim app - tabs tree structure

In what follows let us outline the needed steps to convert the model of Zebra mussel in an application to serve the managers of the reservoir as a tool to assist in decision making processes, allowing him to analyze possible consequences and changes in the population dynamics of the species and the ecosystem, depending on their possible actions to be performed over the system. Further details of the process are provided in [121], and the corresponding files are available at [123].

- Specification of the model in a language understandable by the machine, through the writing of a P-Lingua file.
- Configuration of an application based on MeCoSim adapted to the system under study (customization). The structure of the resulting application will be shown in Fig. 7.30.
- Debugging the model, detecting possible errors, initializing and analyzing the evolution step by step.
- Simulating the model to perform experimental validation.
  After validated with the experts, the application will be ready to assist managers allowing virtual experiments to analyze potential scenarios. For instance, it would be interesting for the managers of the reservoir to know which could happen in case of an external inoculation of larvae produced, an input of mussels due to movement of vessels happened, or a more constant water renewal cycle applied due to a long period of heavy rains.
  For each scenario, the input data must be entered by the managers in the input tables, to set initial populations, environmental conditions (as the area properties, Fig. 7.31) and human actions.
  Once the simulation has finished, the results from the simulation will be shown. Some output tables will provide quantitative data about certain elements of the system, in order to analyze the numbers, as in Fig. 7.32.

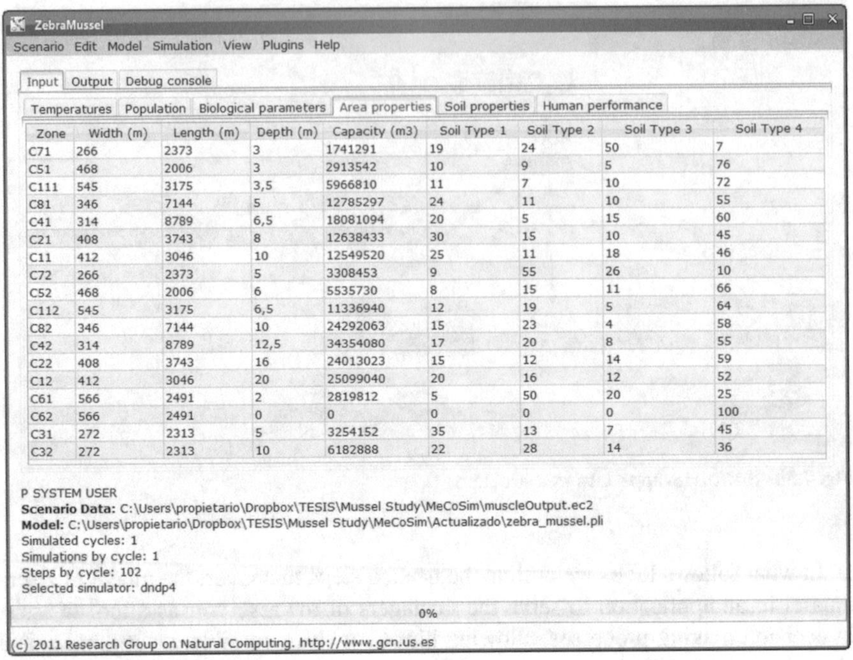

ZebraMussel

Scenario  Edit  Model  Simulation  View  Plugins  Help

Input | Output | Debug console

Temperatures | Population | Biological parameters | Area properties | Soil properties | Human performance

| Zone | Width (m) | Length (m) | Depth (m) | Capacity (m3) | Soil Type 1 | Soil Type 2 | Soil Type 3 | Soil Type 4 |
|---|---|---|---|---|---|---|---|---|
| C71 | 266 | 2373 | 3 | 1741291 | 19 | 24 | 50 | 7 |
| C51 | 468 | 2006 | 3 | 2913542 | 10 | 9 | 5 | 76 |
| C111 | 545 | 3175 | 3,5 | 5966810 | 11 | 7 | 10 | 72 |
| C81 | 346 | 7144 | 5 | 12785297 | 24 | 11 | 10 | 55 |
| C41 | 314 | 8789 | 6,5 | 18081094 | 20 | 5 | 15 | 60 |
| C21 | 408 | 3743 | 8 | 12638433 | 30 | 15 | 10 | 45 |
| C11 | 412 | 3046 | 10 | 12549520 | 25 | 11 | 18 | 46 |
| C72 | 266 | 2373 | 5 | 3308453 | 9 | 55 | 26 | 10 |
| C52 | 468 | 2006 | 6 | 5535730 | 8 | 15 | 11 | 66 |
| C112 | 545 | 3175 | 6,5 | 11336940 | 12 | 19 | 5 | 64 |
| C82 | 346 | 7144 | 10 | 24292063 | 15 | 23 | 4 | 58 |
| C42 | 314 | 8789 | 12,5 | 34354080 | 17 | 20 | 8 | 55 |
| C22 | 408 | 3743 | 16 | 24013023 | 15 | 12 | 14 | 59 |
| C12 | 412 | 3046 | 20 | 25099040 | 20 | 16 | 12 | 52 |
| C61 | 566 | 2491 | 2 | 2819812 | 5 | 50 | 20 | 25 |
| C62 | 566 | 2491 | 0 | 0 | 0 | 0 | 0 | 100 |
| C31 | 272 | 2313 | 5 | 3254152 | 35 | 13 | 7 | 45 |
| C32 | 272 | 2313 | 10 | 6182888 | 22 | 28 | 14 | 36 |

P SYSTEM USER
**Scenario Data:** C:\Users\propietario\Dropbox\TESIS\Mussel Study\MeCoSim\muscleOutput.ec2
**Model:** C:\Users\propietario\Dropbox\TESIS\Mussel Study\MeCoSim\Actualizado\zebra_mussel.pli
Simulated cycles: 1
Simulations by cycle: 1
Steps by cycle: 102
Selected simulator: dndp4

0%

(c) 2011 Research Group on Natural Computing. http://www.gcn.us.es

**Fig. 7.31** MeCoSim app - area properties

Sometimes it will be easier to analyze the qualitative evolution of the species, instead of the exact number. For these cases, some type of graphical evolution by means of charts will be more informative. These charts will be also configurable for the custom simulation app, as in the example shown in Fig. 7.33, showing the evolution of adult mussels per compartment, or in the example shown in Fig. 7.34, allowing the study of the evolution of the amount of larvae in a certain zone of the reservoir.

### 7.4.4  Giant Panda

Now, we present a PDP system based model trying to provide an integrated view of the processes related to Giant panda individuals in the Giant Panda Breeding Base (GPBB for short) of Chengdu, enabling the study the population dynamics of the species under various conditions.

The purposes of a model of this kind can be categorized in the following way:

- Diagnosis: analyze and assess what happened, examining the causes and precursor conditions of some known present or past facts or events.

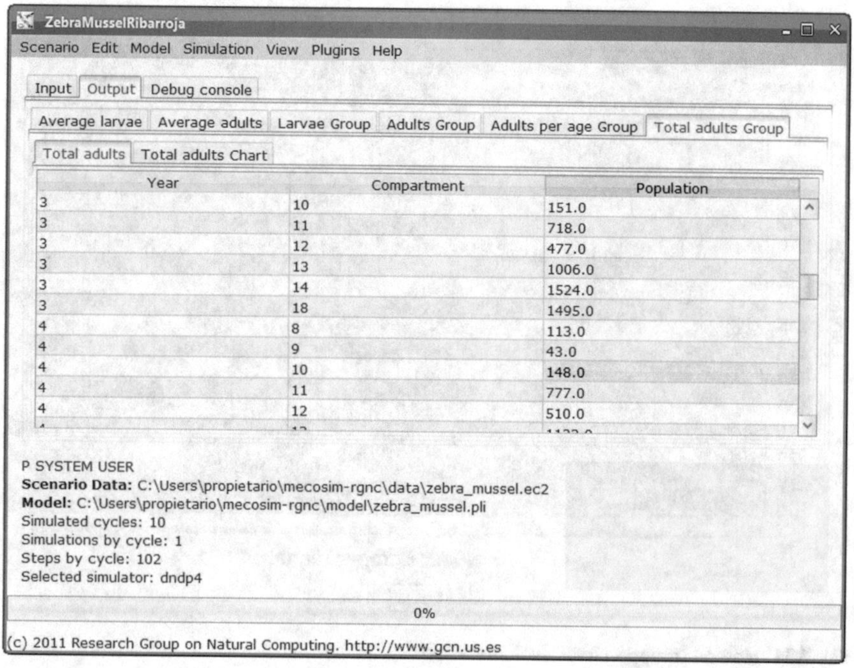

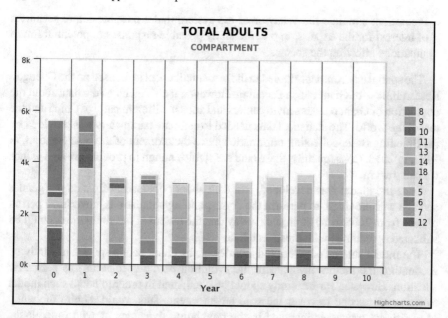

**Fig. 7.32** MeCoSim app - adults output

**Fig. 7.33** MeCoSim app - adults evolution chart

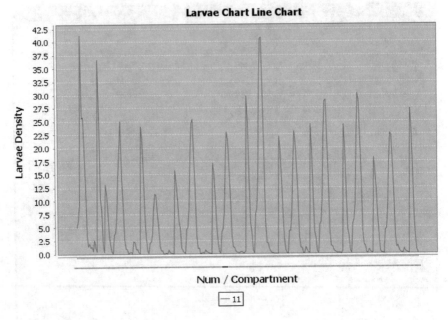

**Fig. 7.34** MeCoSim app - larvae evolution chart

- Prediction: simulate the behavior of the system under possible future scenarios of interest for the experts, in order to study possible responses to potential future situations affecting the species.

The spatial environment studied in the presented model is focused on the Chengdu Research Base of Giant Panda Breeding. However, it also includes the data about the population of Giant pandas sent to centers and zoos in different parts of China and the rest of the world. The statistical data needed to estimate parameters was provided by the Breeding Base, collecting information since the moment of the base creation, in 1987, to 2013. Consequently, these data are stable enough to provide average values we can trust in.

Once the model was developed, real data about population of panda per gender and age was provided, to perform the corresponding simulations, covering a period of time from 2005 to 2013, in such a way that the output of the simulations could be validated by contrast with these real data.

The model studies population dynamics of the Giant panda species in captivity, accounting the factors involved in the evolution of the species in this particular situation. However, further study should be conducted in order to adapt such model to expand its scope to cover the wild environment. This would require the study of additional processes involved in the population dynamics of wild individuals, maybe affected by different threats, feeding problems, competitors, diseases, lower life expectancy, etc.

The main processes/modules included in the present model are as follows:

- **Reproduction**: every year a number of new individuals are born in the Breeding Base and other centers coordinated from there. There exist historical data recording the number of new individuals per year. This number presents a strong variance, from 3 or 4 individuals per sex to a number close to 30. In addition, this number is not related to the number of total individuals in reproductive age, so an average number of males and females is taken as a reference. This fact is probably derived from the controlled nature of the studied system, since in general in wild environmental models for animals, the number of new births is correlated with natural factors as the fertility ratio, the number of individuals in fertile age, the probability to meet depending on the surface, etc. In a later model these or other factors could be considered. The natural growth of the individuals in the population is trivially modeled by increasing the age of the individuals when not affected by diseases or natural facts producing their death.
- **Mortality**: different factors influence the mortality of the individuals in the population, but in a captive environment these are mainly under control. However, as any species, Giant panda has natural factors conditioning its maximum life expectancy, and different mortality ratios depending on the age of the individual. Historical data has been provided, thus permitting statistical data analysis to get curated data for the model.
- **Feeding**: every Giant panda has some feeding needs along the year, mainly bamboo, bamboo shoots and other minor sources of food (i.e. apples, meat and milk). An average need of food is considered per individual, taking into account that these needs are different in different groups of age. The system should provide the necessary amount of food for the living individuals, which seems to be guaranteed in the captive environment in Chengdu Breeding Base and controlled zoos but could not be guaranteed in wild environment.
- **Rescue**: the total number of Giant pandas can evolve not only by the births and deaths of individuals, but can also be increased by bringing in new wild individuals which have been rescued. This number presents a big natural variability among different years, but based on a series of real data from Breeding Base, the average and variance in the total number of captured individuals has been obtained. In addition, there exists a historical proportion of captured individuals per gender, and the age of these captured Giant pandas is not random, but also follows a distribution. All these factors are considered in the model, along with the fact that the lifespan of wild individuals is significantly smaller than in captive individuals (20–25 instead of 34–36), so a proportional increase in the age of the captured individuals is simulated depending on the years the individual lived in the wild environment.

In what follows, the conceptual schema is depicted showing the main processes involved in the model, adding some needed initialization and update modules to the natural processes, in order to synchronize and prepare for the next cycle.

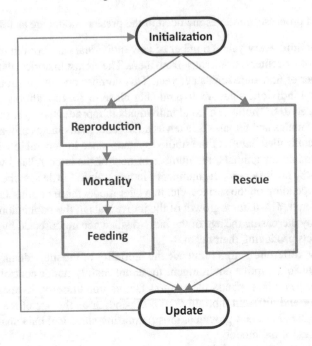

### 7.4.4.1   Model

We consider a PDP system of degree (2, 1)

$$\Pi = (G, \Gamma, \Sigma, T, \mathcal{R}_E, \mu, \mathcal{R}, \{f_{r,1} \mid r \in \mathcal{R}\}, \mathcal{M}_{1,1}, \mathcal{M}_{2,1}, E_1)$$

with one environment and $\Pi_1 = (\Gamma, \mu, \mathcal{M}_{1,1}, \mathcal{M}_{2,1}, \mathcal{R})$ is a P system of degree 2 with (only) two electrical charges (neutral and positive) taking T time units, defined as follows:

- $G$ is an empty graph.
- The working alphabet is
  $\Gamma = \{X_{i,j}, \; Y_{i,j}, \; Z_{i,j}, \; W_{i,j} \; : \; 1 \le i \le 2, \; 0 \le j \le k_{i,5}\} \; \cup$
  $\quad \{ C_{i,j} \; : \; 1 \le i \le 2, \; 0 \le j \le cmaxage\} \cup \{S, \; B, \; O, \; F, \; N, \; A\}$

Symbols $X_{i,j}$ represent individuals of age $j$ before reproduction module, $Y_{i,j}$ represent individuals of age $j$ within mortality module, $Z_{i,j}$ represent surviving individuals of age $j$, $W_{i,j}$ represent individuals of age $j$ after feeding module and $C_{i,j}$ represent rescued individuals of "age" $j$. Objects $S$, $B$, $O$ represents different types of food: bamboo shoots, bamboo, or other food, respectively. Objects $F$, $N$, $A$ are auxiliary symbols that represent the following: $F$ is an object used to generate new quantity of food at the beginning of each time cycle; $N$ is an object used to generate newborns when beginning each time cycle; and $A$ is an object used to trigger the income of rescued individuals.

- The alphabet of the environment is $\Sigma = \emptyset$.
- $T$ represents the number of years to be simulated in the evolution of the ecosystem.
- $\mu = [\ [\ ]_2\ ]_1$ is the membrane structure.
- $\mathcal{M}_{1,1}$ and $\mathcal{M}_{2,1}$ are finite multisets over $\Gamma$ describing the objects initially placed in regions of $\mu$;

    - $\mathcal{M}_{1,1} = \{X_{i,j}^{q_{i,j}} : 1 \le i \le 2,\ 1 \le j \le k_{i,5}\}$ ($q_{i,j}$ represents the initial number of Giant pandas of age $j$).
    - $\mathcal{M}_{2,1} = \{F\ N\ A\}$.

- The set $\mathcal{R}$ of evolution rules of the only P system of $\Pi$ consists of:

– *Initialization rule*

   ★ Generation of objects associated with the food:

   $$r_1 \equiv [\ F\ ]_2^0 \longrightarrow F\ [\ S^{g_3}\ B^{g_4}\ O^{g_5}\ ]_2^+,$$

– *Reproduction rules*

   ★ Rules associated with newborns:

   $$r_2 \equiv [\ N\ ]_2^0 \longrightarrow N\ [\ Y_{1,0}^{g_1}\ Y_{2,0}^{g_2}\ ]_2^+,$$

   ★ Growth rules:

   $$r_3 \equiv [\ X_{i,j}\ ]_2^0 \longrightarrow [\ Y_{i,j}\ ]_2^+,\ \text{for}\ \begin{cases} 1 \le i \le 2 \\ 1 \le j \le k_{i,5} \end{cases}$$

– *Rescued Giant pandas rules*

   ★ Probability to have $c$ rescued individuals:

   $$r_4 \equiv [\ A\ ]_2^0 \xrightarrow{pc_c} A\ C^c\ [\ ]_2^+,\ \text{for}\ cmin \le c \le cmax.$$

   ★ Probability for rescued individuals to have gender $i$:

   $$r_5 \equiv [\ C \xrightarrow{pg_i} C_i\ ]_1^0,\ \text{for}\ 1 \le i \le 2$$

   ★ Probability for rescued individuals to have age $j$:

   $$r_6 \equiv [\ C_i \xrightarrow{pa_j} C_{i,j+1+\lfloor \frac{i}{3} \rfloor}\ ]_1^0,\ \text{for}\ \begin{cases} 1 \le i \le 2 \\ 0 \le j < cmaxage \end{cases}$$

– *Mortality rules*

   ★ Infancy Giant pandas that survive:

$$r_7 \equiv [\, Y_{i,j} \xrightarrow{1-k_{i,6}} Z_{i,j} \,]_2^+, \text{ for } \begin{cases} 1 \le i \le 2 \\ 0 \le j < k_{i,1} \end{cases}$$

★   Infancy Giant pandas that die:

$$r_8 \equiv [\, Y_{i,j} \xrightarrow{k_{i,6}} \lambda \,]_2^+, \text{ for } \begin{cases} 1 \le i \le 2 \\ 0 \le j < k_{i,1} \end{cases}$$

★   (analogous mortality rules for each age interval)

★   Giant pandas which reach the maximum life expectancy:

$$r_{19} \equiv [\, Y_{i,k_{i,5}} \longrightarrow \lambda \,]_2^+, \text{ for } 1 \le i \le 2$$

– *Feeding rules*

★   Feeding process for infancy Giant pandas:

$$r_{20} \equiv [\, Z_{i,j} \; S^{f_{i,1}} \, B^{f_{i,2}} \, O^{f_{i,3}} \,]_2^+ \longrightarrow [\, W_{i,j} \,]_2^0, \text{ for } \begin{cases} 1 \le i \le 2 \\ 0 \le j < k_{i,1} \end{cases}$$

★   (analogous feeding rules for each age interval)

– *Updating rules*

★ Elimination of the remaining food.

$$r_{23} \equiv [\, S \longrightarrow \lambda \,]_2^0$$
$$r_{24} \equiv [\, B \longrightarrow \lambda \,]_2^0$$
$$r_{25} \equiv [\, O \longrightarrow \lambda \,]_2^0$$

★   Preparation for the beginning of a new cycle.

$$r_{26} \equiv [W_{i,j}]_2^0 \longrightarrow X_{i,j+1} [\ ]_2^0, \text{ for } \begin{cases} 1 \le i \le 2 \\ 0 \le j < k_{i,5} \end{cases}$$

$$r_{27} \equiv F [\ ]_2^0 \longrightarrow [F]_2^0,$$

$$r_{28} \equiv C_{i,j} [\ ]_2^0 \longrightarrow X_{i,j+1} [\ ]_2^0, \text{ for } \begin{cases} 1 \le i \le 2 \\ 0 \le j < k_{i,5} \end{cases}$$

$$r_{29} \equiv N [\ ]_2^0 \longrightarrow [N]_2^0,$$

$$r_{30} \equiv A [\ ]_2^0 \longrightarrow [A]_2^0,$$

A list of the constants associated with the rules where the corresponding meanings are specified (for male $i = 1$, for female $i = 2$) is the following:

– $k_{i,j}$: parameters related to mortality ratio for each age interval ($i = 1$ infancy, $i = 2$ subadult, $i = 3$ youth adult, $i = 4$ mid-adult and $i = 5$ elderly), and values for reaching each of those intervals.
– $g_i$: parameters related to number of newborns per year ($i = 1, 2$), and amount of food (bamboo shots, bamboo, and other food) supplied in the GPBB (kg per year).

- $f_{i,j}$: amount of food (kg per year) necessary according to the energetic requirements of the Giant pandas at each age interval.
- $cmin, cmax$: min/max number of rescued Giant pandas per year.
- $cmaxage$: maximum age of rescued Giant pandas.
- $pc_c$: probability to have $c$ rescued individuals.
- $pg_i$: probability for rescued individuals to have gender $i$.
- $pa_j$: probability for rescued individuals to have age $j$.

In order to carry out virtual experiments studying the evolution of the system under different scenarios of interest, the user can interact with the parameters listed above (in the "Notations" section). The evolution of the elements in the system for a single cycle of simulation is as follows:

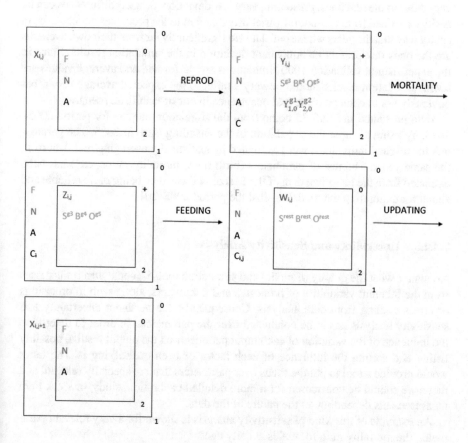

### 7.4.4.2 Experimental Validation

Results of the model have been corroborated by contrasting the output of the simulation against the real data from 2005 to 2013. Simulation data results are found to be

| Natural deviation from average (Average - Annual) difference in births and deaths | | Experimental validation | | | | | | | | |
|---|---|---|---|---|---|---|---|---|---|---|
| | | MALES | | | FEMALES | | | TOTALS | | |
| MALES | FEMALES | Simulation | Real Data | Dev. | Simulation | Real Data | Dev. | Simulation | Real Data | Dev. |
| -3,33% | 1,93% | 84 | 85 | -1,18% | 112 | 111 | 0,90% | 196 | 196 | 0,00% |
| -4,76% | -2,14% | 96 | 98 | -2,04% | 125 | 127 | -1,57% | 221 | 225 | -1,78% |
| -1,40% | -4,54% | 106 | 107 | -0,93% | 137 | 145 | -5,52% | 243 | 252 | -3,57% |
| -4,37% | -0,96% | 117 | 122 | -4,10% | 149 | 150 | -0,67% | 266 | 272 | -2,21% |
| -3,94% | -1,42% | 127 | 131 | -3,05% | 160 | 162 | -1,23% | 287 | 293 | -2,05% |
| -6,16% | 0,49% | 136 | 146 | -6,85% | 171 | 169 | 1,18% | 307 | 315 | -2,54% |
| -1,24% | 0,55% | 146 | 148 | -1,35% | 181 | 178 | 1,69% | 327 | 326 | 0,31% |
| 2,87% | 4,46% | 155 | 151 | 2,65% | 190 | 182 | 4,40% | 345 | 333 | 3,60% |
| | | 967 | 988 | -2,13% | 1225 | 1224 | 0,08% | 2192 | 2212 | -0,90% |

**Fig. 7.35** Experimental validation statistics

very close to the real ones, presenting very low deviation. Smaller differences can be easily explained from the natural variability inherent to the processes under study, in such a way that the years whose real data are significantly far from their own averages are the ones that present an equivalent deviation in the simulation results. Since in the experimental validation 1000 simulations are performed and average values are taken, simulation results should be very similar to the expected average result, but obviously not to each possible real occurrence in our probabilistic scenario.

Data presented in Fig. 7.35 come from the simulation outputs for period 2005–2013, by using as input the population in the Breeding Base in 2005. The parameters for rescued individuals were estimated by statistically averaging real data from the same years. The rest of parameters (birth rates, mortality ratios, etc.) has been extracted from the historical data of the Breeding Base, thus being more reliable and sound for calibrating the model against the period 2005–2013.

### 7.4.4.3 Uncertainty and Sensitivity Analysis

No matter what the quality of model and simulation tools is, some uncertainty rises from the inherent variability of processes and parameters, along with imprecisions or errors coming from data analysis. Consequently, the pertinent uncertainty and sensitivity analysis has to be conducted over the parameters, in order to determine the influence of the variation of each input parameter on the output results, possibly listing and sorting the influence of each factor or even quantifying it. The latter would provide a tool to put the focus over parameters that are specially relevant and therefore should be considered for a more detailed calibration, study, or work field measurements depending on the nature of the data.

An example of this kind of sensitivity analysis is shown for a very relevant parameter, the mortality ratio for males at early ages:

The strong correlation among this ratio and the size of the output population is shown in the fragment of the table and the chart in Fig. 7.36. In order to assure the accuracy of the model, it has been validated for its use in scenarios similar to the one studied, that is, with similar parameter values and environment, such as captive conditions, pairing system, cares, etc.

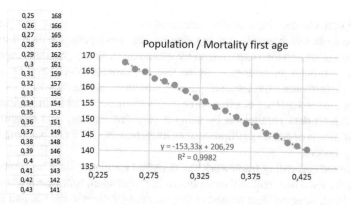

| 0,25 | 168 |
| 0,26 | 166 |
| 0,27 | 165 |
| 0,28 | 163 |
| 0,29 | 162 |
| 0,3 | 161 |
| 0,31 | 159 |
| 0,32 | 157 |
| 0,33 | 156 |
| 0,34 | 154 |
| 0,35 | 153 |
| 0,36 | 151 |
| 0,37 | 149 |
| 0,38 | 148 |
| 0,39 | 146 |
| 0,4 | 145 |
| 0,41 | 143 |
| 0,42 | 142 |
| 0,43 | 141 |

**Fig. 7.36** Correlation between the mortality ratio at early ages (horizontal axis) against the total population at the end of the simulation (vertical axis)

However, both the framework and tools involved in the model design, and the theoretical model itself, allow easy customization since they are parameterized, thus enabling their use in different contexts or even the addition of many other features of interest.

Since the goal of this model is to provide a reliable tool for supporting decision making, further efforts are necessary in terms of parameter estimation, uncertainty, and sensitivity analysis, plus an iterative refinement process involving designers, experts and managers.

## 7.5 Conclusions

In this chapter, a general membrane computing modeling paradigm for both cellular processes and population dynamics of ecological systems, is presented. Among the most important advantages of this new framework with respect to classical models we highlight a high computational potential, modularity (small changes in the system to be modeled entails small changes in the model) and the ability for the modeled process to work in parallel, in any desired order.

In relation to ecological systems and according to [53], our approach (called *Population Dynamics P systems*, PDP systems, for short) provides models able to incorporate spatial distribution among their basic components, to account for adaptation or for a shift in composition of individuals/species, including age structure, and with the possibility of a dynamism in the corresponding parameters associated with the models. Another interesting feature of PDP systems is that modeling elements are "natural" for experts, since individuals can be modeled after single objects, processes like feeding, reproduction, and mortality with rewriting rules and spatial distribution with separate membranes and environments.

PDP systems are (sequentially) simulated by using P-Lingua and MeCoSim, software tools which enable running simulations considering the model under different scenarios and then experimentally validating it. These programs capture PDP semantics through an inference engine that implements DNDP and DCBA algorithms. Nevertheless, the inherent parallel structure associated with PDP systems appoints the convenience of using some parallel hardware architecture. Specifically, GPUs exploit data parallelism by using a very fast memory and simplifying the cores by means of SIMD multiprocessors interconnected to a fast bus with the main memory system.

Presented computing framework and designed simulators have been successfully applied to model different real ecosystems. In this chapter, four case studies are presented. First, an ecosystem related to three scavenger birds in the Catalan Pyrenees, where one of them, the Bearded vulture, is an endangered species. The goal is to find out what is the most adequate management of the ecosystem for the sustainable development of species.

The second case study is a complex fluvial ecosystem, characterized by a composition of dynamic habitats with many variables that interact simultaneously. Specifically, the Riba-roja reservoir (Spain) occupied by an exotic invasive species, the zebra mussel (Dreissena polymorpha).

The aim of the zebra mussel modeling is the design of a management strategy for the reservoir in order to control or eradicate, if possible, the this species which causes serious environmental and economic damages.

The third case study is the Pyrenean chamois dynamics in the Catalan Pyrenees (Spain). This is an emblematic species that provides significant economic contributions in the area and constitutes an important food resource for obligate and facultative scavengers. In recent years, several diseases have caused a drastic decrease in the number of individuals with the population suffering from several infectious epidemics. More recently, a new disease has appeared associated to some pestivirus. Due to the importance of the species and the diseases impact on the population, it is very interesting to provide a model in order to facilitate the management of their ecosystems. The PDP model designed has been assessed by ecologists resulting in very encouraging and promising results.

Finally, a PDP model to study the ecosystem related to Giant panda individuals in the Giant Panda Breeding Base of Chengdu (China) has been presented. The model enables to study the population dynamics of the species in captivity under different conditions.

In the four cases studied, PDP models prove to be very useful tools to model complex, partially desynchronized, processes working in parallel. It is worth pointing out that the same framework proves to be highly scalable, since the number of considered elements significantly differ.

# References

1. Ackerman, J.D., B. Sim, S.J. Nichols, and R. Claudi. 1994. A review of the early life history of zebra mussels (dreissena polymorpha): comparisons with marine bivalves. *Canadian Journal of Zoology* 72: 1169–1179.

2. Arkin, A., J. Ross, and H.H. McAdams. 1998. Stochastic kinetic analysis of developmental pathway bifurcation in phage lambda-infected Escherichia coli cells. *Genetics* 149: 1633–1648.

3. Bearded Vulture C++ ad-hoc simulator. http://www.gcn.us.es/?q=node/338.

4. Ben-Ari, M. 2008. *Principles of the Spin model checker*. London: Springer.

5. Blakes, J., J. Twycross, S. Konur, F.J. Romero-Campero, N. Krasnogor, and M. Gheorghe. 2014. Infobiotics Workbench: A P systems based tool for systems and synthetic biology. In *Applications of Membrane Computing in Systems and Synthetic Biology*. Emergence, Complexity and Computation, vol. 7, ed. P. Frisco, M. Gheorghe, and M.J. Pérez-Jiménez, 1–41. Springer International Publishing (Chapter 1).

6. Bower, J., and H. Bolouri. 2001. *Computational modeling of genetic and biochemical networks*. Cambridge: MIT Press.

7. Brown, C.J. 1997. Population dynamics of the Bearded Vulture Gypaetus barbatus in southern Africa. *African Journal of Ecology* 35: 53–63.

8. Cardelli, L. 2005. Brane calculi: interactions of biological membranes. *Lecture Notes in Bioinformatics* 3082: 257–278.

9. Cardona, M., M.A. Colomer, M.J. Pérez-Jiménez, D. Sanuy, and A. Margalida. 2009. Modeling ecosystems using P systems: The bearded vulture, a case study. In *Membrane Computing, 9th International Workshop, WMC 2008, Edinburgh, UK, July 28–31, 2008, Revised Selected and Invited Papers*. Lecture Notes in Computer Science, vol. 5391, ed. D.W. Corne, P. Frisco, Gh. Păun, G. Rozenberg, and A. Salomaa, 137–156.

10. Cardona, M., M.A. Colomer, A. Margalida, I, Pérez-Hurtado, M.J. Pérez-Jiménez, and D. Sanuy. 2010. A P system based model of an ecosystem of some scavenger birds. In *Membrane Computing, 10th International Workshop, WMC 2009, Curtea de Arges, Romania, August 24-27, 2009, Revised Selected and Invited Papers*. Lecture Notes in Computer Science, vol. 5957, ed. Gh. Păun, M.J. Pérez-Jiménez, A. Riscos-Núñez, G. Rozenberg, and A. Salomaa, 182–195.

11. Cardona, M., M.A. Colomer, A. Margalida, A. Palau, I. Pérez-Hurtado, M.J. Pérez-Jiménez, and D. Sanuy. 2011. A computational modeling for real ecosystems based on P systems. *Natural Computing* 10 (1): 39–53.

12. Caughley, G., and A.R.E. Synclair. 1994. *Wildlife Ecology and Management*. Oxford: Blackwell Science.

13. Claudi, R., and G.L. Mackie. 1994. *Practical Manual for Zebra Mussel Monitoring and Control*. London: Lewis Publishers.

14. Colomer, M.A., S. Lavín, I. Marco, A. Margalida, I. Pérez-Hurtado, M.J. Pérez-Jiménez, D. Sanuy, E. Serrano, and L. Valencia-Cabrera. 2011. Modeling population growth of Pyrenean Chamois (Rupicapra p. pyrenaica) by using P systems. In *Membrane Computing, 11th International Conference, CMC 2010, Jena, Germany, August 24-27, 2010, Revised Selected Papers*. Lecture Notes in Computer Science, vol. 6501, ed. M. Gheorghe, T. Hinze, Gh. Păun, G. Rozenberg, and A. Salomaa,144–159.

15. Colomer, M.A., A. Margalida, D. Sanuy, and M.J. Pérez-Jiménez. 2011. A bio-inspired computing model as a new tool for modeling ecosystems: The avian scavengers as a case study. *Ecological modelling* 222 (1): 33–47.

16. Colomer, M.A., I. Pérez-Hurtado, M.J. Pérez-Jiménez, and A. Riscos-Núñez. 2012. Comparing simulation algorithms for multienvironment probabilistic P system over a standard virtual ecosystem. *Natural Computing* 11: 369–379.

17. Colomer, M.A., A. Margalida, and M.J. Pérez-Jiménez. 2013. Population Dynamics P System (PDP) Models: A Standardized Protocol for Describing and Applying Novel Bio-Inspired Computing Tools. *PLOS ONE* 8 (4): e60698. doi:10.1371/journal.pone.0060698.

18. Colomer, M.A., A. Margalida, L. Valencia, and A. Palau. 2014. Application of a computational model for complex fluvial ecosystems: The population dynamics of zebra mussel Dreissena polymorpha as a case study. *Ecological Complexity* 20: 116–126.
19. Colomer, M.A., M. García-Quismondo, L.F. Macías, M.A. Martínez-del-Amor, I. Perez-Hurtado, M.J. Pérez-Jiménez, A. Riscos-Núñez, and L. Valencia-Cabrera. 2014. Membrane System-Based Models for Specifying Dynamical Population Systems. In *Applications of Membrane Computing in Systems and Synthetic Biology*. Emergence, Complexity and Computation, vol. 7, ed. P. Frisco, M. Gheorghe, and M.J. Pérez-Jiménez, 97–132. Springer (Chapter 4).
20. Cormen, T.H., C.E. Leiserson, and R.L. Rivest. 1994. *An Introduction to Algorithms*. Cambridge: The MIT Press.
21. Crampe, J.P., J.M. Gaillard, and A. Loison. 2002. L'enneigement hivernal: un facteur de variation du recrutement chez l'isard (Rupicapra pyrenaica pyrenaica). *Canadian Journal of Zoology* 80: 306–1312.
22. Da Silva Peixoto, M., L. Carvalho de Barros, and R. Bassanezi. 2008. Predator-prey fuzzy model. *Ecological Modelling* 214: 39–44.
23. Danos, V., and C. Laneve. 2004. Formal molecular biology. *Theoretical Computer Science* 325 (1): 69–110.
24. Díaz-Pernil, D., M.A. Gutiérrez-Naranjo, M.J. Pérez-Jiménez, and A. Riscos-Núñez. 2007. A linear-time tissue P system based solution for the 3-coloring problem. *Electronic Notes in Theoretical Computer Science* 171: 81–93.
25. Díaz-Pernil, D., I. Pérez-Hurtado, M.J. Pérez-Jiménez, and A. Riscos-Núñez. 2008. P-lingua: A programming language for membrane computing. In *Sixth Brainstorming Week on Membrane Computing*, vol. II, ed. D. Díaz, C. Graciani, M.A. Gutiérrez, Gh. Păun, I. Pérez-Hurtado, and A. Riscos, 135–155. Sevilla: Fénix Editora.
26. Díaz-Pernil, D., I. Pérez-Hurtado, M.J. Pérez-Jiménez, and A. Riscos-Núñez. 2009. A P-lingua Programming Environment for Membrane Computing. In *Membrane Computing 9th International Workshop, WMC 2008, Edinburgh, UK, July 28-31, 2008, Revised Selected and Invited Papers*. Lecture Notes in Computer Science, vol. 5391, ed. D. Corne, P. Frisco, G. Păun, G. Rozenberg, and A. Salomaa, 187–203.
27. Donázar, J.A. 1993. *Los buitres ibéricos: biología y conservación*, ed. J.M. Reyero.
28. Ernst, M.D., J.H. Perkins, P.J. Guo, S. McCamant, C. Pacheco, M.S. Tschantz, and C. Xiao. 2007. The daikon system for dynamic detection of likely invariants. *Science of Computer Programming* 69 (1–3): 35–45.
29. Fontana, F., L. Bianco, and V. Manca. 2005. P Systems and the Modelling of Biochemical Oscillations. In *Membrane Computing, Sixth International Workshop, WMC6, Vienna, Austria*. Lecture Notes in Computer Science, vol. 3850, ed. R. Freund, Gh. Păun, G. Rozenberg, and A. Salomaa, 199–208.
30. García-Quismondo, M., R. Gutiérrez, M.A. Martínez-del-Amor, E. Orejuela.-Pinedo, and I. Pérez-Hurtado. 2009. P-Lingua 2.0: A software framework for cell-like P systems. *International Journal of Computers, Communications and Control* 4, 3:234–243.
31. García-Quismondo, M., R. Gutiérrez-Escudero, I. Pérez-Hurtado, M.J. Pérez-Jiménez, and A. Riscos-Núñez. 2010. An overview of P-Lingua 2.0. In *Membrane Computing*. Lecture Notes in Computer Science, vol. 5957, ed. Gh. Păun, M.J. Pérez-Jiménez, A. Riscos-Núñez, G. Rozenberg, and A. Salomaa 264–288.
32. Gheorghe, M., F. Ipate, and C. Dragomir. 2012. Kernel P Systems. In *Tenth Brainstorming Week on Membrane Computing*, ed. M.A. Martínez-del-Amor, Gh Păun, and F.J. Romero-Campero, 153–170. Sevilla: Fénix Editora.
33. Gheorghe, M., F. Ipate, R. Lefticaru, M.J. Pérez-Jiménez, A. Turcanu, L. Valencia Cabrera, M. García-Quismondo, and L. Mierla. 2013. 3-col problem modelling using simple kernel P systems. *International Journal of Computer Mathematics* 90 (4): 816–830.
34. Gillespie, D.T. 1976. A general method for numerically simulating the stochastic time evolution of coupled chemical reactions. *Journal of Computational Physics* 22: 403–434.

35. Gillespie, D.T. 1977. Exact stochastic simulation of coupled chemical reactions. *The Journal of Physical Chemistry* 81: 2340–2361.
36. Gillespie, D.T. 1992. A rigorous derivation of the chemical master equation. *Physica A* 188: 404–425.
37. Gillespie, D.T. 2001. Approximate accelerated stochastic simulation of chemically reacting systems. *The Journal of Physical Chemistry* 115: 1716–1733.
38. Gillespie, D.T., and L. Petzold. 2003. Improved leap-size selection for accelerated stochastic simulation. *The Journal of Physical Chemistry* 119: 8229–8234.
39. GPL license. http://www.gnu.org/copyleft/gpl.html.
40. Goss, P.J.E., and J. Peccoud. 1998. Quantitative modelling of stochastic system in molecular biology by using stochastic Petri nets. *Proceedings of the National Academy of Sciences of USA* 95: 6750–6755.
41. Harel, D. 1987. Statecharts: A visual formalism for Complex Systems. *Science of Computer Programming* 8 (3): 231–274.
42. Hallstan, S., U. Grandin, and W. Goedkoop. 2010. Current and modeled potential distribution of the zebra mussel (Dreissena polymorpha) in Sweden. *Biological Invasions* 12: 285–296.
43. Herrero, J., I. Garin, C. Prada, and A. García-Serrano. 2010. Inter-agency coordination fosters the recovery of the Pyrenean chamois Rupicapra pyrenaica pyrenaica at its western limit. *Fauna & Flora International, Oryx* 44 (4): 529–532.
44. Higgins, S.N., and M.P. Vander Zanden. 2010. What a difference a species makes: a meta-analysis of dreissenid mussel impacts on freshwater ecosystems. *Ecological Monographs* 80 (2010): 179–186.
45. Holcombe, M., M. Gheorghe, and N. Talbot. 2003. A hybrid machine model of rice blast fungus. *Magnaphorte Grisea. BioSystems* 68 (2–3): 223–228.
46. Holzmann, G.J. 2003. *The SPIN Model Checker: Primer and Reference Manual*, 1st ed. Reading: Addison-Wesley Professional.
47. Ipate, F., R. Lefticaru, and C. Tudose. 2011. Formal verification of P systems using Spin. *International Journal of Foundations of Computer Science* 22 (1): 133–142.
48. Lefticaru, R., C. Tudose, and F. Ipate. 2011. Towards automated verification of P systems using Spin. *International Journal of Natural Computing Research* 2 (3): 1–12.
49. Lefticaru, R., F. Ipate, L. Valencia Cabrera, A. Ţurcanu, C. Tudose, M. Gheorghe, M.J. Pérez-Jiménez, I.M. Niculescu, and C. Dragomir. 2012. Towards an integrated approach for model simulation, property extraction and verification of P systems. *Tenth Brainstorming Week on Membrane Computing*, 291–318.
50. Ionescu, M., Gh. Păun, and T. Yokomori. 2006. Spiking Neural P systems. *Fundamenta Informaticae* 71 (2–3): 279–308.
51. Ito, M., C. Martín-Vide, and Gh. Păun. 2001. A Characterization of Parikh Sets of ET0L Languages in Terms of P systems. In *Words, Semigroups, and Transductions*, ed. M. Ito, Gh. Păun, and S. Yu, 239–253. World Scientific.
52. Jenner, H.A., J.W. Whitehouse, C.J.L. Taylor, and M. Khalanski. 1998. Cooling water management in European power stations: biology and control of fouling. *Hydroécologie Appliquée* 10 (1–2): 1–225.
53. Jørgensen, S.E. 2009. *Ecological Modelling. An introduction*. Southampton: WIT press.
54. Kachitvichyanukul, V., and B.W. Schmeiser. 1988. Binomial random variate generation. *Communications of the ACM* 31 (2): 216–222.
55. Karatayev, A.Y., L.E. Burlakova, and D.K. Padilla. 2002. Impacts of zebra mussels on aquatic communities and their role as ecosystem engineers. In *Invasive Aquatic Species of Europe - Distribution, Impacts and Management*, ed. E. Leppäkoski, S. Gollasch, and S. Olenin, 433–446. Dorchecht: Kluwer Academic Publishers.
56. Macías-Ramos, L.F., I, Pérez-Hurtado, M. García-Quismondo, L. Valencia-Cabrera, M.J. Pérez-Jiménez, and A. Riscos-Núñez. 2012. A P-Lingua based simulator for Spiking Neural P systems. In *Membrane Computing, 12th International Conference, CMC 2011, Fontainebleau, France, August 23–26, 2011, Revised Selected Papers*. Lecture Notes in Computer Science, vol. 7184, ed. M. Gheorghe, Gh. Păun, G. Rozenberg, A. Salomaa, and S. Verlan, 257–281.

57. Macías-Ramos, L.F., and M.J. Pérez-Jiménez. 2012. On recent developments in P-lingua based simulators for Spiking Neural P systems. In *Pre-proceedings of Asian Conference on Membrane Computing (ACMC 2012)*, ed. L. Pan, Gh Păun, and T. Song, 14–29. Wuhan: Huazhong University of Science and Technology.

58. Macías-Ramos, L.F., and M.J. Pérez-Jiménez. 2013. Spiking Neural P systems with functional astrocytes. In *Membrane Computing - 13th International Conference CMC 2012, Budapest, Hungary, Revised Selected Papers*. Lecture Notes in Computer Science, vol. 7762, ed. E. Csuhaj-Varjú, M. Gheorghe, G. Rozenberg, A. Salomaa, and G. Vaszil, 228–242.

59. Mackie, G.L., W.N. Gibbons, B.W. Muncaster, and I.M. Gray. 1989. *The zebra mussel, Dreissena polymorpha, a synthesis of European experiences and a preview for North America*, Queen's Printer for Ontario.

60. MacMahon, R.F., and J.L. Tsou. 1990. Impact of European zebra mussel infestation to the electric power industry. *Annual Meeting of the American Power Conference*, Chicago (USA), 10.

61. Margalef, R. 1977. *Ecología*, ed. Omega. Barcelona, Spain.

62. Margalida, A., D. García, and A. Cortés-Avizanda. 2007. Factors influencing the breeding density of Bearded Vultures, Egyptian Vultures and Eurasian Griffon Vultures in Catalonia (NE Spain): management implications. *Animal Biodiversity and Conservation* 30 (2): 189–200.

63. Margalida, A., J. Bertran, and R. Heredia. 2009. Diet and food preferences of the endangered Bearded vulture Gypaetus barbatus: a basis for their conservation. *Ibis* 151: 235–243.

64. Martín-Vide, C., Gh. Păun, J. Pazos, and A. Rodríguez-Patón. 2003. Tissue P systems. *Theoretical Computer Science* 296 (2): 295–326.

65. Martínez-del-Amor, M.A. 2013. *Accelerating Membrane Systems Simulators using High Performance Computing with GPU*, Ph.D. thesis, University of Seville.

66. Martínez-del-Amor, M.A., I. Pérez-Hurtado, M.J. Pérez-Jiménez, A. Riscos-Núñez, and M.A. Colomer. 2010. In *IEEE Fifth International Conference on Bio-inspired Computing: Theories and Applications (BIC-TA 2010)*, vol. 1, ed. K. Li, Z. Tang, R. Li, A.K. Nagar, R. Thamburaj, 59–68.

67. Martínez-del-Amor, M.A., I. Pérez-Hurtado, M.J. Pérez-Jiménez, and A. Riscos-Núñez. 2010. A P-Lingua based simulator for tissue P systems. *The Journal of Logic and Algebraic Programming* 79 (6): 374–382.

68. Martínez-del-Amor, M.A., I. Pérez-Hurtado, M.J. Pérez-Jiménez, A. Riscos-Núñez, and F. Sancho-Caparrini. 2011. A simulation algorithm for multienvironment probabilistic P systems: A formal verification. *International Journal of Foundations of Computer Science* 22 (1): 107–118.

69. Martínez-del-Amor, M.A., I. Pérez-Hurtado, A. Gastalver-Rubio, A.C. Elster, and M.J. Pérez-Jiménez. 2012. Population Dynamics P systems on CUDA. In *10th Conference on Computational Methods in Systems Biology, CMSB2012, London, UK, October 3-5, 2012*. Proceedings Lecture Notes in Computer Science, vol. 7605, ed. D. Gilbert, and M. Heiner, 247–266.

70. Martínez-del-Amor, M.A., I. Karlin, R.E. Jensen, M.J. Pérez-Jiménez, and A.C. Elster. 2012. Parallel simulation of probabilistic P systems on multicore platforms. In *Tenth Brainstorming Week on Membrane Computing*, vol. II, ed. M. García-Quismondo, L.F. Macías-Ramos, Gh. Păun, and L. Valencia-Cabrera, 17–26. Sevilla: Fénix Editora.

71. Martínez-del-Amor, M.A., I. Pérez-Hurtado, M. García-Quismondo, L.F. Macías-Ramos, L. Valencia-Cabrera, A. Romero-Jiménez, C. Graciani-Díaz, A. Riscos-Núñez, M.A. Colomer, and M.J. Pérez-Jiménez. 2013. DCBA: Simulating Population Dynamics P systems with proportional object distribution. In *Membrane Computing- 13th International Conference CMC 2012*, Budapest, Hungary, Revised Selected Papers. Lecture Notes in Computer Science, vol. 7762, ed. E. Csuhaj-Varú, M. Gheorghe, G. Rozenberg, A. Salomaa, and G. Vaszil, 257–276.

72. Martínez-del-Amor, M.A., M. García-Quismondo, L.F. Macías-Ramos, L. Valencia-Cabrera, A. Riscos-Núñez, and M.J. Pérez-Jiménez. 2015. Simulating P systems on GPU devices: a survey. *Fundamenta Informaticae* 136 (3): 269–284.

73. Minchin, D., F. Lucy, and M. Sullivan. 2005. Ireland: a new frontier for the zebra mussel Dreissena polymorpha (Pallas). *Oceanological and Hydrobiological Studies* 34: 19–30.
74. Morales, Y., L.J. Weber, A. Mynett, and E. Newton. 2006. Mussel dynamics model: a hydroinformatics tool for analyzing the effects of different stressors on the dynamics of freshwater mussel communities. *Ecological Modelling* 197: 448–460.
75. M.P.A. Group. Daikon web page.
76. M.P.A. Group. 2010. The daikon invariant detector user manual.
77. Mullon, G., P. Cury, and L. Shannon. 2004. Viability model of trophic interactions in marine ecosystems. *Natural Resource Modeling* 17 (2004): 71–102.
78. Murray, J.D. 2002. *Mathematical Biology: An Introduction.* New York: Springer.
79. Mutyam, M., and K. Krithivasan. 2001. P systems with membrane creation: Universality and efficiency. In *Proceedings of the Third International Conference on Machines, Computations, and Universality, MCU '01*, ed. M. Margenstern, and Y. Rogozhin, 276–287. London, UK.
80. Nguyen, V., D. Kearney, and G. Gioiosa. 2009. An algorithm for non-deterministic object distribution in p systems and its implementation in hardware. In *Membrane Computing, 9th International Workshop, WMC 2008, Edinburgh, UK, July 28-31, 2008, Revised Selected and Invited Papers*. Lecture Notes in Computer Science, vol. 5391, ed. D.W. Corne, P. Frisco, Gh. Păun, G. Rozenberg, and A. Salomaa, 325–354.
81. *NVIDIA CUDA website.* 2014. https://developer.nvidia.com/cuda-zone.
82. Palau, A., I. Cía, D. Fargas, M. Bardina, and S. Massuti. 2003. *Resultados preliminares sobre ecología básica y distribución del mejillón cebra en el embalse de Riba-Roja (río Ebro).* Monografía de Endesa, Dirección de Medio Ambiente y Desarrollo Sostenible, Endesa, Lleida.
83. Pan, L., and Gh. Păun. 2009. Spiking Neural P systems with anti-spikes. *International Journal of Computers, Communications and Control* 4 (3): 273–282.
84. Pan, L., and M.J. Pérez-Jiménez. 2010. Computational complexity of tissue-like P systems. *Journal of Complexity* 26 (3): 296–315.
85. Pan, L., Gh Păun, and M.J. Pérez-Jiménez. 2011. Spiking Neural P systems with neuron division and budding. *Science China Information Sciences* 54 (8): 1596–1607.
86. Pan, L., J. Wang, and H.J. Hoogeboom. 2011. Asynchronous Extended Spiking Neural P systems with Astrocytes. In *International Conference on Membrane Computing*. Lecture Notes in Computer Science, vol. 7184, ed. M. Gheorghe, Gh. Păun, G. Rozenberg, A. Salomaa, and S. Verlan, 243–256.
87. Păun, A., and Gh Păun. 2002. The power of communication: P systems with symport/antiport. *New Generation Computing* 20 (3): 295–305.
88. Păun, Gh. 1998. Computing with membranes. *Journal of Computer and System Sciences* 61: 108–143.
89. Păun, Gh. 1999. P systems with active membranes: Attacking NP complete problems. *Journal of Automata, Languages and Combinatorics* 6: 75–90.
90. Păun, Gh., M.J. Pérez-Jiménez, and A. Riscos-Núñez. 2008. Tissue P systems with Cell Division. *International Journal of Computers, Communications & Control* 3 (3): 295–303.
91. Pérez-Hurtado, I., L. Valencia-Cabrera, M.J. Pérez-Jiménez, M.A. Colomer, and A. Riscos-Núñez. 2010. MecoSim: A General purpose software tool for simulating biological phenomena by means of P systems. In *IEEE Fifth International Conference on Bio-inspired Computing: Theories and Applications (BIC-TA 2010)*, vol. 1, ed. K. Li, Z. Tang, R. Li, A.K. Nagar, and R. Thamburaj, 637–643.
92. Pérez-Hurtado, I., L. Valencia-Cabrera, J.M. Chacón, A. Riscos-Núñez, and M.J. Pérez-Jiménez. 2014. A P-Lingua based Simulator for Tissue P Systems with Cell Separation. *Romanian Journal of Information Science and Technology* 17: 89–102.
93. Pérez-Jiménez, M.J. and F.J. Romero-Campero. 2006. P Systems, a new computational modelling tool for systems biology. In *Transactions on Computational Systems Biology VI*. Lecture Notes in Bioinformatics, vol. 4220, ed. C. Priami, and G. Plotkin, 176–197.
94. Pescini, D., D. Besozzi, G. Mauri, and C. Zandron. 2006. Dynamical probabilistic P systems. *International Journal of Foundations of Computer Science* 17 (1): 183–195.

95. Pioz, M., A. Loison, P. Gibert, D. Dubray, P. Menaut, B. Le Tallec, M. Artois, and E. Gilot-Fromont. 2007. Transmission of a pestivirus infection in a population of Pyrenean chamois. *Veterinary Microbiology* 119: 19–30.
96. Regev, A., and E. Shapiro. 2004. The π-calculus as an abstraction for biomolecular systems. In *Modelling in Molecular Biology*, ed. G. Ciobanu, and G. Rozenberg, 219–266. Berlin: Springer.
97. Regev, A., E.M. Panina, W. Silvermann, L. Cardelli, and E. Shapiro. 2004. BioAmbients: an abstraction for biological compartments. *Theoretical Computer Science* 325: 141–167.
98. Romero-Campero, F.J., and M.J. Pérez-Jiménez. 2008. Modelling gene expression control using P systems: The Lac Operon, a case study. *BioSystems* 91 (3): 438–457.
99. Romero, F.J., and M.J. Pérez-Jiménez. 2008. A model of the Quorum Sensing System in Vibrio Fischeri using P systems. *Artificial Life* 14 (1): 95–109.
100. Russell, J.C., V. Lecomte, Y. Dumont, and M. Le Corre. 2009. Intraguild predation and mesopredator release effect on long-lived prey. *Ecological Modelling* 220: 1098–1104.
101. Sakanoue, S. 2007. Extended logistic model for growth of single-species populations. *Ecological Modelling* 205: 159–168.
102. Sakanoue, S. 2009. A resource-based approach to modelling the dynamics of interacting populations. *Ecological Modelling* 220: 1383–1394.
103. Sanz-Ronda, F.J., S. López, S. San Martín, and A. Palau. 2014. Physical habitat of zebra mussel (Dreissena polymorpha) in the lower Ebro River (Northeastern Spain) Influence of hydraulic parameters in their distribution. *Hydrobiologia* 735 (1): 137–147.
104. Schneider, D.W., C.D. Ellis, and K.S. Cummings. 1998. A transportation model assessment of the risk to native mussel communities from zebra mussel spread. *Conservation Biology* 12: 788–800.
105. Shaffer, M.L. 1983. Determining minimum viable population sizes for the grizzly bear. *Bears: Their Biology and Management*, vol. 5, *A Selection of Papers from the Fifth International Conference on Bear Research and Management*, Madison, Wisconsin, USA, February 1980, 133–139.
106. Soulé, M.E. (ed.). 1987. *Viable Populations for Conservation*. Cambridge: Cambridge University Press.
107. Spin web site. http://www.spinroot.com/.
108. Strayer, D., and L. Smith. 1996. Relationships between zebra mussels (dreissena polymorpha) and unionid clams during the early stages of the zebra mussel invasion the hudson river. *Freshwater Biology* 36 (3): 771–779.
109. Strayer, D.L. 2009. Twenty years of zebra mussels: lessons from the mollusk that made headlines. *Frontiers in Ecology and the Environment* 7: 135–141.
110. Strayer, D.L., N. Cid, and H.M. Malcom. 2010. Long-term changes in a population of an invasive bivalve and its effects. *Oecologia* 165: 1063–1072.
111. Suzuki, Y., and H. Tanaka. 2000. Chemical evolution among artificial proto-cells. In *Proceedings of the Seventh International Conference on Artificial Life*, ed. M.A. Bedau J. McCaskill, N.H. Packard, and S. Rasmussen, 54–63. MIT.
112. Suzuki, Y., and H. Tanaka. 2000. Computational living systems based on an abstract chemical system. In *Proceedings of the 2000 Congress on Evolutionary Computation, CECO*, La Jolla, California, 1369–1376.
113. Suzuki, Y., and H. Tanaka. 2000. A new molecular computing model, artificial cell systems. In *Proceedings of the Genetic and Evolutionary Computation Conference, GECCO 2000*, ed. L. Darrell Whitley, 833–840. Morgan Kaufman Publishers.
114. Suzuki, Y., and H. Tanaka. 2003. Abstract rewriting systems on multisets and their application to modelling complex behaviours. In *Proceedings of the First Brainstorming Week on Membrane Computing, February 5–11*, ed. M. Cavaliere, C. Martín-Vide, and Gh Păun, 313–331. Tarragona, Spain.
115. The GNU GPL Website. http://www.gnu.org/copyleft/gpl.html.
116. The Java Website. https://www.java.com/.
117. The MeCoSim Web Site. http://www.p-lingua.org/mecosim/.

118. *The PMCGPU project*. 2013. http://sourceforge.net/p/pmcgpu.
119. The P-Lingua Website. http://www.p-lingua.org/.
120. Timar, L., and D.J. Phaneuf. 2009. Modeling the human-induced spread of an aquatic invasive: the case of the zebra mussel. *Ecological Economics* 68: 3060–3071.
121. Valencia-Cabrera, L. 2015. *An environment for virtual experimentation with computational models based on P systems*, Ph.D. thesis, University of Seville.
122. Vayttaden, S., S. Ajay, and U. Bhalla. 2004. A spectrum of models of signalling pathways. *ChemBioChem* 5: 1365–1374.
123. Zebra mussel model on MeCoSim site. http://www.p-lingua.org/mecosim/doc/case_studies/multienvironment/zebramussel.html.

Printed in the United States
By Bookmasters